Adventures in MicroStation 3D

Samir Haque, Sam Hendrick, and Scott Williams

Adventures in MicroStation 3D

Samir Haque, Sam Hendrick, and Scott Williams

Published by
OnWord Press
2530 Camino Entrada
Santa Fe, NM 87505-4835 USA

LPA, Inc. Architects provided master planning, architecture, landscape architecture, and interior design services for the San Marcos Town Center project.

SAN 694-0269
First Edition, 1996
10 9 8 7 6 5 4 3 2 1

Printed in the United States of America

Library of Congress Cataloging-in-Publication Data
Samir Haque, Sam Hendrick, and Scott Williams
Adventures in MicroStation 3D
p. cm.
Includes index
ISBN 1-56690-068-9
1. Computer-aided design. 2. MicroStation
TA174.A416 1995
604.2'0285'53--dc20 95-33181
CIP

Trademarks

Warning and Disclaimer

This book is designed to provide information about MicroStation. Every effort has been made to make this book complete and as accurate as possible; however, no warranty or fitness is implied.

The information is provided on an "as-is" basis. OnWord Press shall have neither liability nor responsibility to any person or entity with respect to any loss or damages in connection with or rising from the information contained in this book.

About the Authors

Samir Haque is an Engineer-Scientist, with degrees in Biology, Electrical Engineering, and Physics. He has been working with computers for 15 years and with MicroStation for five years. In addition to loving MicroStation, he also spends time looking at stars and playing his violin, sometimes at the same time.

Sam Hendrick has a background in Electro-Mechanical design and has done training for CADAM, AutoCAD, and MicroStation. He is currently a Corporate Training Specialist for a rapidly growing software company. Aside from his other interests in flying and two-man beach volleyball, his true passion is freeing the enslaved CAD users of the world from mediocrity.

Scott Williams studied Architecture and practiced in the field for five years. He has a Bachelors degree in Organizational Leadership from Biola University. Presently he works for Earth Tech, a geological-environmental company, as the corporate CAD manager. He has eight years of MicroStation experience. While not evangelizing the world with MicroStation, he enjoys mountain biking and computer animation.

Acknowledgments

We wish to extend a special thanks to Carlos Soria, Brian Conor, Beth Mullons, Jim Wirick, and Bob Cupper of LPA for their assistance.

Thanks to Gary Cochrane of Bentley Systems, Inc., and Bryan Bergstrand for their technical reviews and suggestions.

Special thanks to Karen Williams for her illustrations and Kim Hatch of Earth Tech for his tremendous support.

Many thanks to William Eamigh for fa-fooming.

Thanks to Dennis Preston for sending the authors to the quietest, most peaceful cabin in California.

OnWord Press

OnWord Press is dedicated to the fine art of professional documentation.

Thanks to the following people and the other members of the OnWord Press team who contributed to the production and distribution of this book.

Dan Raker, Publisher
Gary Lange, Vice President
David Talbott, Acquisitions Editor
Carol Leyba, Production Manager
Margaret Burns, Project Editor
Lynne Egensteiner, Cover Designer
Kate Bemis, Indexer

How To Use the Bonus CD-ROM

The CD-ROM that is bound in the back of this book is compatible with DOS, Macintosh, and UNIX systems. Included on the CD-ROM are all the drawings to complete the exercises in the book. In addition, the CD-ROM has many additional example files, including images, animations, and resource files. Please note all files on the CD-ROM are read only and cannot be edited. However, they can be copied to your hard drive for editing.

The CD-ROM also ships with a specific workspace for the book. The workspace is called *Inside3D* and appears as a pulldown menu next to Palettes. Each member of the pulldown menu is a chapter of the book. Each chapter has a submenu that opens several different files that are needed for the chapter exercises.

Please read the *README.TXT* file that is on each partition.

For DOS systems use the *LOADIM3D.BAT* file to copy the necessary files to your hard drive. It is assumed that you have MicroStation installed in the USTATION directory. Please consult the *README.TXT* file for instructions on how to install the files individually. Use the *IM3D.BAT* file to start MicroStation. This will automatically load the workspace. Otherwise, it can be manually selected from inside MicroStation after the appropriate files have been copied to the local hard drive. Please consult your MicroStation manual on how to use workspaces (better yet, buy *The MicroStation 95 Productivity Book* to learn all the workspace tricks). The workspace is intended to help you be more productive with the book; however, if you wish to work on individual files, feel free to copy them to your local drive.

The CD-ROM has lots of other material on it, please explore and enjoy!

Dedication

I dedicate this book to my parents, Zia and Farrakh, who have always inspired me to do more, through their own actions.

Samir Haque

I dedicate this book to my loving children, Christyn and Curtis. Their contribution to this book was to give up their Dad for the summer. I also dedicate this book to you for buying it, so that my children did not suffer in vain. Many thanks to all of you.

Sam Hendrick

I wish to dedicate my portion of this book to my loving and patient wife Karen, Jesus Christ who is the strength of my life, and my friends Sam and Samir.

Scott Williams

Contents

Introduction

Our journey begins by planning our travels in the 3D wilderness. You will see how 3D objects are created and manipulated. You probably have worked on 2D representations or sections of 3D objects, such as site plans, elevations, or cross-sections. In this chapter you will see how these 2D drawings relate to the 3D world and how to generate the 3D drawings from the 2D information.

The Art Gallery

An imaginary Art Gallery is used to demonstrate the basics of the 3D world. The gallery is examined as a 3D object and the 2D sections that make it up. You will then see the relationship between the sections and the actual object. The 3D Art Gallery is full of objects you might meet on your 3D expedition, and as you travel from room to room in our gallery, you will chart the nature of the 3D adventure you are about to embark upon.

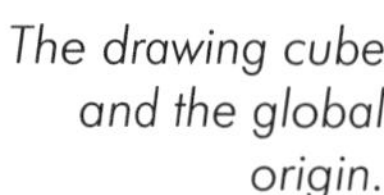

The drawing cube and the global origin.

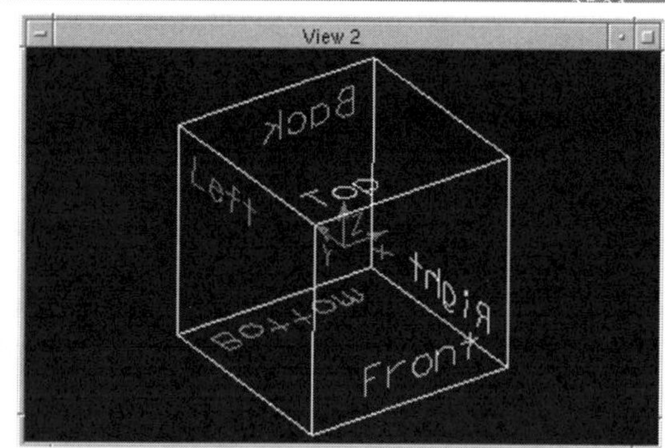

The design cube is where you will draw all of your elements; no elements can be drawn outside it. The Global Origin is in the exact center of the design cube. The drawing cube is used to show the faces of the design cube but is much smaller than the design cube, but the both have the same center, the global origin.

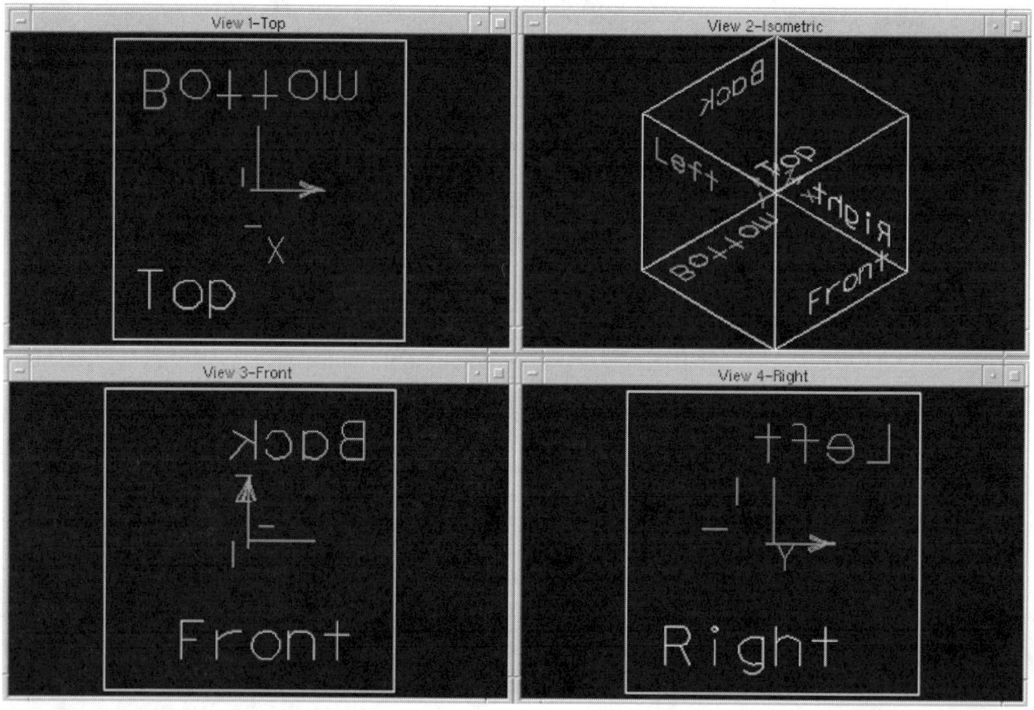

The standard views in MicroStation.

The standard views are Top, Iso, Front, and Right. These views display a volume of the design cube. The Top view shows the X,Y plane, the Front view shows the X,Z plane, and the Right view shows the Y,Z plane.

1. Select the command **Inside3D → Chapter1.**

Plan (Top) view of the 3D Art Gallery.

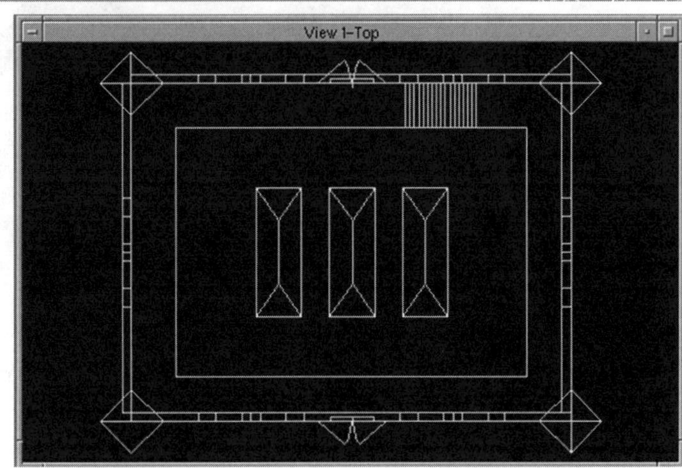

The 3D Art Gallery floor plan is a 2D representation of the overhead view of the gallery. The floor plan shows us the location, walls, doors, and windows. The lower left corner of the gallery is at the global origin (0,0,0). The dimensions of the floor are also given; the total dimensions of the building are 100 feet wide (X) by 75 feet deep (Y) by 30 feet tall (Z). The floor plan is in the Top view. In your mind, try to build a picture of the gallery by starting with the floor plan. After you can picture the floor plan, mentally add the following 2D planes, one at a time. All the drawing done on the floor is done in this view, at the active depth of the floor where Z=0 (hereafter referred to as the global Z=0 plane). Elements are created in a 2D plane as they were before.

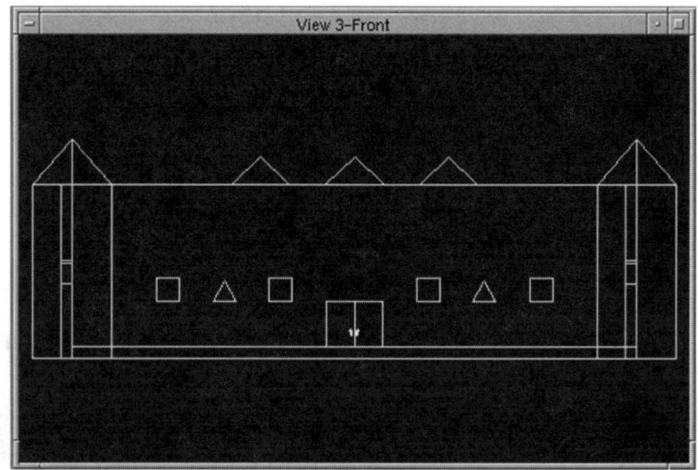

Front elevation of the 3D Art Gallery.

The front elevation shows the front facade of the building. You are looking at the front door. The elevation and floor plan are orthographic to each other. You will see later in this chapter how to project one plane onto another to get information about the 3D model. The Front view is used to display the front elevation. All work on this front facade is done in this view, at the depth of the front wall (global Y=0 plane).

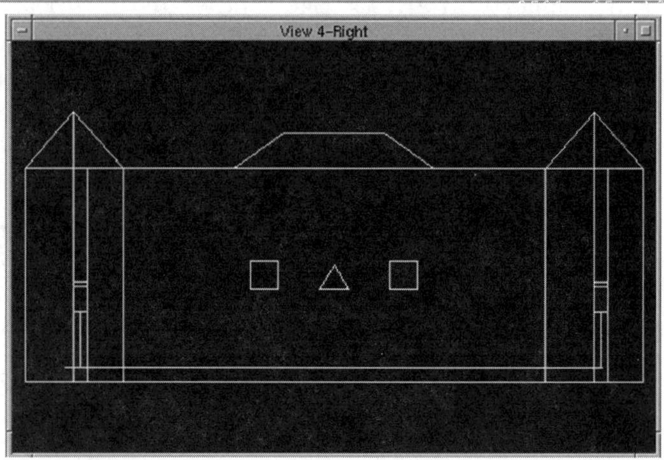

Right side elevation of the gallery.

Another view of the gallery shows the right side elevation. The Right view is used to display the right elevation. You should be mentally positioning these walls together to create the completed gallery in your mind. All the work done on the right wall is done in this view at the view depth of the right wall (global X=100 plane).

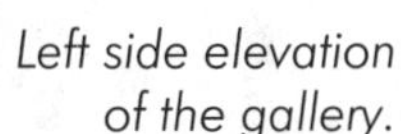

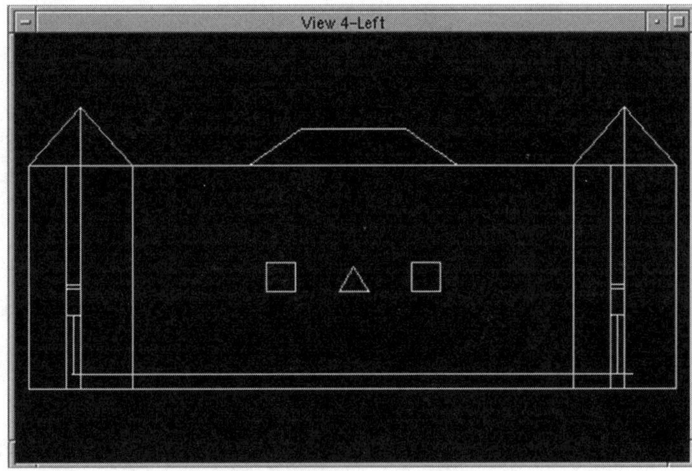

Left side elevation of the gallery.

The left side elevation is the opposite wall to the right side elevation. As you might expect, the Left view is used to display the left elevation. All work done on the left wall is done in this view at the depth of the left wall (global X=0 plane).

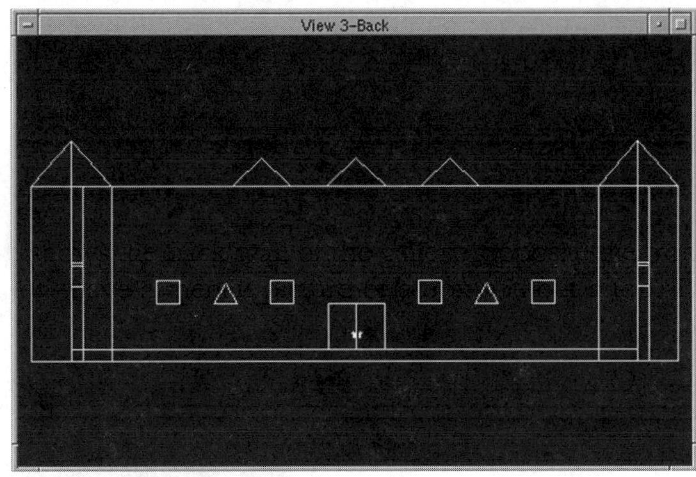

Back side elevation of the gallery.

The previous figure shows the back wall of the gallery, opposite the front wall. You should now have a mental picture of a box without a top. To display the back elevation, the Back view is chosen. The back wall is drawn in this view at the depth of the back wall (global Y=75 plane).

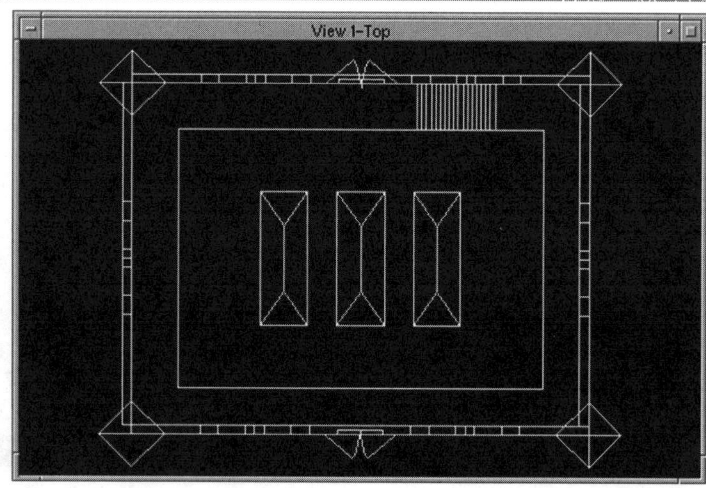

The roof of the 3D Art Gallery.

The roof of the gallery can be seen in the Top view also. The box now has a top, and you can look at the complete building using the Iso view. All work on the roof is also done in the Top view; however, the active depth is now set to the roof height (global Z=30 plane). Notice how each face was created at its own active depth, that is, we locked one dimension at a specific value, and drew 2D elements at that depth. By knowing the dimensions of the building, you know where to lock your dimensions and where to create your elements.

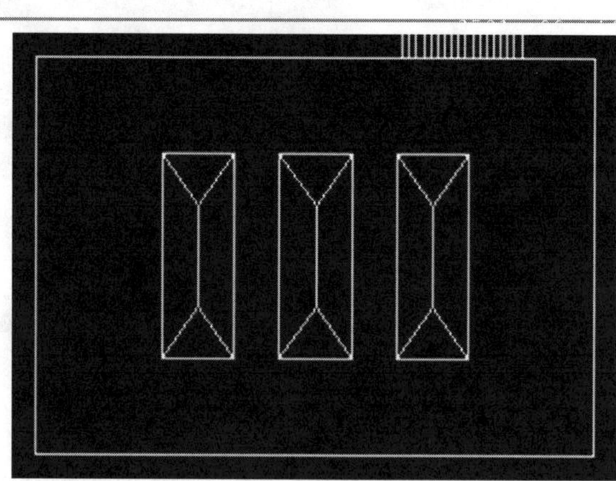

Adding skylights for the roof.

Skylights were added with precision input and tentative points in the Top view. The top of the skylight was placed at global Z=35 feet.

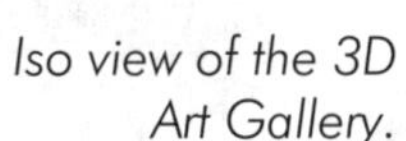

Iso view of the 3D Art Gallery.

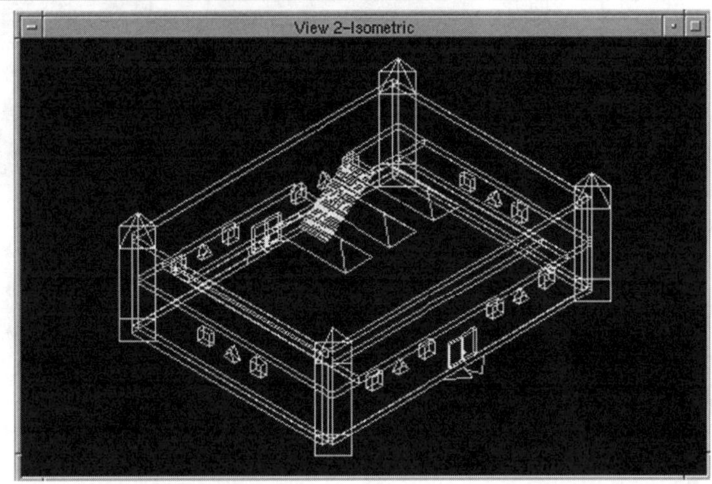

The Iso view shows us the gallery at an angle that is equally inclined for all axes. Basically, you are positioned above and to the left of the gallery looking down.

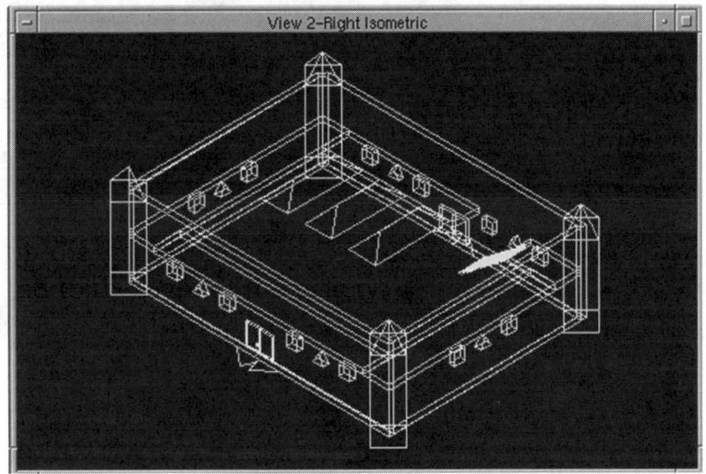

Right Iso view of the gallery.

This view shows us the gallery from the opposite side. You are now positioned above and to the right of the gallery.

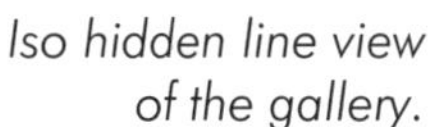
Iso hidden line view of the gallery.

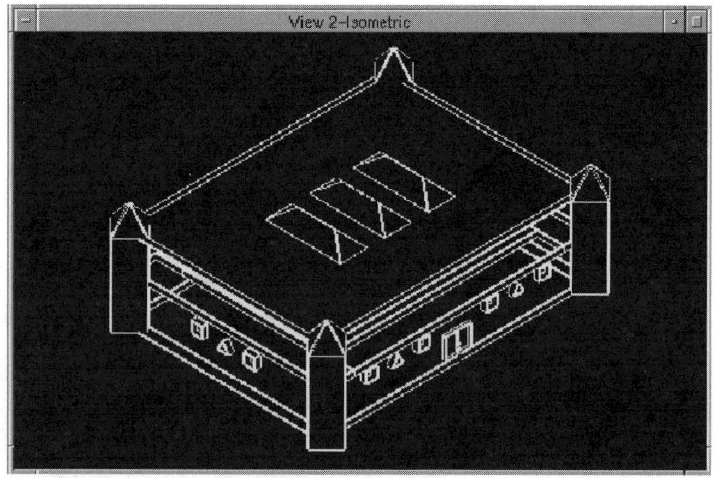

Finally, a view with the lines that are not normally visible are now hidden from view.

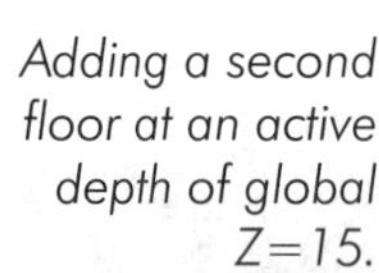
Adding a second floor at an active depth of global Z=15.

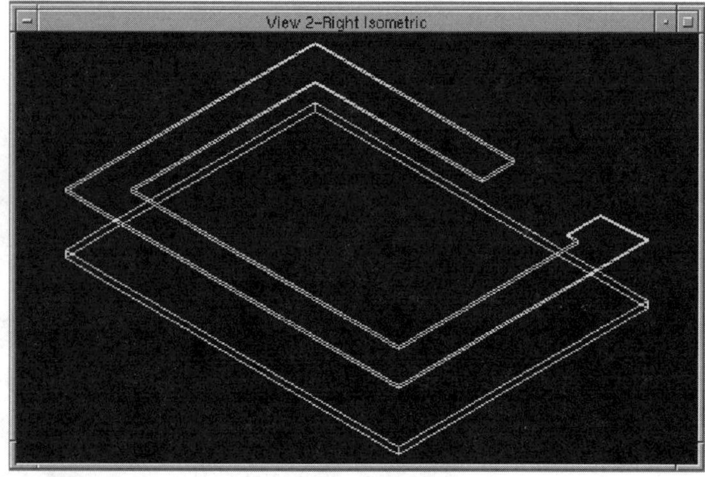

The gallery has a second floor 15 feet above the ground floor. The second floor is drawn in the Top view at an active depth of global Z=15 feet. A little room was left to put a staircase in.

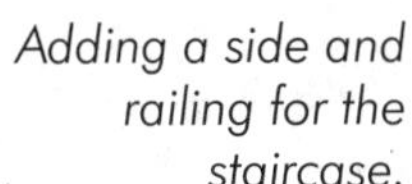

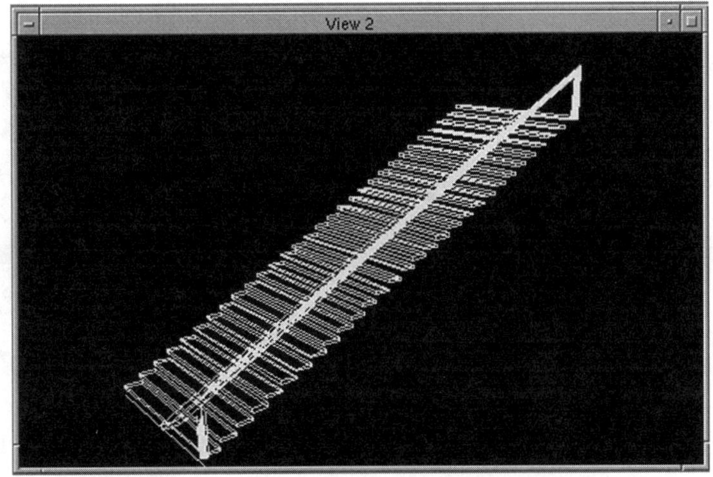

Adding a side and railing for the staircase.

The railing and staircase side is drawn in the Front view. At an active depth of the global Y=65 foot plane (10 feet in front of the back wall), a railing and staircase side is drawn. Now you are ready to place each stair.

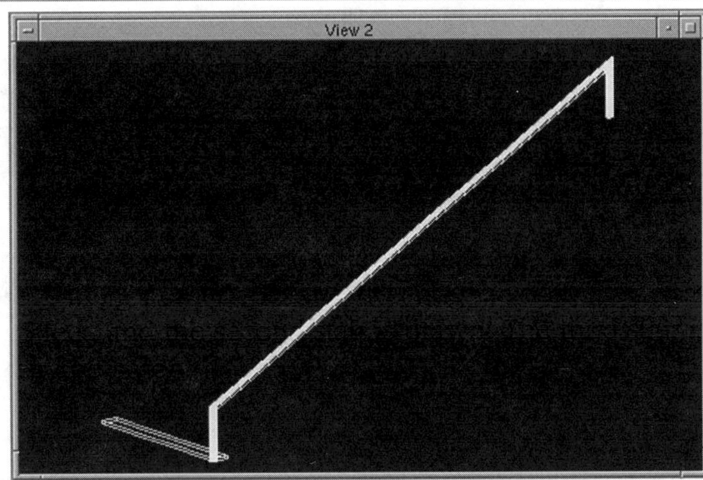

Placing a stair in the staircase.

The first stair is placed in the Top view at an active depth of global Z=0.75 feet, or 9 inches above the ground floor. After creating the stair, the active depth is set to Z=1.5 feet, and the second stair is drawn. The third stair is drawn at Z=2.25 feet, and so on.

A sculpture of the 3D design cube.

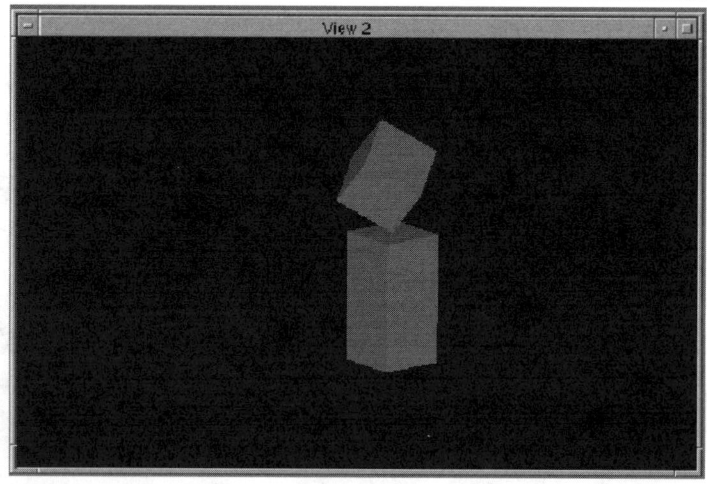

The design cube within which you will work is seen as sculpture in the gallery. The cube was created using precision input. Because the cube is equal in each axis, it can be drawn in any orthogonal view.

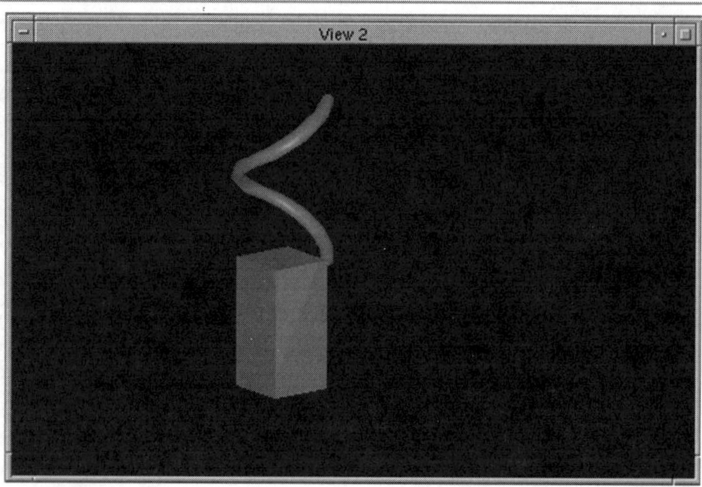

A 3D spiral sculpture.

The 3D spiral sculpture was created in the Top view (at global Z=0 feet), and its height was set in the Front view (at global Z=5 feet).

A 3D sculpture.

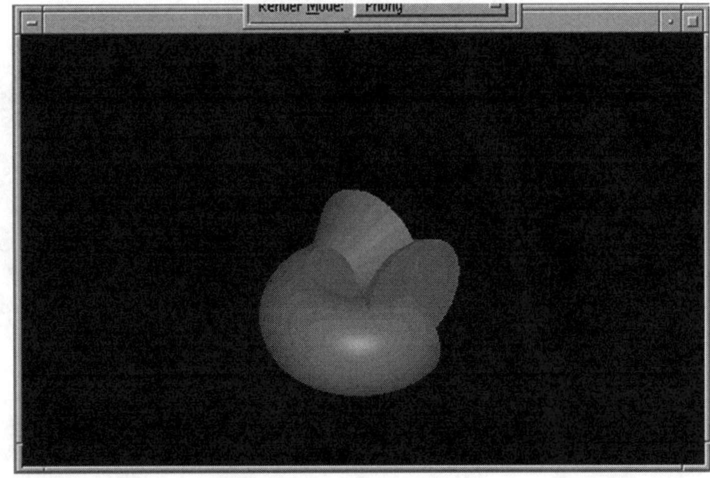

Using other 3D tools, sculptures can be created. Here a combination of a torus (the doughnut) and a truncated cone make up the sculpture. All distances were keyed in.

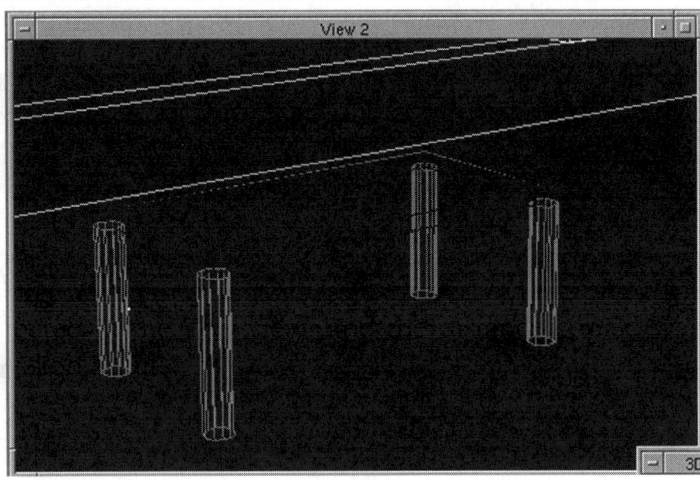

Creating a bench by first creating cylinders for feet.

Creating a cylinder from global Z=0 feet to Z=1.5 feet lets you create a leg of the bench. Copying the leg three times creates the other legs. The *Copy* command was executed in the Top view, and the distance copied was keyed in using precision input.

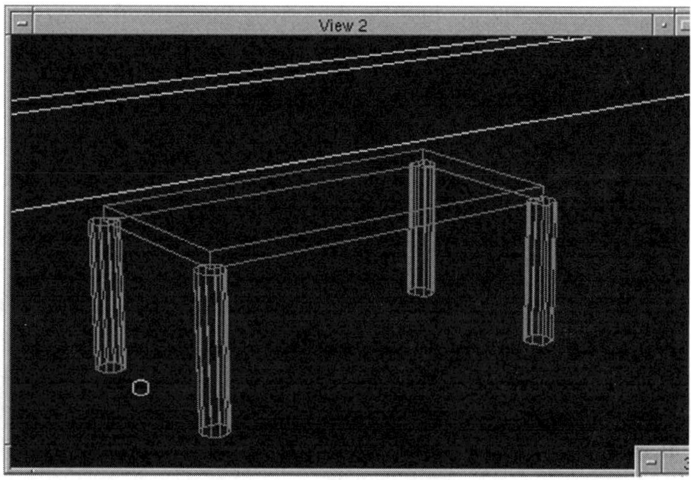

Adding a slab to create a bench seat.

The bench seat is created by creating a slab in the Top view at an active depth of global Z=1.5 feet. The bench thickness is 0.25 feet (3 inches) and is specified by a precision input keyin.

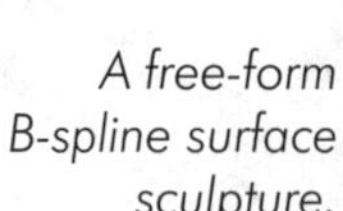

A free-form B-spline surface sculpture.

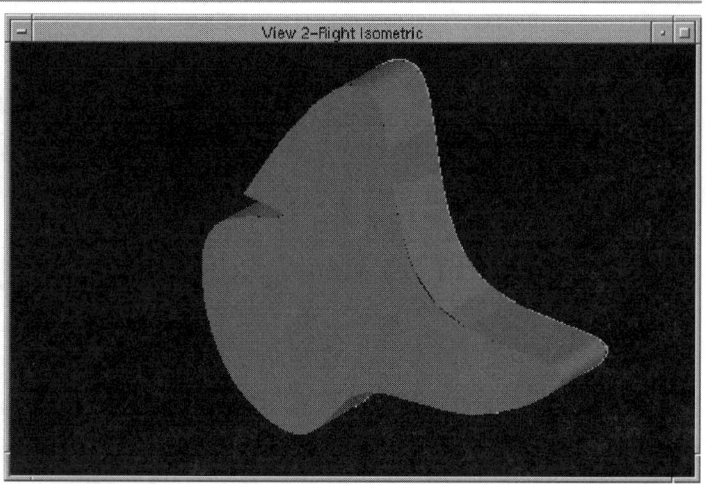

Another sculpture created using B-spline surfaces. The sculpture was started at global Z=0.

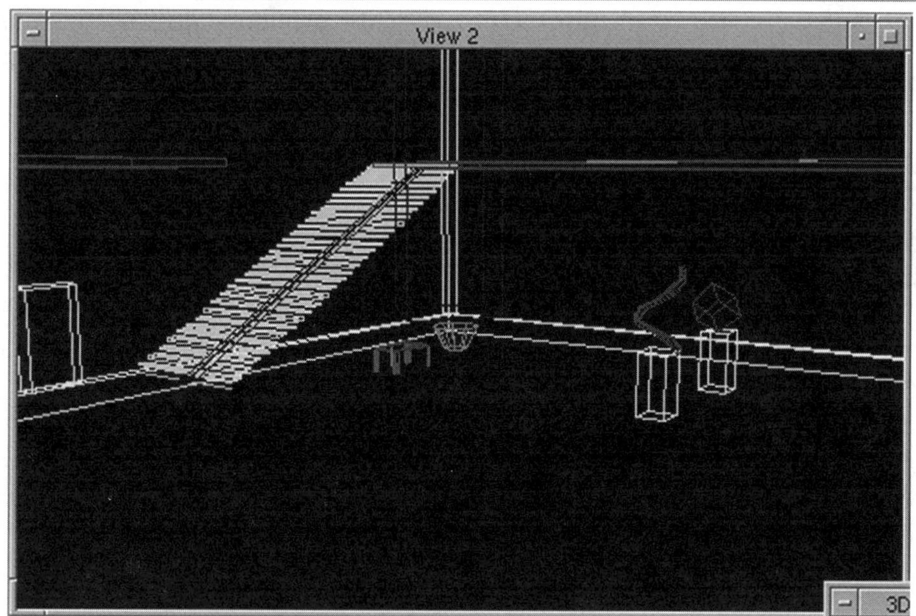

A view of the lower part of the gallery.

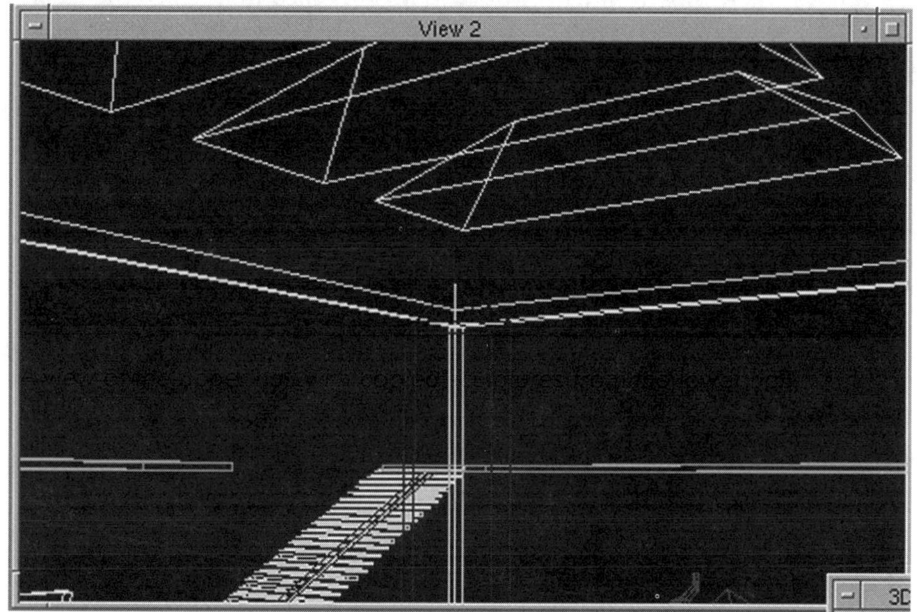

A view of the upper hall with copied sculptures from the lower hall.

Sculptures are copied from the Front view by specifying a distance using a precision input keyin.

Summary

The 3D expedition has begun! To proceed, you must understand how the 2D and 3D worlds relate. The 2D representations such as sections and elevations can help you draw accurate and precise 3D objects, which enables you to create more comprehensive drawings and eases the construction of the object. Using orthogonal projections, you can see how a 3D object looks from different views. You must be able to look at a 3D object and imagine what it would look like from the left side or top side, etc. Orthographic projection will aid you in constructing all the surfaces of an object and is a vital tool for a safe 3D expedition.

When your 3D model is complete, all the plans can be generated from this model, and any modifications are made to one drawing and then the plans are regenerated. A further benefit is that we can take the 3D model and render and animate it to produce realistic color images of the design, thus eliminating the need to construct a 3D model out of wood and plastic.

The modeling process is aided by locking a dimension at a specific depth and drawing 2D elements at that depth. Selecting the appropriate view greatly aids the work flow.

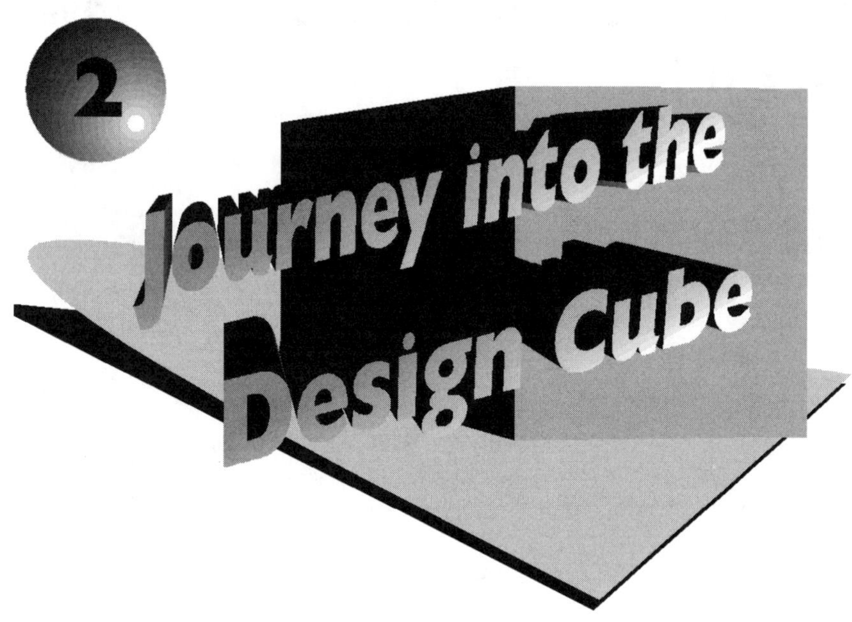

Introduction

Our journey will take us to places that have never been seen before, new objects will be created and seen for the first time, and it all happens inside the design cube. Safari Sam knows that unfamiliar settings can be frightening, so follow closely and don't stray too far from the path.

The design cube is the volume within which you will create all of your geometry. No element can be created outside the design cube. The definition of the design cube is covered by understanding *working units* in 3D. You will learn how to merge 2D information into 3D objects in the design cube. You will also learn to use the appropriate seed file in your application and to customize seed files. The use of display and active depth will help you create and modify elements. More basic 3D tools such as view volume, depth lock, 3D tentative point, and clipping planes are introduced. Finally, you will cover 3D precision input and how keyin syntax changes in 3D.

What the Design Cube Is and How To Use It

The design cube and the global origin.

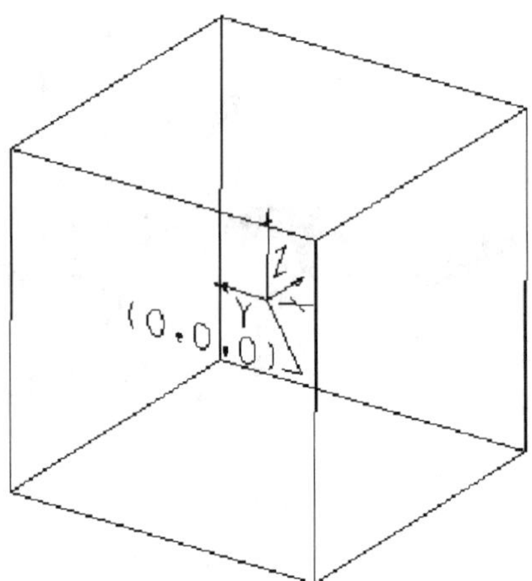

The design cube is MicroStation's default space within which you can draw. This space is, as the name implies, a cube. The cube is defined by X-, Y-, and Z-coordinates (Cartesian coordinates) and represents the total volume of the design file. The coordinate system can change to spherical, cylindrical, and user-defined systems. You can place elements within the design cube, but not outside of it. The global origin (0,0,0) is in the exact center of the design cube. The global origin can be redefined using the `GO=x,y,z` keyin.

Redefining the global origin can be useful if you want to have all positive numbers or want to match map coordinates. The changes should be saved in a seed file and use of that seed file should be restricted to that application.

Working units in 3D act the same as in 2D. MicroStation uses positional units for drawing and lets you use feet and inches or meters as your working units. You define the Master and Sub Units and then define the Working Resolution, which is the Positional Units per Sub Unit. Coordinate keyins are in the MU:SU:PU format, for example, 12:10:20 means 12 master units, 10 subunits, and 20 positional units. Each data point entered is saved as a 32-bit integer, so there are 2^{32} data points in each dimension (X, Y, and Z). The distance between two of these points is defined as the positional unit or a single unit of resolution (UOR).

Your working volume (or design cube) is then found from the working area and working length (the length of one edge of the cube):

> (2^{32} * Sub Unit/Positional Unit * Master Unit/Sub Unit) = Working Length
>
> 2^{32} is 4,294,967,296.

Working Units setting dialog box.

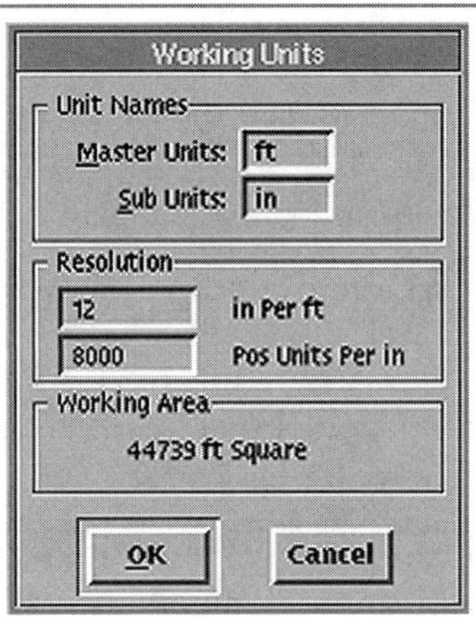

Exercise 1: Setting the Working Units

1. Open the *Inside3D → Chapter2 → Architecture* file.

   For most architectural applications:
   Master Units = ′
   Sub Units = ″
   12″ per ′
   Positional Units/Sub Unit = 8000

NOTE: *The use of ′ or feet will affect the coordinate style used in dimensions. For example, using ′, the dimension text for 5 feet would read 5′.*

Therefore,

(4,294,967,296 * [1/8000] * [1/12]) = 44,739 feet.

So the Working Area is:

$(44{,}739)^2$ = 2,001,599,834 square feet of area on each face of the design cube.

To get the Working Volume cube 44,739 feet:

$(44{,}739 \text{ ft})^3$ = 8.9548 E 13 cubic feet.

For a mapping application we would use the following working units:

Master Units = ft
Sub Units = th (tenths of a foot)
10 th per feet
100 Positional Units per Sub Unit

2. See if you get the same answer:

   Working Length: 4,294,967.3 ft
   Working Area: 1.8447 E 13 square ft
   Working Volume: 7.9228 E 19 cubic ft

Creating and using consistent working units is essential to working easily in 3D. You can draw elements at real size, without taking the scale into consideration. When you plot, you scale the drawing as needed.

Basic File Concepts

Creating files is important in the 3D wilderness; the process must be followed consistently. Seed files contain all of the settings and view configurations; they are our templates. Customizing the seed file settings saves you from setting them each time you start a new drawing with that seed file. MicroStation provides discipline-specific seed files, in addition to the generic seed file, *seed3d.dgn.* All your working unit information is contained in your seed files, so consistent use of seed files will enable you to maintain standards for all your drawings. The following list of seed files comes with MicroStation; the seed files are located in subdirectories in the \ustation\wsmod directory:

File Name	Discipline
ARCH\SEED\SDARCH3D.DGN	Architecture
CIVIL\SEED\SDCIV3D.DGN	Civil Engineering
DEFAULT\SEED\SEED3D.DGN	Default
MAPPING\SEED\MAP3D.DGN	Mapping
MECHDRFT\SEED\SDMECH3D.DGN	Mechanical

These are the basic 3D seed files available in each discipline; many more seed files are available.

A good exercise would be to open the seed files individually and determine the settings for each file. Many of you will be creating drawings of the same type (civil, architectural, mechanical, and others) and will want to customize the settings and view configurations for your own discipline. The following exercise shows you how to customize a 3D seed file:

Exercise 2: Customizing a Seed File

1. In DOS, go to the directory of the seed file you wish to modify and make a copy of it, for example:
   ```
   copy sdarch3d.dgn myarch3d.dgn
   ```
2. Start MicroStation and open *myarch3d.dgn.*
3. Check the working units and adjust them as necessary.
4. Add any graphical elements if necessary, such as a cube (or sphere or cylinder, depending on your coordinate system; see Chapter 8) or axes.
5. When you have finished modifying the drawing, select the *Compress Design* and *Save Settings* commands.

Compress Design permanently removes deleted elements from the file, whereas *Save Settings* saves the view configuration, and those settings controlled through settings boxes.

Display Depth and Active Depth

Being able to see clearly in the 3D wilderness is very important to successful adventuring. You can't go anywhere if you can't see the way. The display volume is bounded by the view area and the front and back clipping planes. The front and back clipping planes are set by the *Set Display Depth* command.

Set Display Depth icon.

The *Set Display Depth* command is found in the 3D View Control subpalette of the 3D palette. Elements in front of the front plane and elements behind the back plane are not visible, even though they might be in the view area.

To set the display depth graphically you must have at least two views open, and it is helpful if you use the standard views. Try the following exercise to learn how to use the *Set Display Depth* command.

Exercise 3: Setting Display Depth

1. Open the *Inside3D* → *Chapter 2* → *Display Depth* file.
2. Make sure all four standard views are open (Top, Iso, Front, and Right). Select the view in which you want to set the Display Depth (1). Once you select the view whose Display Depth you want to change, you will see "marching ants" around the perimeter of the view, and a dynamic box appears in the Iso view.

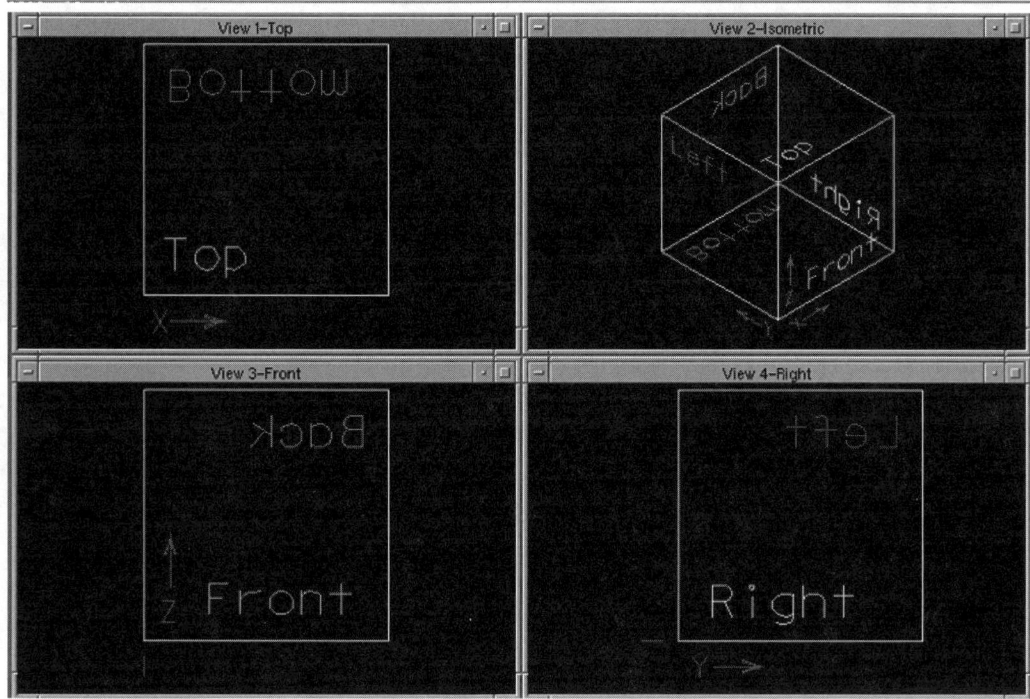

The standard views in MicroStation.

3. First, define the front clipping plane. The front clipping plane is defined by entering a data point in another view, preferably an orthogonal view. You could first enter a tentative point to an element and then a data point to accept. The clipping plane will be at the same depth as the element. In your Iso view you can watch the clipping planes move dynamically, before entering a data point. Enter a data point in the Front view above the line (2).

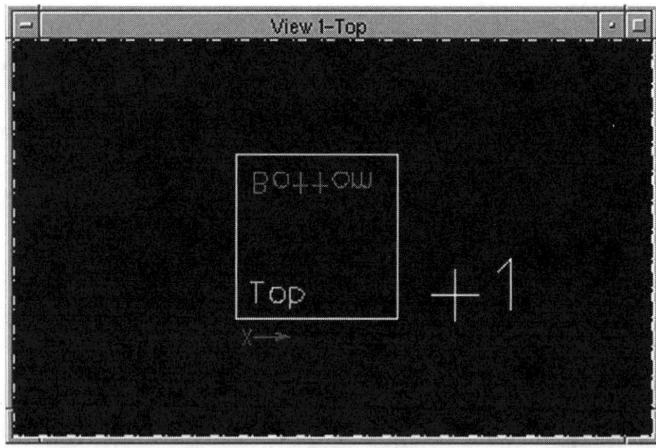

Data points entered to set display depth.

4. Now, in a view other than the one you selected for the display depth, select another data point to set the back clipping planes (3). If you choose the same depth as the front clipping plane, the display depth is not changed. Enter a data point below the line. Now your display depth has been set.

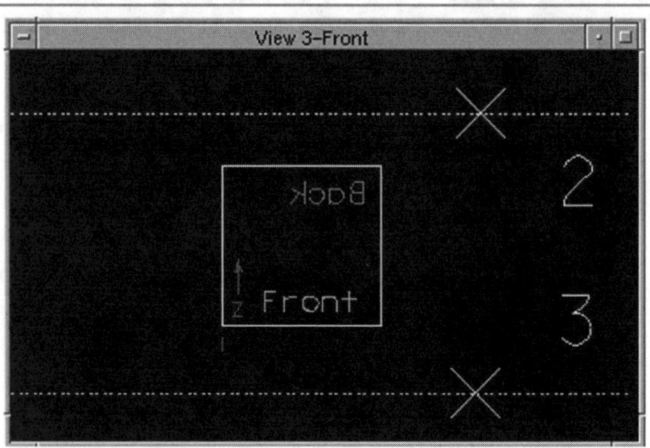

Dynamic display in Front view of Top view's display volume.

The display depth can also be set by using keyin commands. To set the display depth use the following keyin:

```
DP=FRONT,BACK
```

where *front* and *back* are distances along the view's Z-axis from the (0,0,0) point (global origin), to the actual location of the front and back clipping planes. Finally, select the view within which you would like the display depth set.

You can move the clipping planes after they have been set by using the following keyin command:

```
DD=FRONT,BACK
```

Once again, front and back refer to the relative distance to move the front and back clipping planes.

> **NOTE:** *Remember, to "get back to Kansas" (to return to a complete view of all elements) use the Fit Active Design or Fit All command. The View→Previous command will return you to your last view.*

A related command is *Show Display Depth*, which is used to show a view's display depth.

Show Display Depth icon.

After selecting the command from the 3D View Control palette, you then select a view to show its display depth. The display depth is shown in the Command Window. For example:

```
View 1: -0.00050,0.7950
```

This example was for a Top view. The numbers represent the Z-depth of the front and back clipping planes. The first number is the location on the Z-axis of the back clipping plane, and the second number is the location of the front clipping plane. Picture a piece of paper parallel to the screen that is 0.7950 working units in front of the screen and another piece of paper 0.0050 working units behind the screen. Anything in front of the front piece of paper is not visible, and anything behind the back piece of paper is also not visible. However, anything in between is visible. You can also use the `DP=?` keyin command to get a view's display depth.

Once you have chosen your display depth, you can then set an active depth.

Set Active Depth icon.

The *Set Active Depth* command is found on the 3D View Control subpalette of the 3D palette. When you set the active depth you will be constrained to drawing elements at that depth. The active depth is a plane parallel to the screen in which data points can be entered. No data points can be entered in front of or behind the active depth. Measurement of the active depth is along the view's Z-axis, and the active depth plane is always perpendicular to the view's Z-axis. Hence, some people refer to it as the *Active Z-Depth.*

NOTE: *A view's active depth is always within the display depth.*

Exercise 4: Setting Active Depth

1. Open the *Inside3D → Chapter 2 → Set Active Depth* file.
2. Make sure the standard views are open and select the Set Active Depth icon.
3. Select the view in which you would like to set the active depth. For example, if you are working on a multistory building and you want to work on details on each floor, you would select the Top view and set your active depth to the floor. If you are working on walls, then you would select the Front or Right views.
4. In the Iso view you can watch the display depth and a dynamic rectangular shape that represents the active depth move up or down inside the display depth, indicating the depth at that point.

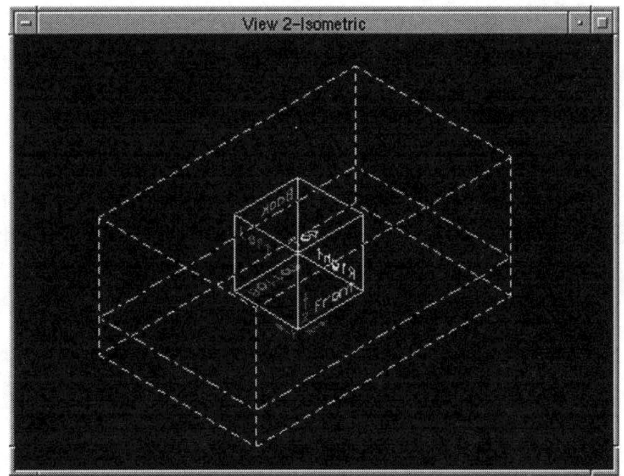

To set the active depth in the Top view, select the Top view first, then tentative point to an element in another view at the desired depth and select a data point to accept the tentative point.

5. In a separate view you can enter a data point when you are satisfied that the active depth is set correctly.
6. It is useful to select a tentative point at a point whose depth you know, and then data point to accept. For example, with the floor you would select a tentative point at a point on the floor and set your active depth, thus ensuring that all future elements you draw will be at the same depth as the floor.

The active depth can also be set by using the following keyin command.

```
AZ=DEPTH
```

where *depth* is the distance along the view's Z-axis.

NOTE: *If you set the active depth outside the display depth (or set the display depth outside the active depth), the active depth is changed to the nearest clipping plane.*

To move the active depth, use the following keyin:

```
DZ=DEPTH
```

where *depth* is the distance along the view's Z-axis.

Show Active Depth icon.

The *Show Active Depth* command is used to get a view's active depth. The command is found on the 3D View Controls subpalette, and by selecting a view, you can get its active depth. The keyin command is:

```
AZ=?
```

Exercise 5: Setting Display Depth and Active Depth

1. Open the *Inside3D → Chapter 2 → Show Active Depth* file if it isn't already open. Use the *Save As* command to save the file as `chap2.001`
2. Place a Block in the Top view, whose origin is at 0,0,0. Select the Place Block icon, and enter the following:
   `xy=0,0,0`
3. Then, for the opposite corner in the Top view, tentative to any point in the Top view and then enter:
   `dx=1,1`

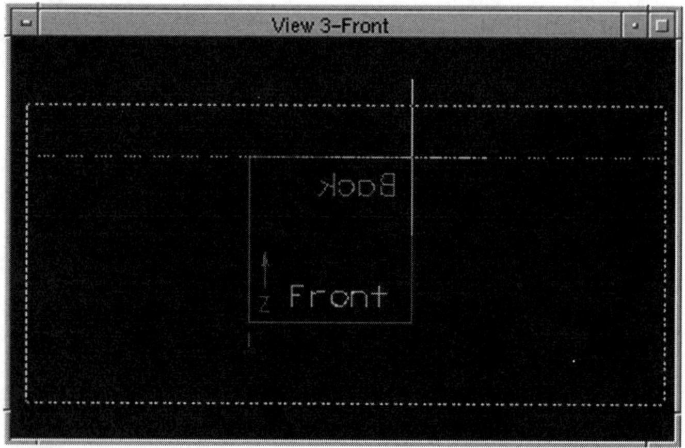

Place Block for display depth and active depth exercise.

4. Select the display depth icon, and then select (place a data point in) the Top view.

5. In the Front (or Right) view, enter another data point (1) above the line and another data point below the line (2).
6. Now select the *Set Active Depth* icon. Select the Top view; once you do so you will see "marching ants" around the inside perimeter of the view. Now tentative point to any point on the block in the Front view (the front view is just a line), and data point to accept. Congratulations! You have now set the active depth in the Top view.

Placing Points Away from the Active Depth Plane

The 3D data and tentative points are used for placing data points away from the active depth of a view, without having to constantly reset the active depth. The 3D tentative and data points are set from the *User:Button Assignment* command. These commands are really holdovers from previous versions of MicroStation. The regular tentative point will work just fine in place of the 3D tentative point, so long as *Depth Lock* is turned off.

The Settings: Locks Toggles dialog box.

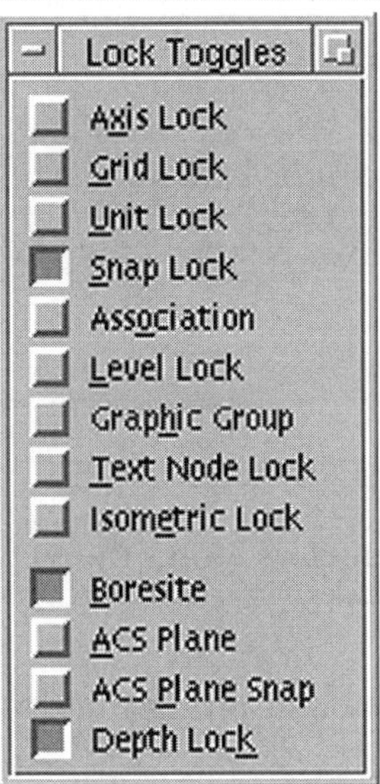

Depth Lock is found on the *Settings Locks Toggles* dialog box. If the command is on, it allows each tentative point to be placed on the active depth, and no where else. If the *Depth Lock* is turned off, then you can tentative point to any point on or off the active depth. A combination of Depth Locks and the regular tentative point allows you to do the same thing a 3D tentative point does and is the recommended way of proceeding.

A related command is the *Boresite Lock*, also found on the Locks Toggle dialog box. If the *Boresite Lock* is on, you can identify elements at any depth. If the *Boresite Lock* is off, then you may only identify elements at or near the active depth. A tentative point will nullify the effects of the *Boresite Lock*. Snapping to an element and then selecting a data point to accept a manipulation will change the element's current depth to the view's active depth.

Precision Input by Keyin

Precision input is necessary to draw accurate and precise designs. Using the keyboard we can enter exact values for data points. Precision input works almost the same way as in 2D; however, you must also enter a depth (Z) coordinate.

> **NOTE:** *If you do not enter a depth coordinate (or any other coordinate) the coordinate will be regarded as 0.*

The keyins remains the same for 3D as well.

```
XY=x,y,z
```

where X, Y, and Z are absolute, numerical coordinates.

```
DX=x,y,z
```

for relative coordinates along the view's axes.

```
DL=x,y,z
```

for relative coordinates along the design plane axes.

The following keyin syntax is also correct:

```
DX=1,,1
```

This means, place a data point 1 working unit in the X- and Z-directions, and `y=0`, relative to the last data point entered along the view axes.

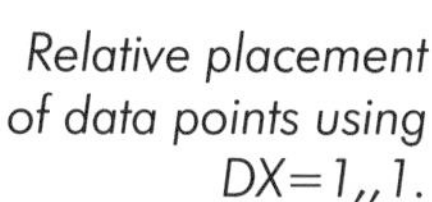

Relative placement of data points using DX=1,,1.

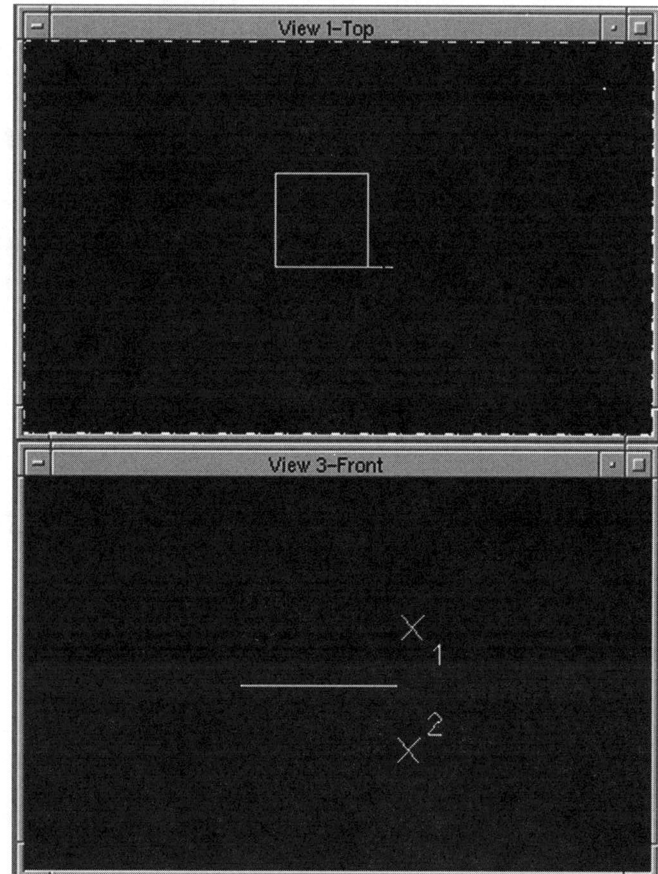


```
XY=,1,1
```

This means place a data point at the absolute coordinate (0,1,1).

Summary

The design cube is a place where we will be spending a lot of time in the future. You must be able to define it with working units, view it properly with display depth, and move where you want to with active depth. These commands are the heart and soul of 3D control and successful element creation and editing.

Introduction

Attention Tenderfoots and Trailblazers, prepare to seek higher ground! Now that you have ventured into the design cube, let's discover how to view the environment. Remember, the design cube represents the total 3D volume in MicroStation. The small cube that is visible when you start is just a drawing aid and not your design cube. In this chapter you will get your bearings in the display cube and learn how to move inside the cube and view the 3D elements you have created.

The ability to find your bearings and manipulate the view you have of the cube is essential to proper element creation. Two-dimensional viewing procedures are reviewed and compared with 3D procedures. View manipulation and rotation are covered to see how 3D views change with orientation. Finally, you will perform basic operations involving Z-axis manipulation and its effect on 3D views.

Display Volume

The *display volume* is the volume displayed in a 3D view. This volume is bounded by the window area and the display depth. In most cases, only a small volume of the design cube is viewed. Each of MicroStation's standard views displays a separate part of the design cube. This feature helps you control exactly how you draw in 3D.

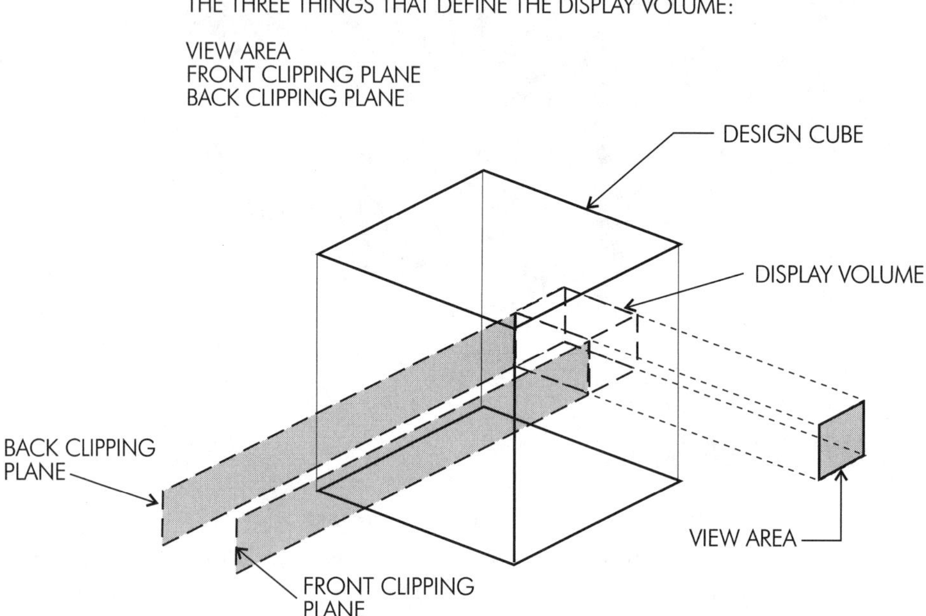

The display volume is defined by the front and back clipping planes and the window size.

From Chapter 2, you learned that the display depth is bounded by the front and back clipping planes. Only elements in between the planes are visible, helping you to control the amount of geometry displayed. The extent of the view in the X- and Y-directions is controlled by the window area (you control the window area by resizing the window or zooming in or out), and the Z-direction (depth) is controlled by the front and back clipping planes. The combination of X, Y, and Z determines the display volume.

✔ **TIP:** *When working in a 3D file, you will rarely need to have all of the design cube displayed at one time. Set your window area and display depth according to the geometry on which you are working or the amount of space you need to create geometry.*

As you create or alter geometry you may need to change the display volume to enclose the new geometry. The best way to do this is to use the *Fit Active Design* command. Remember, no geometry is visible outside the display volume. However, as you execute the *Fit Active Design* command in each view, that view's display volume changes (by moving the front and back clipping planes and resizing the window) to enclose all the active geometry, but the display volume changes for that view only.

The fact that you have independent display volumes can cause some problems, such as invisible or partially visible elements, but the way to "get back to Kansas" is the *Fit Active Design* or *Fit All* command, if you have reference files attached. When you use *Fit Active Design* in every view, the display volumes are equal in size, but each keeps its orientation.

You create your display volume in a specific view to choose the proper clipping planes and window size, for instance, the Top view. To ensure that your other views are displaying similar volumes, use the keyin command `align`. The `align` keyin matches the display volume in your destination view with your source view.

3D Versus 2D Viewing Procedures

Many view control procedures work in 3D the same way they do in 2D. The commands that do not change in 3D include *Update View*, *Window Center*, *Zoom In*, and *Zoom Out*.

The View Control palette.

If an element is not inside the display volume, it is not visible, no matter how far you zoom out (zooming in or out only changes the Window size, letting you see the display volume from a distance, but will not redefine the clipping planes or the display volume initially created). You can change a view area using these commands, without changing the display volume.

The *Window Area* command has subtle, but important differences in 3D. In 2D you create your window area in one view (source view) and then select another view (destination view) for displaying. For instance, you might use the *Window Area* command in View 1 and select View 2 to see the windowed area. The difference in the 3D environment is that the source view's active depth and display depth are transferred to the destination view. The destinations view's active depth and display depth are now the same as the source view's. Using the *Window Area* command in this fashion may cause elements to be created at the wrong depth, because the active depth might change.

> **NOTE:** *Many an adventurer has slipped and fallen on the Window Area command. You must remember that the attributes of the display volume are transferred to the destination view. To avoid this, make your source and destination views the same view.*

The *Fit Active Design* and *Fit All* commands are different in the 3D world. The *Fit All* and *Fit Active Design* commands change the display volume by altering both the front and back clipping planes and the window area, so that all of the geometry is inside the display volume. Therefore, all the geometry is visible.

A 3D object partially hidden because it is larger than the display volume.

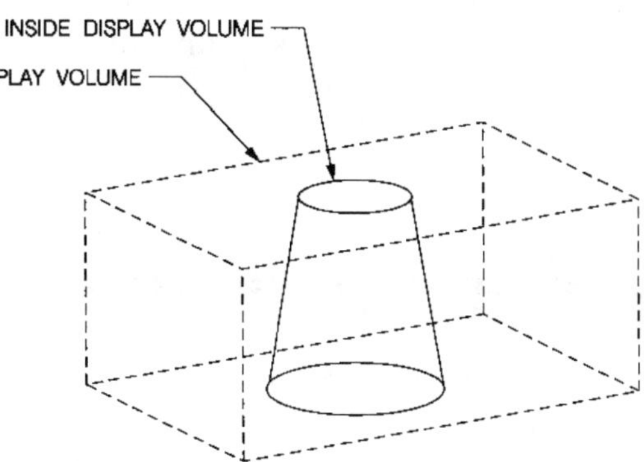

After using Fit Active Design, all of the 3D object is now visible, because the front and back clipping planes and window's size have been changed.

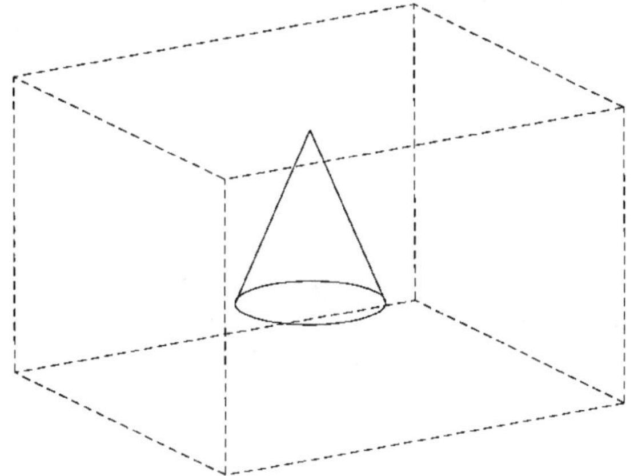

➾ **NOTE:** *Using Fit Active Design and Fit All will move your clipping planes and readjust your window area.*

✔ **TIP:** *To restore your last display volume, use the VIEW→ PREVIOUS command.*

The *pan* function allows you to move the window area without changing the view magnification. Dynamic panning is activated the same way in 3D; hold down the <Shift> key and the Data button and move the cursor in the direction you wish to pan. Panning will not change your display volume.

Setting Up Views and View Rotation

There are eight views in the drawing environment; each can have a different orientation. The views are orthographic to each other, except for the iso and custom views. The four default views (Top, Iso, Front, and Right) are the best ones to work with.

Create and open a new design file using the *seed3d.dgn* seed file and call it *views.dgn*. Remember, the following are the default views:

View 1 is top
View 2 is iso
View 3 is front
View 4 is right

In a 3D file you notice that things look different; the orientation may not be visually sensible, or what you expect. View Rotation enables you to manipulate the view so the elements are oriented in a way that is visually understandable. In the 3D world you can rotate around three axes, as opposed to one (Z-axis) in the 2D world.

Orthogonal Views and Standard Views

There are six basic orthogonal (*orthogonal* means they are perpendicular to each other) views in a 3D file; Front, Back, Left, Right, Top, and Bottom. These views correspond to the six sides of the design cube and are referred to as standard views. In the Top view, the X- and Y-axes are parallel to the screen. In the Front view the X- and Z-axes are parallel to the screen. The Right view has the Y- and Z-axes parallel to the screen.

In addition to the orthogonal views, there are two other standard views, the Right and Left Isometric views. Each of the faces of the design cube is rotated so that it is equally inclined from the screen surface. These views rotate the X-axis down 30 degrees and Y-axis +/- 30 degrees.

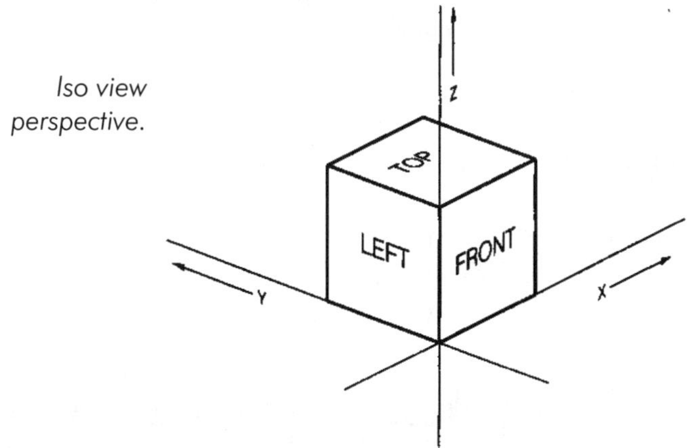

Iso view perspective.

The standard views offer quick and accurate viewing of the display volume. These views can be accessed through the View Rotation palette, which is found on the *View* → *Rotation* → *Settings* menu.

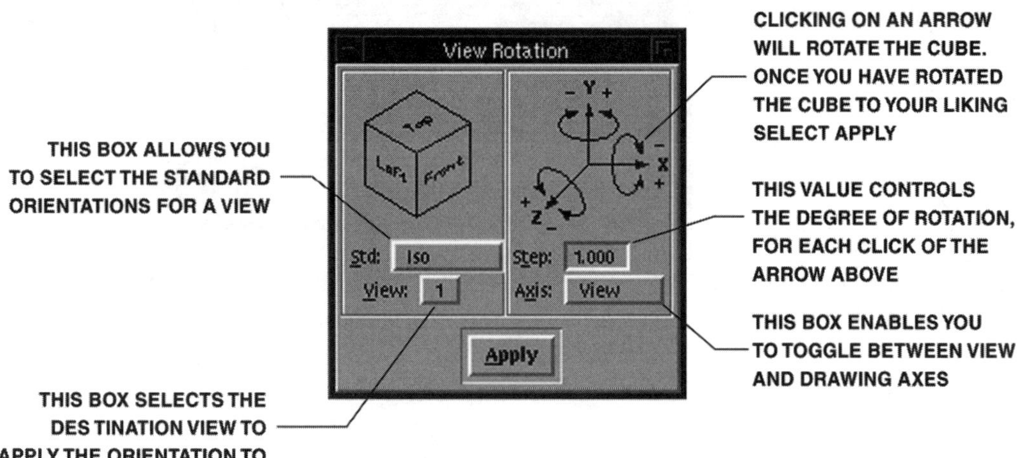

The View Rotation dialog box.

The left side of the palette represents the orientation of the view, and the right side of the palette allows the orientation of the display volume to be changed.

Once the desired rotation is achieved, the rotation can be applied to the view. The standard views can be accessed under the STD button and the destination view can be changed from the View button of the dialog box.

Saved Views

A saved view is a view that the user names and saves with the design file. First, a source view is created by opening a view and setting its rotation and then selecting *View → Saved.* A dialog box for Saved Views is opened. The saved view can be attached to a destination view.

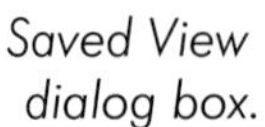

Saved View dialog box.

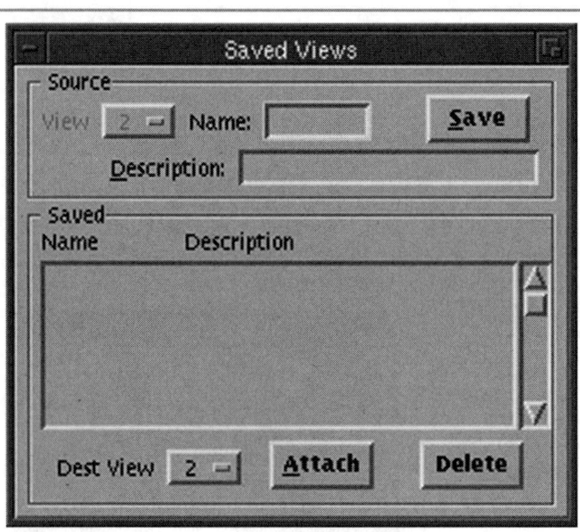

Like 2D saved views, 3D saved views remember all of the view's attributes, such as level information and symbology. In addition, the display volume and active depth are also saved. The saved views enable you to save orthogonal views that look 2D, which can be helpful in modeling.

Z-Axis Operations

Many operations such as *Copy*, *Move*, *Scale*, and others can affect the visibility of elements. The best way to ensure correct element manipulation is to have one view that is updated by *Fit Active Design*. You should also use *Fit Active Design* after you manipulate an element. The first three rules for view control are *Fit Active Design*, *Fit Active Design*, and *Fit Active Design*. Of course, the active depth and display depth have to be reset to draw the next elements at the correct depth. If you feel uneasy about changing active depth and display depth, review the exercises in Chapter 2.

Summary

View control is important to proper modeling. Many commands affect view control and display volumes. Remember the *View* → *Previous* command if you lose geometry. The most important concepts are those of the front and back clipping planes and the window size. When you choose a display volume, zooming in or out will only change your *view* of the display volume and not the display volume itself. Traveling in the 3D world can be confusing. To get a clear view of the landscape, use *Fit Active Design* or *Fit All* to see all of your geometry, and then use the *Set Display Depth* command to reset your clipping planes so you can focus on the elements you need to work on.

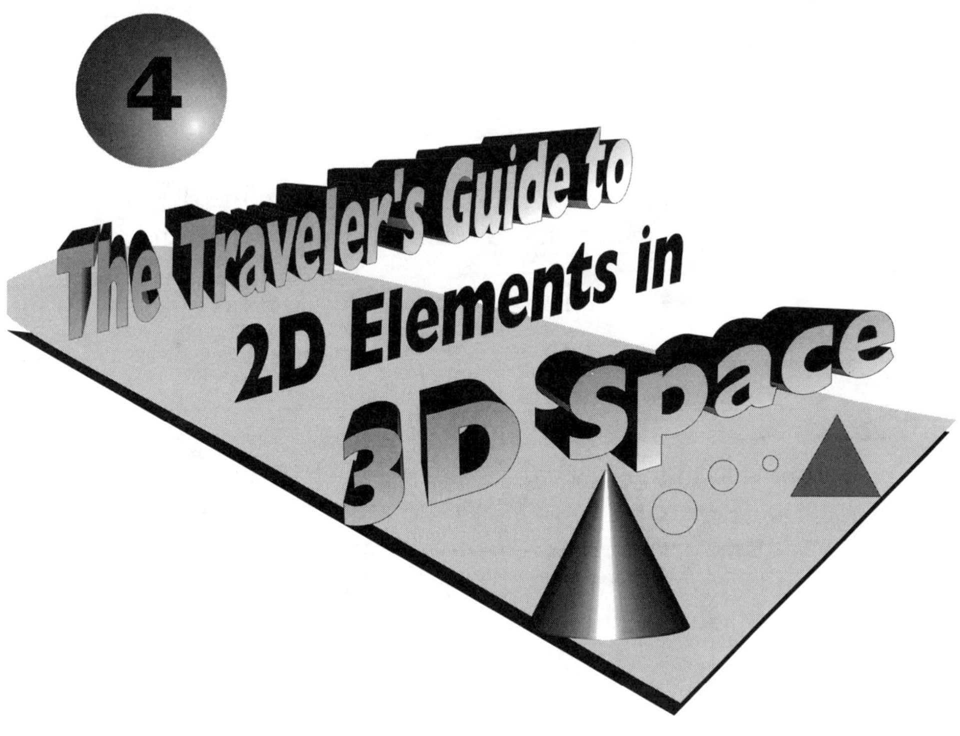

Introduction

As a seasoned 2D MicroStation adventurer, you are familiar with the basic element creation and manipulation commands. To make you more comfortable, I will first reintroduce you to your old 2D friends. Basically, 2D commands work the same in 3D as they do in 2D. All you have to do is remember that you are working with three axes. As we discussed in Chapter 2, you need to remember to set your active depth when creating geometry. When you created geometry in a 2D design file, you were working on the X,Y plane because a 2D design file only has one plane for you to create geometry on. This chapter is divided into two sections. The first section covers element creation, and the second section covers element manipulation. So let's get started on our tour of 2D elements in 3D space.

Element Creation

Exercise 1: Place Line

In this section you will create 2D elements while controlling the depth at which they are created. You will practice setting your active depth and using precision input.

The Place Line icon.

The *Place Line* command in a 3D design file is not much different than in a 2D design file. In this exercise, you will set the active depth for View 1 (the Top view) and place several lines. Then you will change your active depth for View 1 and create more lines. This will help to illustrate how to use the active depth to create 2D or 3D geometry at differing depths in a 3D design file.

1. Open the **Inside3D** → **Chapter 4** → **Place Line** file.

NOTE: *You can see that there are four views open. They are*
View 1 = Top
View 2 = Isometric
View 3 = Front
View 4 = Right

Having these four particular views open and in this order is a good place to start. Notice that Views 1, 3, and 4 are orthographic from each other. This makes this view arrangement a natural one.

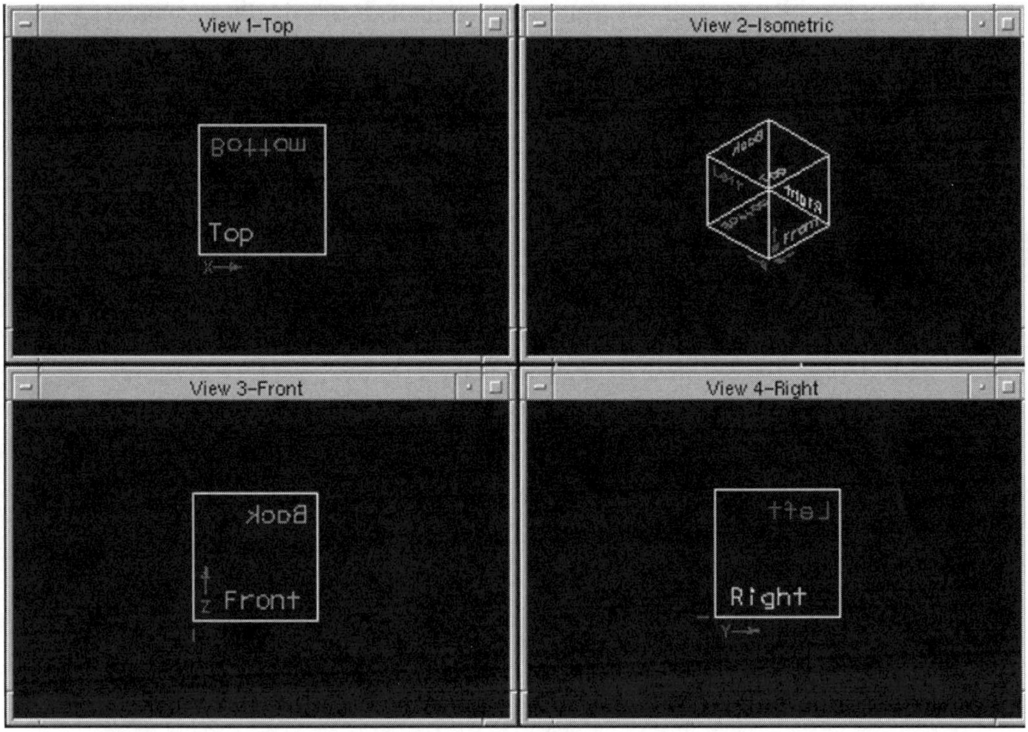

The four views in a 3D file.

2. Open the 3D palette **(Palettes → 3D → 3D)**. Move the palette so it is off to the side.
3. From the 3D palette, pull off the 3D View Control subpalette and select the *Set Active Depth* tool.

The 3D palette and the 3D View Control subpalette.

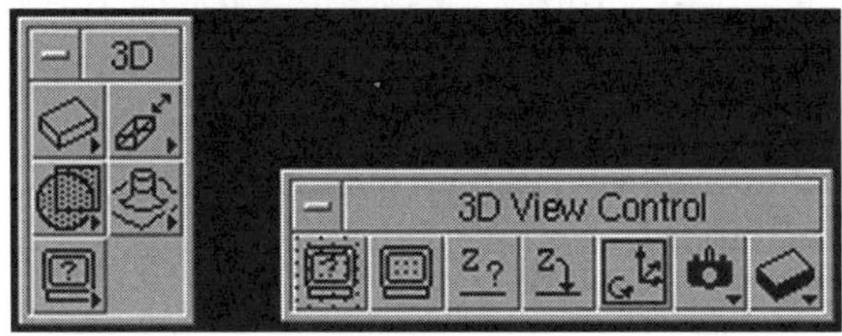

4. Select View 1 as the view in which you want to set the active depth by selecting a data point in that view. Move the cursor down into

View 3, and you will see the active depth dynamically displayed as a horizontal line moving up and down with the movement of the cursor.

NOTE: *MicroStation will not let you set the active depth outside of the display depth. The display depth in this file has been set so that you will not see the extents of the display depth in Views 1, 3, and 4. In View 2 (the isometric view), you will see the extents of the display depth in the X,Y plane.*

5. Tentative point to the middle of the cube in View 3 and accept with a data point. You have now set the active depth for View 1 by identifying a point on the Z-axis in View 3.
6. Select the *Place Line* tool and in View 1 create several line segments in and around the cube. As you create the line segments, notice how they appear in the other views. In Views 3 and 4, you are seeing them from an edge view. View 2 is a better view of what is really happening.
7. You are now going to reset the active depth for View 1. Select the *Set Active Depth* tool and data point in View 1. Move the cursor down into View 3 and select a tentative and a data point on the top of the cube. The active depth for View 1 is set to the top of the cube.
8. Select the *Place Line* tool and in View 1 create several lines again in and around the cube. Notice in the other views how the lines are appearing at a different depth than the first set of lines you created. In View 2 (the isometric view), you may not be able to clearly see the depth difference of the lines you created.

Congratulations! You have just created 2D lines in 3D space. Now, that wasn't so difficult, was it?

Exercise 2: Place Line String

The Place Line String icon.

Stay in the same file. There are two types of *Linestrings* in a 3D design file. First is the default planar linestring. Second is the non-planar linestring. To create a non-planar linestring, you must turn on the check button for Non-planar from the popdown area for the Place Line subpalette or the Tool Settings palette. The difference between these two linestrings is actually quite straightforward, if the name doesn't give it away for you. A planar linestring can only exist on one plane, such as the X,Y or Z,X plane. This means if you started a linestring in View 1 (the X,Y plane), placing at least three points, you would not be able to define a point in the Z-axis either by keyin or a data point in View 3 or 4. Views 1, 3, and 4 represent three different planes.

View 1 = XY (Top)
View 2 = ZX (Front)
View 3 = ZY (Right)

This illustration shows the three different planes represented by Views 1, 3, and 4.

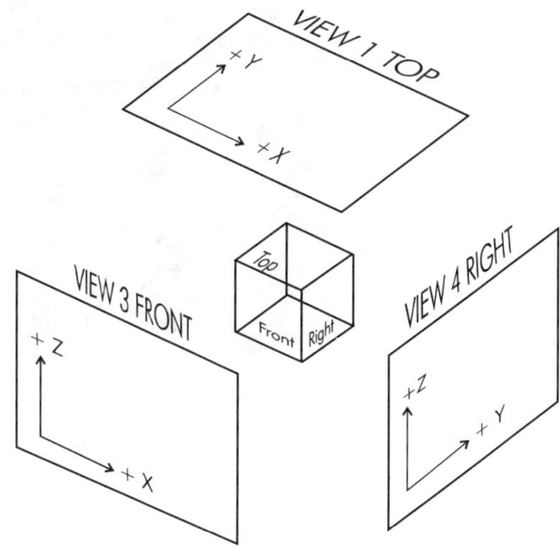

This exercise has two parts. First, we will place a planar linestring, and then a non-planar linestring.

Part 1: Planar Linestring

1. To ensure that we are both working with the same design file, stay in the same file.
2. Select the *Place Line String* tool from the Line subpalette. Set your active color to one (blue).
3. Make sure that the non-planar check button on the Tool Settings palette is off. You are going to create a linestring as shown in the following figure. After you have placed several points of the linestring, and before you press Reset to end the command, I want you to move your cursor into View 3 and notice that your linestring remains planar or flat. Your cursor is there, but no line is dynamically dragging behind it. Move your cursor back into View 1 and complete the linestring.

In View 1, your linestring should look somewhat like this.

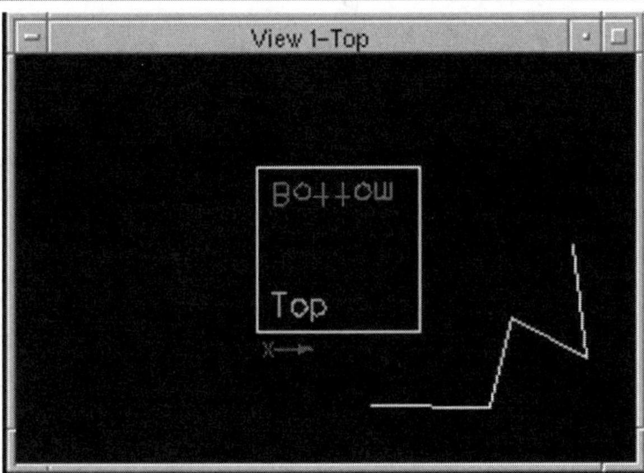

4. Select the *Place Line String* command. Set your active color to 2 (green). Create another linestring in View 3 as shown in the following figure.

Your linestring in View 3 should look like this.

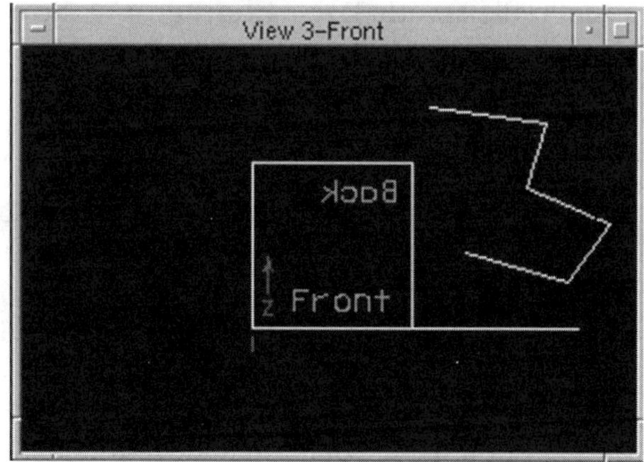

5. Set your active color to 3 (red). Create one last linestring in View 4 as shown in the following figure.

Your linestring in View 4 should look like this.

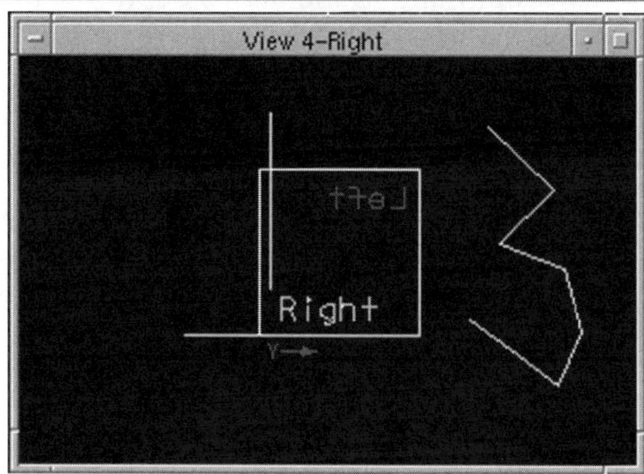

6. Take a step back and look at what you have drawn. The colors should help you to see which line is which. The isometric view is sometimes helpful, but can also add to the confusion.

Part 2: Non-Planar Linestring

Next, you are going to create non-planar linestrings. In this part of the exercise, you are going to start your linestring in one view (or plane), and then move to a different view (or plane) to complete the linestring.

1. Open file **Inside3D → Chapter 4 → Non-Planar Linestring**.
2. Select the *Place Line String* tool. Set your active color to 1 (blue).
3. From the Tool Settings palette, click on the check button for Non-planar. This will allow you to create Non-planar linestrings.
4. In View 1, start to create a linestring by placing three segments of the linestring without pressing Reset to end the command. Move the cursor into View 3 and notice that you can still see the line dynamically dragging from the cursor. In View 3, place three more segments. Then move to View 4 to place three more segments before pressing Reset to end the command. Notice that you can see in all of the views that the linestring exists in more than one plane.
5. If you were watching carefully as you created the linestring in step 4, you might have noticed that as you started to create line segments in Views 3 and 4, the active depth for those views defined what depth the line segments were created at. Try repeating step 3 and watching how each of the view's active depths controls the depth at which the line segments are created.

✔ **TIP:** *The active depth for a view can be set while in the middle of a command. Just hit Reset after you have set the active depth and you will return to where you were in the command. It works like the view controls.*

Exercise 3: Place Point Curve

The Place Point Curve icon.

As with the linestring, the *Point Curve* can be either planar or Non-planar. The point curve can be a lot of fun to create, but is not used that often in 3D. Its highly evolved cousin, the B-spline, is more mathematically correct, and therefore used more in 3D. This exercise is similar to the previous exercise, except that you will place point curves instead of linestrings. There are two parts to this exercise.

Part 1

1. Open file **Inside3D → Chapter 4 → Place Point Curve.**
2. Select the *Place Point Curve* tool. Set your active color to 2 (green).
3. Make sure that the check button on the Tool Setting palette is *not* set for Non-planar.
4. In View 1, to the right of the cube, start placing several points for the point curve as shown in the following figure. Also, notice how it appears in the other views.

Your point curve should look something like this.

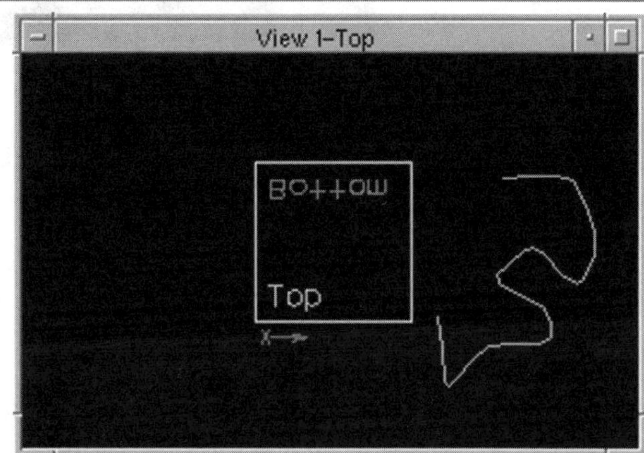

5. Now you are going to change the active depth for View 1. From the 3D View Control subpalette, select the *Set Active Depth* tool. If you can't find the 3D View Control subpalette, refer to step 2 of the Place Line exercise.
6. With a data point, select View 1 as the view for which you want to change the active depth. Move your cursor into View 3 and data point approximately in the middle of the cube.
7. Set your active color to 1 (blue).
8. Select the *Place Point Curve* tool. In View 1, to the left of the cube, place several points for the point curve as shown in the following figure.

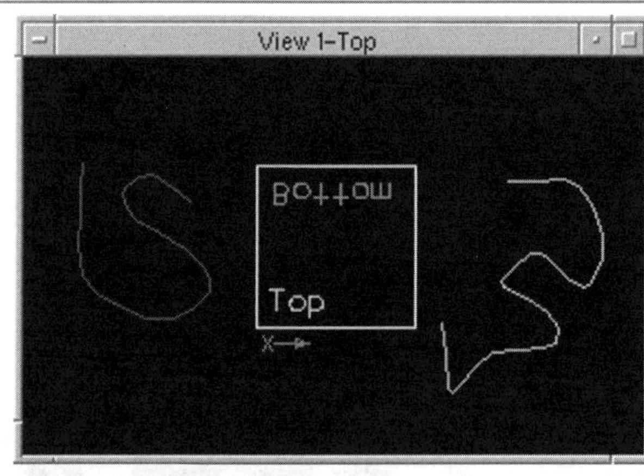

Your point curve should look something like this.

Part 2

In part 2, you will create a Non-planar point curve. To do this, you will have to start your point curve in View 1 and finish it in View 3.

1. Open file *Inside3DChapter 4Non-Planar Point Curve.*
2. Select the *Place Point Curve* tool. Set your active color to 1 (blue).
3. From the Tool Settings palette, turn on the check button for Non-planar.
4. In View 1, to the right of the cube, place several points for the point curve without pressing Reset to end the command.
5. Move your cursor into View 3. Notice how the point curve follows your cursor. This is a Non-planar point curve.
6. In View 3, place several more points to define the point curve. Notice how the points are created on the active depth for View 3.
7. Look at View 1 and move your cursor left and right. You can see how the point curve is being created on the active depth of View 3.
8. Press Reset to end the command.

Exercise 4: Place Block

The Place Block icon.

Next on your guided tour is the *Place Block* command. Let's look at the block element and see what is special about it in 3D space. For one thing, the block is a planar element (that means it is flat, but you know that by now, right?). As in 2D, the block is a closed shape. Being a closed shape in 3D makes it a surface. A surface, when rendered, appears to be solid and hides whatever is behind it. In this exercise, you will create blocks and practice setting the active depth.

1. Open file **Inside3D → Chapter 4 → Place Block**.
2. Select the *Place Block* tool from the Polygon subpalette. Set your active color to 2 (green).
3. Select the *Set Active Depth* tool. Data point in View 1 to select it as the view for which you want to set the active depth. This time you are going to set the active depth from View 2 (the isometric view). Move your cursor into View 2 and you will see the display depth and the active depth dynamically moving up and down inside of the display depth.
4. Tentative point on any part of the top of the cube. Verify that you have snapped to the top of the cube by noticing the highlighted color of the block on top of the cube. Look at Views 3 and 4 and you will see where the active depth is. Press the data point once you are sure that you have the top of the cube.
5. In View 1, place a block on the right side of the cube as shown in the following figure.

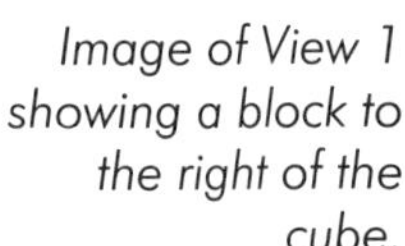

Image of View 1 showing a block to the right of the cube.

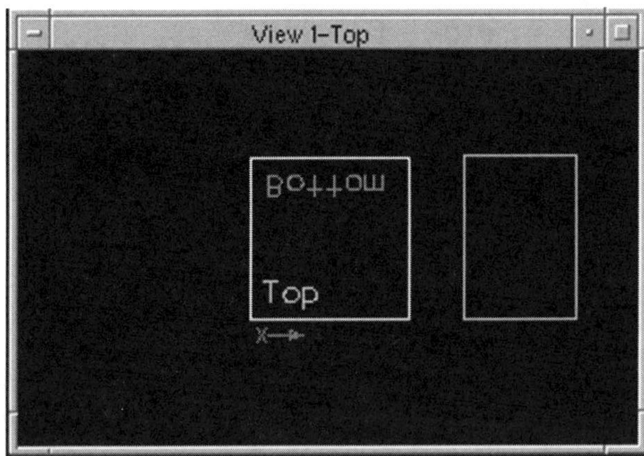

6. You are now going to change the active depth for View 3. Select the *Set Active Depth* tool and data point in View 3. Move your cursor into View 1 and notice the active depth dynamically moving up and down. Tentative point in the middle of one of the sides of the cube to set the active depth for View 3.

✔ **TIP:** *After you become comfortable with the Set Active Depth command, it's always a safe practice to check the active depth once you have set it. You can do this by placing a line or block and looking at the other views to see if it is at the correct depth, or use the Precision Input dialog box to check your value.*

7. Place a block in View 3 just to the right of the cube as shown in the following figure.

View 3 should look like this.

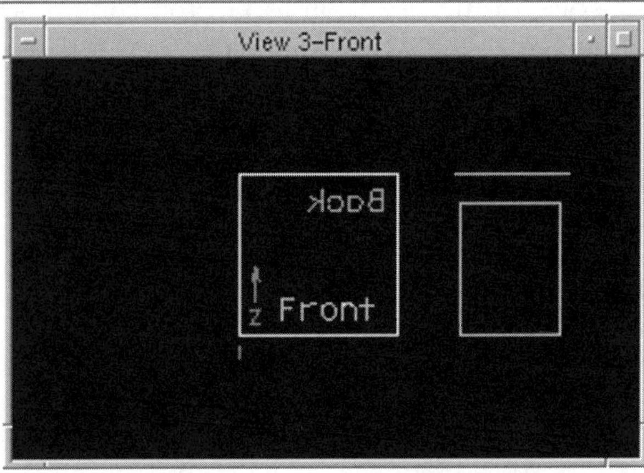

Your isometric view (View 2) should show both of the blocks that you have created. View 2 should look like the following figure.

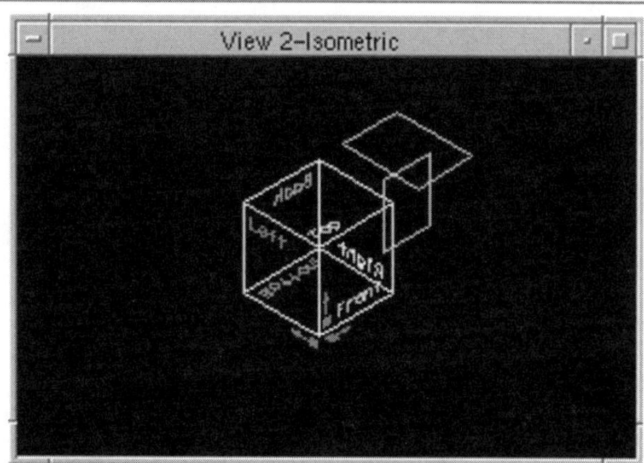

Your isometric view (View 2) should look like this.

Exercise 5: Place Shape

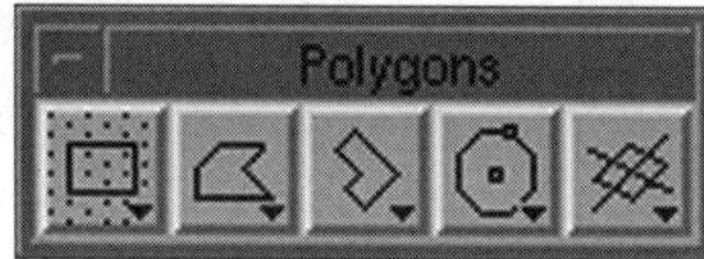

The Place Shape icon.

The *Place Shape* element, as you know, is a close cousin of the *Place Block* element. It, too, is a planar element and a closed shape. In this exercise you will practice some of your newly acquired skills from Chapter 2. You are going to place shapes using precision inputs. You will begin with absolute coordinates (XY=), and then complete the shape with relative (DL=) keyins.

1. Open file **Inside3D → Chapter 4 → Place Shape.**
2. Select the *Place Shape* tool. Set your active color to 7 (cyan).
3. Key in the following inputs and look at each view and see how the shape is created.

NOTE: *The* DL= *input is relative to the last point defined. So if you make a typing error in the middle of the following inputs, start over.*

By selecting Undo, you will only undo the last line segment you created, not the point that it defined. Trust me, it's easier this way.

```
XY=1
DL=1
DL=,,.75
DL=-.5
DL=,,-.5
DL=-.25
DL=,,-.25
```

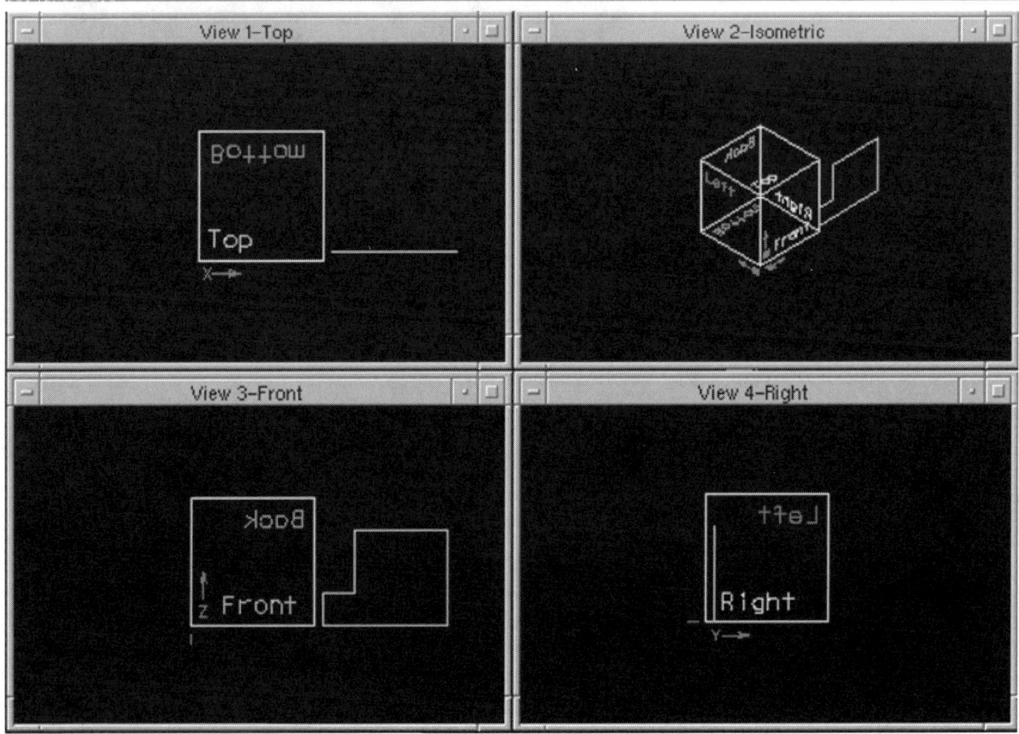

Your shape should look like this in all four views.

Review the keyins you input. Try to visualize what each one will do before you key it in. I can't emphasize enough how important it is to develop a mental image of what it is you are creating in 3D space. It's a major concern that all the members of our expedition are capable of navigating in the

3D world. Practice and hard work are the only ways to become proficient. Enough with the pep talk; let's move on.

Exercise 6: Place Circle

The Place Circle icon.

Well, explorer, so far you have met the challenge. Granted, these 2D elements should not be completely foreign to you. Each of these exercises you complete will help you to master this brave new 3D world. In Chapter 2 you were introduced to the lock setting, the *Depth Lock*. This setting can be of great assistance as you work in 3D. Allow me to refresh your memory on what this lock setting does. The depth lock locks the creation of geometry to the active depth, regardless of whether you snap to other geometry at differing depths to create it. Also, you will set the active depth by using a precision keyin.

1. Open file **Inside3D → Chapter 4 → Depth Lock.**
2. Select the *Set Active Depth* tool. Data point in View 1 to identify it as the view you want change the active depth for.
3. This time you are going to set the active depth for View 1 by precision keyin. Key in the following:
   `XY=,,1.5`
4. Set your active color to 1 (blue).
5. Select the *Place Circle By Center* tool. Set the radius to `.75`.
6. In View 1, tentative point on the cube until you see the top of the cube highlight in View 2 (the isometric view). Then data point to accept. Notice that the circle was not created on the active depth for View 1. The placement was overridden when you snapped to an existing element.
7. Open the Lock Toggles palette **(Settings → Locks → Toggles)**. Notice the Depth Lock check button at the bottom. Turn the Depth Lock setting on.

8. Select the *Place Circle By Center* tool. Now you are going to place the circle with Depth Lock on.
9. In View 1, tentative point on the cube until you see the top of the cube highlight in View 2 (the isometric view). Then data point to accept. This time the circle was locked to the active depth by the Depth Lock.
10. Notice the location of the circle in the other views. With Depth Lock on, it does not matter what you snap to. The element will always be created on the active depth.

Element Manipulation

In this part of the chapter, we will discuss and practice how to manipulate your old 2D friends in 3D space. The commands will be familiar to you, but how they function in 3D may surprise you. As your guide, I suggest you stay close and don't stray too far from the trail. If you are observant, you will discover new ways to manipulate 2D elements.

Exercise 7: Copy/Move Element

The Place Block icon.

Since you are not a novice to MicroStation, and you know how to use the *Copy* and *Move Element* commands in 2D space, I will combine these two commands into one exercise. So from this point on, I will refer only to the *Copy* command, so as not to waste valuable time with two identical exercises. In the following exercise, you will copy a block by using arbitrary placement and precision input.

1. Open **Inside3D → Chapter 4 → Copy/Move Element.**

➾ **NOTE:** *This file contains geometry that you are going to manipulate in the following exercise. The geometry is on preset levels. You will be instructed to turn certain levels on or off to display the necessary geometry for a particular exercise.*

2. Select the *Copy Element* tool. Notice that a blue block is already displayed in the file.
3. In View 1, tentative and data point on the top right corner of the blue block.
4. You are going to copy the block two units in the Z-axis. Key in the following input:

   `DL=,,2`
5. Press Reset to end the *Copy* command. Notice the location of the new block in the other views.
6. Now you are going to make a copy of the new block. This time you are going to work in View 2. Tentative and data point to the upper right corner of the new block in View 2. (The upper right corner is the corner that is closest to the top of the screen.)
7. This time you are going to copy the block two units in the Y-axis. Key in the following:

   `DL=,2`
8. Press Reset to end the command. Again, look at the other views to see the results.

When doing an exercise like this, it's easy just to read the exercise and key in the numbers, without really understanding what you have done. Please take a moment (if necessary) and study the views and digest the concepts.

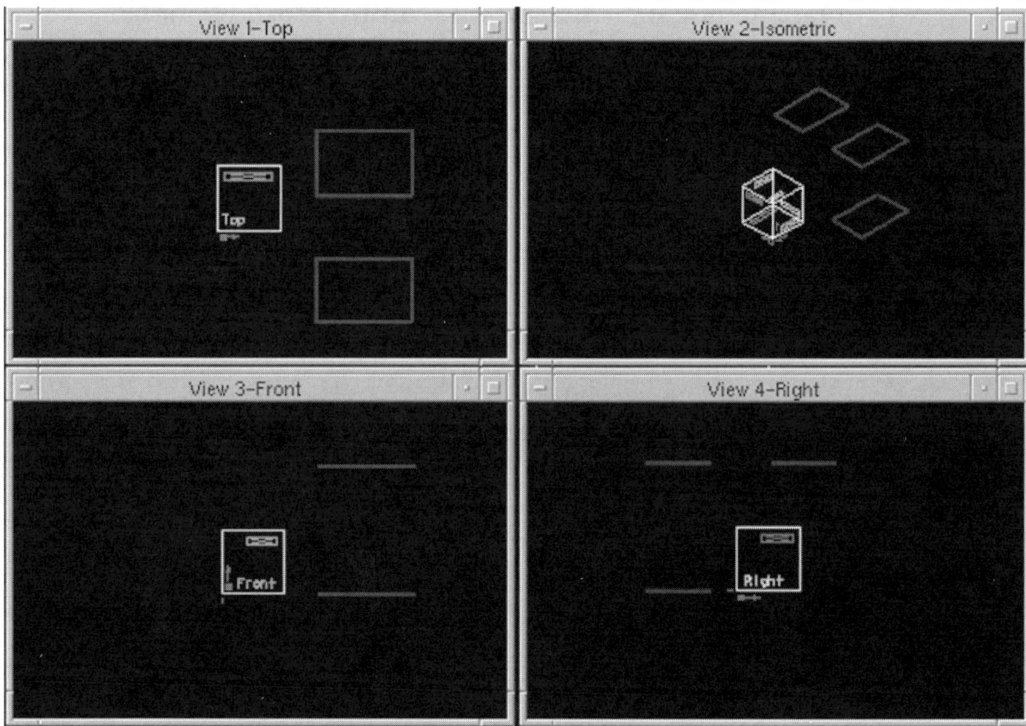

Your four views of the copied blocks should look like this.

9. Now you are going to use two views again to copy and place an element. But this time you are going to turn on the Axis Lock to control the placement of the copy. Turn on Axis Lock from the Toggles palette.
10. In View 1, data point on any part of the last block you created.
11. Move your cursor to the left of the cube in View 1 and data point to place the block. Press Reset to end the command.

Hang in there, traveler. One last lesson with the block. This time I want you to use the `DX=` keyin to copy the block in View 1.

12. Select the *Copy Element* tool. In View 1, data point on the last block you created (that would be the uppermost left block in View 1). Key in the following:

    `DX=,-2`
13. Press Reset to end the command.

➾ **NOTE:** *Remember that the X, Y, Z of the DX= keyin is relative to the view axes and not the design file axes. Refer to Chapter 2 for further review.*

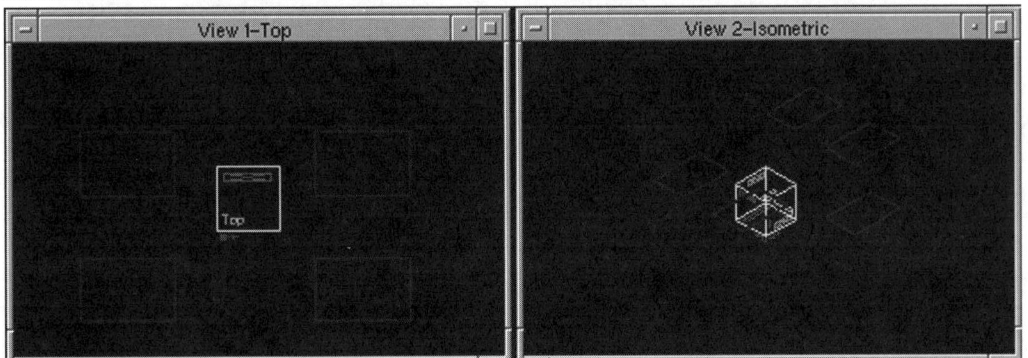

Your Views 1 and 2 with the blocks copied using the DX= keyin should look like this.

14. That about covers the exercises for the *Copy Element* command.

Exercise 8: Copy/Move Parallel

The Copy/Move Parallel icon

This single tool, through the use of the Tool Settings popdown, is actually four different commands. We sure got our money's worth on this tool. Again, I will combine the *Copy* and *Move Parallel* commands into one exercise. I will only refer to the *Copy Parallel* command from this point on. These commands work the same in 3D as they did in 2D, with the exception that you will be able to copy some elements in two planes and not just one. The planar elements such as circles, blocks, and shapes can only be copied parallel on the plane they exist on, whereas elements like lines and non-planar linestrings can be copied in two axes. In the following exercises you will *Copy Parallel* planar and non-planar elements.

1. Open file **Inside3D → Chapter 4 → Copy/Move Parallel.**

2. Turn on level 11 for all views.
3. Select the *Copy Parallel* tool. From the Tool Setting popdown window, turn on the check button for Distance and Make Copy. You are now in the *Copy Parallel by Keyin* command. Key in the following distance: `.25`
4. In the file you will see two lines, two blocks, one circle, and the cube.
5. In View 1, select the line to the right of the cube.
6. Move your cursor to the left and right of the selected line and notice a copy appearing. So far, it's just like 2D, right?
7. Now move your cursor into View 3 and notice the difference. See how the copy of the line is now moving in the axes. Lines are not planar in regard to the direction you can copy them.
8. In View 3, data point below the selected line and press Reset.
9. Next, in View 3, select the circle with a data point. Move your cursor outside of the selected circle and data point to create the copy. Press Reset to end the command.
10. Notice how the copy of the circle is confined to the plane the original circle was created on.
11. In View 1, select the block that is to the right of the cube.
12. Move your cursor into View 2. Move the cursor inside the selected block and data point to create the copy. Reset to end the command.

Your View 2 should look like this.

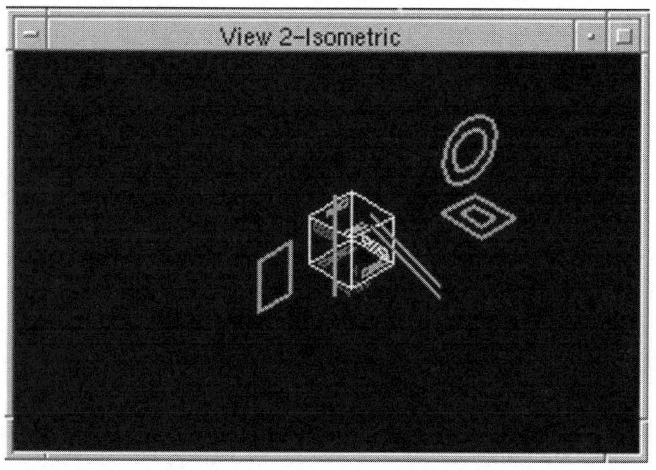

13. Take some time to experiment on your own with the other elements in the file.

Exercise 9: Scale Element

The Scale Element icon.

The *Scale Element* command, like the other 2D commands we have looked at, works much the same in 3D as it does in 2D. But, beware; the combined effects of the active depth and display depth can produce unexpected results. When you selected an element to scale in 2D, the point by which you selected the element becomes the point from which that element is scaled. When in 3D, this operation takes on a new dimension (no pun intended). Your cursor works from the active depth. So, if the active depth is above or below the element you selected, it is as if you have moved the cursor away from the selected element. This *may* cause the element to move out of the display depth. This is the unexpected result.

If the element moves outside of the display depth, it will not be visible on your screen. You will get to experience this in one of the following exercises. The thing to keep in mind is that the cursor can be moved away from the element not only in the X- and Y-axes, but also the Z-axis. This exercise will be in two parts. In the first part, you will scale and copy an element. In the second part you will see how the combination of the active depth and display depth affect this command.

Part 1

1. Open file **Inside3D → Chapter 4 → Scale Element**.
2. Turn on level 12 for all views. You will see the word "scale" appear as text.
3. Select the *Scale Element* tool and turn on the Make Copy check button.
4. Change the active scale for all three axes to 2.

Your Scale Element (Copy) Tool Palette should look like this.

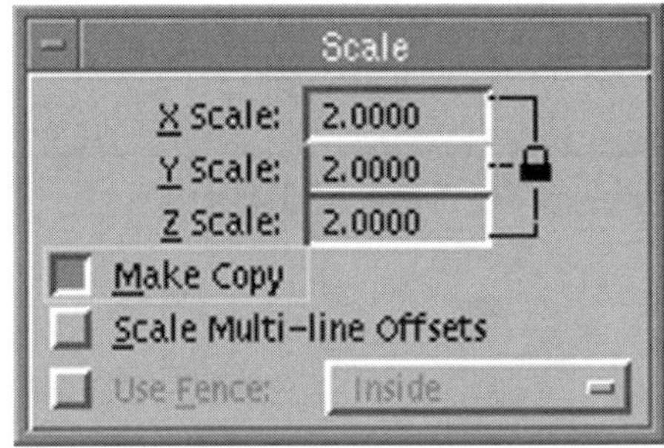

5. In View 1, select the text element using a data point.
6. Move your cursor around in View 1 and notice how the text moves in the other views. It is moving around on a plane.
7. Now move the cursor into View 3. This is where the third axis comes into play.
8. Move the cursor over the text as a starting point. Now move the cursor above or below the text. Notice that the text is moving up and down on the Z-axis.
9. Move the cursor in View 3 below the selected text so that the new scaled text is about even with the top of the cube to the left. Press a data point and then Reset.

Part 2

In this exercise, you will open an existing file. This file has a preset display depth setting that is different from the files you have been working on. You had better bring a snack along; this could be a long exercise.

1. Open **Inside 3D → Chapter 4 → Scale Elements → Part 2.**
2. Select the *Set Display Depth* tool from the 3D View Control subpalette.
3. Select View 1. Look at Views 2 and 4 and see the dynamic display of the current display depth for View 1. Notice how the front clipping plane is at the top of the cube and the back clipping plane is at the bottom of the cube.

4. Press Reset now. We didn't want you to change the display depth, just to look at it.
5. Select the *Scale Element* tool. As you will notice, I have ventured ahead and set the active scale to 2 for you. This is what a guide gets paid for.
6. In View 1, select the word "Scale."
7. Move your cursor into View 3, just below the selected text. Notice in View 1 how the text is visible. It is still inside the display depth for View 1.
8. Now move your cursor further below the selected text. Watch in View 1 as the new scaled text moves out of the display depth. Without Safari Sam here to explain these strange goings on, one might become bewildered and abandon the quest.
9. In View 3, try moving your cursor above the selected text. The same effect will occur in View 3. Again, the new scaled text moves outside of the display depth and is not visible in View 1.
10. Press Rest now.
11. Select the *Set Active Depth* tool. Select View 1 and move your cursor into View 3. Tentative/data point to the middle of the cube. To help you visualize where the active depth is for View 1, reselect the *Set Active Depth* tool. Select View 1 again and notice the location of the new active depth relative to the display depth in Views 2 and 4. Press Reset now.

NOTE: *The word "scale" no longer lies on the active depth for View 1. The importance of this fact will become obvious in the next step. So pay close attention.*

12. Select the *Scale Element* tool.
13. In View 1, select the word "scale." Notice that in View 1 the new scaled text does not appear. If you look in View 3, you will notice that the new scaled text appears. This is where the active depth for View 1 comes into play. Carefully look at View 3 and notice the distance between the new scaled text and the selected text. The distance between the new scaled text and the selected text is the same as the distance between the selected text and the active

depth. You may want to find a comfortable spot on the trail and sit down for awhile. Think about these concepts until they sink in.

14. Press Reset now.

Exercise 10: Rotate Element

The Rotate Element icon.

When rotating 2D elements in 3D, you need to know which axes you are rotating them on. For example, if you select a 2D element that lies on the X,Y plane, then you are rotating it about the Z-axis. When rotating elements, it is wise to do it in an orthographic view. The Top view is an example of a standard orthographic view. Avoid trying to rotate elements from isometric or custom views. This will rotate the element relative to the view, and not the design. In the following figure, notice the block that exists on the X,Y plane. It was rotated from View 1 (Top) about the Z-axis.

The dashed line represents the sweep of the rotation.

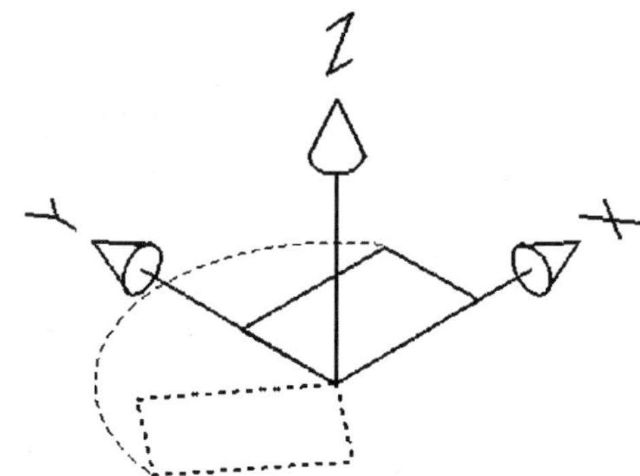

1. Open file **Inside3D → Chapter 4 → Rotate Element.**
2. Turn on level 10 for all views.
3. Select the *Rotate Element* tool.
4. In View 1, select the block. Note that the block exists on the X,Y plane.

5. In View 1, define the pivot point by selecting a tentative/data point on the upper left corner of the block.
6. Move your cursor around in View 1 and see how it spins about the pivot point. You are now rotating the block around the Z-axis.
7. Press Reset, and you are again prompted to select the pivot point.
8. This time move your cursor into View 3. Tentative/data point on the left end of the block as it appears on the edge in this view.
9. In View 3, move the cursor around the pivot point. Notice in View 2 how the Y-axis is the axis of rotation.
10. Press Reset twice.

Exercise 11: Mirror Element

The Mirror Element icon.

The *Mirror Element* command is another command that, as long you stay in one view, works just as it did in 2D. Once you move your cursor into a view that has a different plane, the command functions differently. As you know, there three ways to *Mirror Element.* There is horizontal, vertical, and about a line. Let's just look at horizontal for a moment. What is horizontal to an element in the top view is not the same in the front view. We will investigate this concept in the following exercise.

1. Open file **Inside3D → Chapter 4 → Mirror Element.**
2. Turn on level 10 for all views.
3. Select the *Mirror Element* tool.
4. Turn on the Make Copy check button and select Mirror About Horizontal from the option menu.
5. In View 1, select the block anywhere on the top half. Moving your cursor around, you will notice that this functions the same as it does in 2D.

6. Now move your cursor into View 3. You will see that it is now horizontal to the element as that element is seen from the front or X- and Z-axes.
7. With your cursor in View 3, tentative/data point in the middle of the cube. Press Reset.
8. Change the Mirror About option to vertical.
9. In View 1, select the block anywhere on the top half.
10. In View 1, move your cursor around to see where vertical is relative to the block.
11. Now move your cursor into View 4. It now dynamically displays the new block, but vertical to the element in View 4.
12. In View 4, move your cursor over the lower left corner of the cube and press a data point and then Reset. In View 1, you will see that the new block appears above the original block. You now see that vertical and horizontal are relative to the view and not just the element.

Exercise 12: Array Element

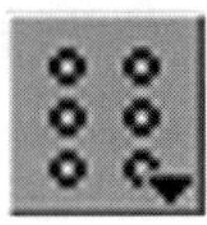

The Array Element icon.

The *Array Element* command is like the *Mirror Element* command in that it too is relative to the view you are working in. The *Array Element* command works with rows and columns. These rows and columns are relative to the view, that is, rows are horizontal and columns are vertical.

1. Open file **Inside3D → Chapter 4 → Array Element.**
2. Turn on level 13 for all views.
3. Select the *Array* tool.
4. Set the Array options to match those of the following figure.

Your Array options should match these.

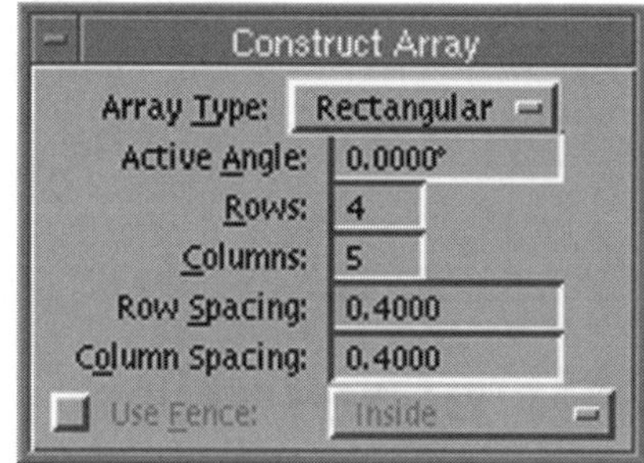

5. In View 1, select the circle as the element to array and accept. You can see how the array was done relative to View 1.
6. In View 3, from the edge view, select the last circle on the right and accept it as the element to array. You can see how the *Array* command works relative to the view that you are working in.

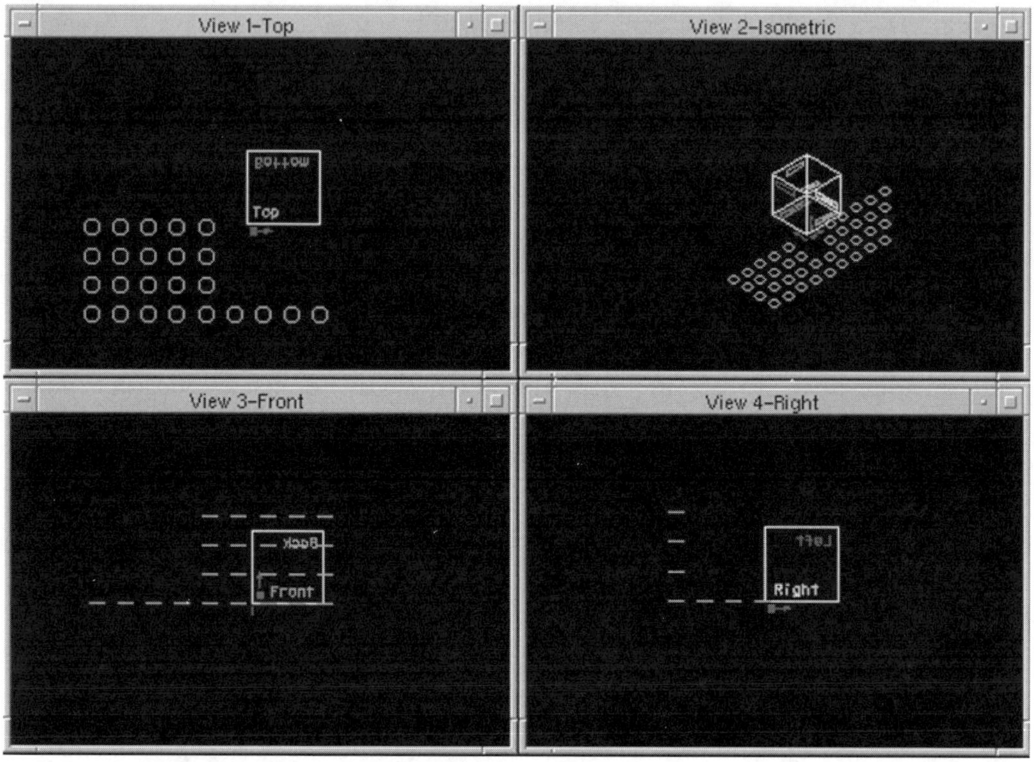

All of your views should look like this after applying the Array command.

✔ **TIP:** *When selecting an element from an edge view, it is always handy to have an isometric view open. This way you can see the element highlight.*

Fence Manipulation

Now, if you are the curious type, you may have been wondering how fence operations differ in 3D. For the most part, they don't change at all. That's not to say that there aren't some things to understand about using fences in 3D.

❒ A fence has no depth constraints. If the fence mode is set to Overlap, and the fence crosses an element, regardless of that element's depth, it will be selected by the fence.

❒ The fence, if set to Inside or Void mode, will only select elements that are totally in the display depth.

❒ If you place a fence, and then in that same view zoom in so that the fence is not visible, the fence will be turned off.

Element Modification

Well, traveler, how are you feeling? When I say "active depth," you don't break into a cold sweat, do you? This term and others will become a second language to you by the end of this journey. Now it's time to see how some of the modify element commands work in 3D. The common theme in this chapter has been "remember that you have three axes." This section is no different. Depending on how you use the modify commands in 3D, they can be just like 2D. All you need to do is be vigilant of which plane you are working in, and what is your active depth.

Exercise 13: Modify Element

The Modify Element icon.

The *Modify Element* command can do everything in 3D that it does in 2D and more. The *Modify* command is generally used to modify the vertex or radius of an element. In 2D you could only modify elements on the X,Y plane. In 3D, for example, you can modify the vertex of a planar block. By moving the vertex to a different depth, you have changed it into a non-planar block. In the following exercise, you will see how that can happen.

1. Open file **Inside3D → Chapter 4 → Modify Element.**
2. Turn on level 10 for all views.
3. You are now going to check the active depth for View 1. Select the *Set Active Depth* tool.
4. Select View 1 and observe the active depth dynamically displayed in the other views. Then press Reset to terminate the command.
5. First, you will modify the block that exists on the active depth.
6. In View 1, select the upper right corner of the lower block.
7. Then, move your cursor to the right, making the block into a rectangular shape. Then press Reset. The block remained planar because it exists on the active depth.
8. In View 1, select the lower right corner of the upper block. Notice how the vertex of the block is forced down to the active depth for this view.
9. In View 1, tentative/data point to the lower right corner of the lower block. Then press Reset to end the command. You just changed the block to a non-planar block. Take a moment to digest how the block was modified.

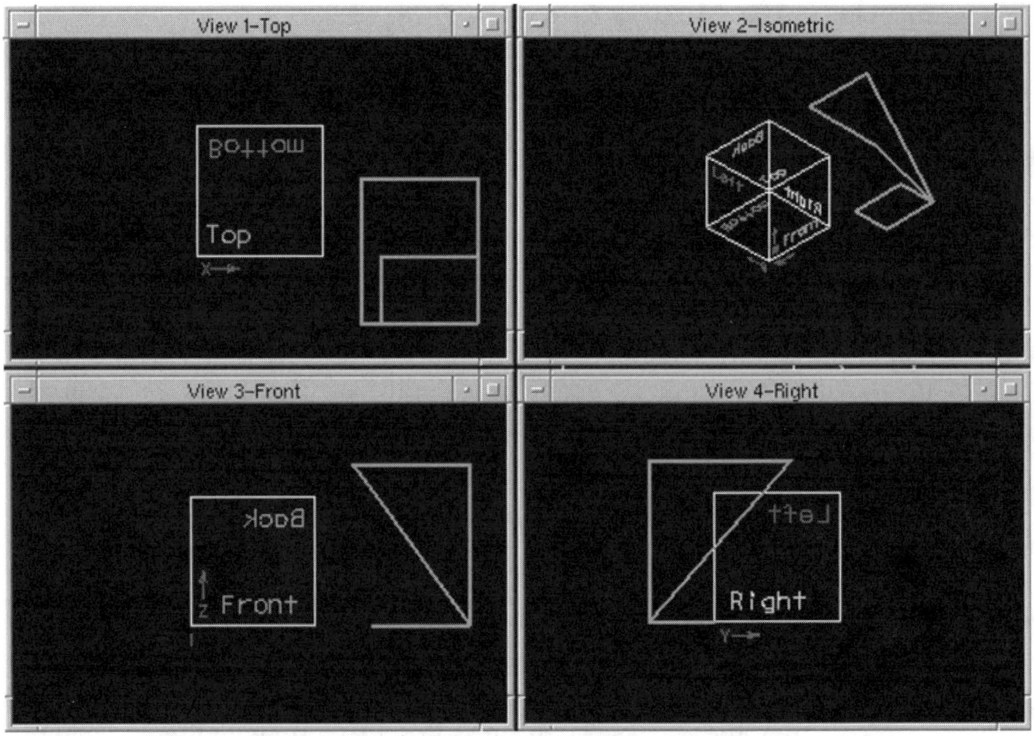

This is what your modified blocks should look like in all four views.

Exercise 14: Extend Line

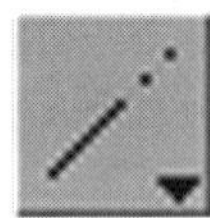

The Extend Line icon.

You may be surprised by how the *Extend Line* command works in 3D. The *Extend Line* command does not require that the line to be extended, and the point to extend to, be at the same depth. This feature makes this command simple to operate, as you will see in the following exercise. In this exercise you will extend a line to a point defined by another that is at a different depth.

1. Open file **Inside3D → Chapter 4 → Extend Line**.
2. Turn on level 11 for all views. Notice the lines and how they are at differing depths.
3. Select the *Extend Line* tool.
4. In View 1, select the vertical line.
5. In View 1, tentative/data point on the horizontal line. The vertical line was extended to the horizontal line, even though it was at a different depth. If I know you, and I think I do by now, you are already thinking of ways to use this command.

Exercise 15: Extend 2 Elements to Intersection

Extend 2 Elements to Intersection icon.

In the 2D world, the *Extend 2 Elements to Intersection* command would physically connect the selected elements. In this strange new world of 3D, things are not always as they look. When using this command in 3D, what appears to be an intersection in one view, may not be an intersection at all. To understand this command, you need to know that the intersection that is created is relative to the view, and not the elements. In the following exercise, you will see what I'm saying.

1. Open file **Inside3D → Chapter 4 → Extend 2 Elements to Intesect.**
2. Turn on level 12 for all views. Notice the elements and how they are at differing depths.
3. Select the *Extend 2 Elements to Intersection* tool.
4. In View 1, select the vertical line.
5. In View 1, select a point to the left of the arc on the horizontal line. Data point to initiate intersection. In View 1, it appears that an intersection was made between the two lines, but a look at View

2 tells another story. The intersection that was created was an intersection relative to View 1.

6. In View 1, select the arc. Then select the vertical line below the intersection of the arc. Data point to initiate intersection. Again, everything looks fine in View 1. View 2 shows us that these elements are not all at the same depth.
7. In View 1, complete the little pie wedge by selecting the arc and then the horizontal line. Data point to initiate intersection.

This exercise helps to illustrate that even if everything looks good in one view, another view can act as a reality check.

Summary

You have done well, traveler. You have created and modified planar and non-planar elements without hurting yourself or others. If you are sore between the ears, don't be alarmed. It's a shared feeling with many adventurers who have escaped from the flatlands of 2D. As you become acclimated to the altitude, you will see that it was well worth the effort. Now it's time to meet some of the natives, as we have a rendezvous with the 3D primitives.

Introduction

Three-dimensional primitives and surfaces are the basic building blocks of the 3D world. They form the desert plains, the mountains, and the valleys you will become familiar with on your adventure. Don't worry, no mountain is too high, no valley too deep, no desert too wide for you to conquer, not with Safari Sam along.

Most elementary shapes can be created using one or a combination of 3D primitives. The primitives are blocks, spheres, cones, cylinders, etc. Once created, they can be distorted or manipulated to represent most objects. Three-dimensional surfaces are used to create those objects that are a bit more difficult to climb over, such as rocky cliffs and slippery slopes. The four main types are surfaces of projection and revolution, tubular surfaces, and skin surfaces.

The flexible and easy methods of creation make the primitives and surfaces the core of most 3D applications. In architecture they are extremely useful in the design process. They are used for massing and the study of proportions, as well as quick renderings. In the mechanical discipline they are used as the base shapes for die casts and molds. By manipulation and Boolean operations (explained in Chapter 10) accurate, complicated mechanical designs can be created and rendered. The specific applications of these primitives will be explained in Chapters 14 through 16.

3D Primitives

The 3D Primitives palette *(Palettes→ 3D→ Primitives)* contains the tools used to create primitives.

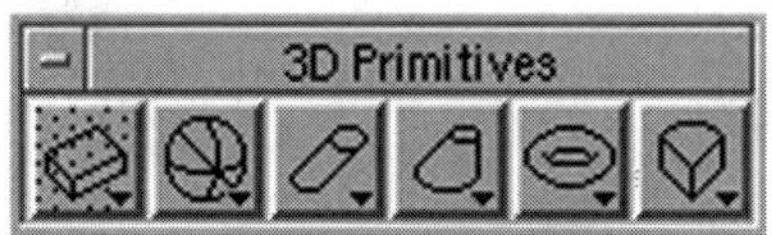

The 3D Primitives palette.

The Slab

The Slab icon.

The first primitive is the *slab*. This element is a true 3D cube with six sides. It is defined by an X-, Y-, and Z-value. These values are specific to the element, not the view or the design cube. The slab can be started in any view, that is, Top, Right, or Front. The first data point determines the starting view. A line will appear on the screen following the cursor. The second data point determines the length and rotation of the slab. The third data point defines width. The first three points are placed in the starting view. The last data point determines the height and is placed in a different view.

There are many useful options to help in placing slabs. These options appear on the 3D Primitives palette or the tool settings palette when the *Place Slab* command is activated.

The Place Slab palette.

The type option controls whether or not the element will be a surface (hollow) or solid (capped with a top and bottom). Orthogonal will force the sides of the slab to be at 90-degree angles. Axis constrains the direction the object is drawn. Screen X, Screen Y, and Screen Z set the element's axis in relationship with the screen. Drawing X, Drawing Y, and Drawing Z set the element's axis in relationship with the drawing. Points allow the element's axis to be placed randomly by precision input or data points.

> **NOTE:** *Use the axis lock when you wish to constrain the cursor. It allows you to constrain it in all three dimensions at once.*

The length, weight, and height can be constrained on the palette. When these options are activated, exact values can be input.

The Sphere

The Sphere icon.

The next primitive is the *sphere*. A sphere is a 3D volume with a constant radius and circular cross-section. It is the basic ball shape. A sphere is defined with two values, the center point and the radius.

The Place Sphere tool settings dialog box.

The radius can be input manually or by exact values. The method is controlled in the dialog area. In addition, the axis of the radius can be controlled in the same area. With the combination of these options, spheres can be created quickly and with great accuracy.

If the radius value is set in the palette, when the tool is activated and the center is defined, the sphere will appear on the screen, dynamically following the cursor. MicroStation is just waiting for you to define the axis of the sphere. When creating a sphere manually, place the center point with a data point. The sphere will display and dynamically update as the cursor moves around the screen. The axis control, on the palette, can be activated to restrain the movement of the cursor. At this point you can complete the sphere by placing a data point tentative to an existing element or using precision input.

Take a rest; you deserve it. Have a drink from your water bottle, but notice that the bottle is a cylinder. A cylinder is defined by two identical circles in parallel planes. The volume of the bottle is contained between the top and bottom circles that make the bottle.

The Cylinder

The Cylinder icon.

The *cylinder* has a number of useful setting that help in creation. The Type setting controls whether or not the cylinder is a surface (hollow) or solid (capped with a top and bottom). Axis determines the orientation. If the orthogonal setting is on, a right cylinder can be created. If off, the cylinder can be skewed. Radius and height can also be constrained with exact values.

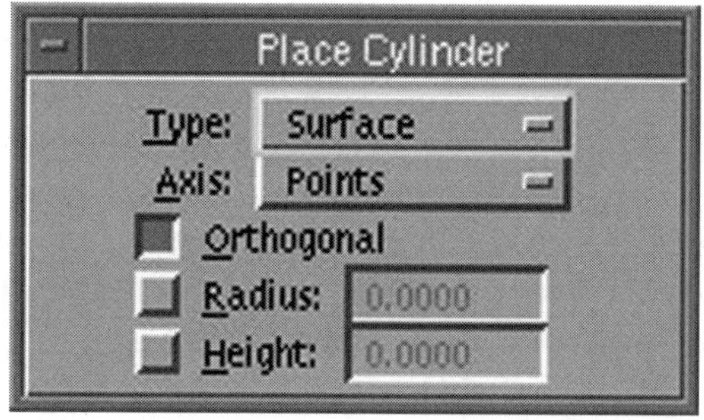

The Place Cylinder tool settings dialog box.

The Cone

The Cone icon.

A *cone* is basically a cylinder, but the top and bottom circles have different radii. The one addition to the tools of the cylinder is the separate constraints for the top and bottom of the cone.

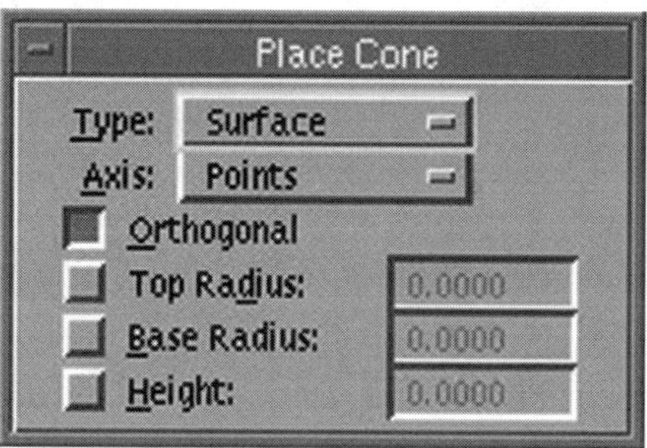

The Place Cone tool settings dialog box.

Exercise 1: Exploring 3D Terrain

How about getting your feet wet with an exercise? Now you can explore some real 3D terrain, slabs, spheres, and cylinders.

1. Open the **Inside3D** → **Chapter 5** → **3D Terrain** file.

NOTE: *You can see that there are the four following views open:*
View 1 = Top
View 2 = Isometric
View 3 = Front
View 4 = Right

Having these four particular views open and in this order is a good place to start. You notice that Views 1, 3, and 4 are orthographic from each other. This makes this view arrangement a natural one.

2. Set the color to yellow (`co=4`).
3. Activate the 3D primitives palette **(Palettes → 3D → Primitives)**. Select the *Place Slab* tool. Make sure type is set to solid and the orthogonal setting is activated.
4. Place the first data point at 0,0 using `xy=0,0,0` Place the second data point by inputting `DL=20,0,0`. Place the third point with `DL=0,20,0`. Place the fourth point with `DL=0,0,-2.5`.
5. Set active color to red (`co=3`).
6. Select the *Place Sphere* command from the 3D Primitives palette. Activate the Radius button and input 5 into the dialog area. Tentative point on the upper right corner of the slab, in the top view. You will now see the sphere floating on the screen. Tentative point on the upper left corner of the slab and data point to set the axis of the sphere. Now turn off the radius constraint. Move the cursor to the front view and tentative point to the lower left corner of the slab and data point. In the command window, type
`DL= 2.5,0,0.`

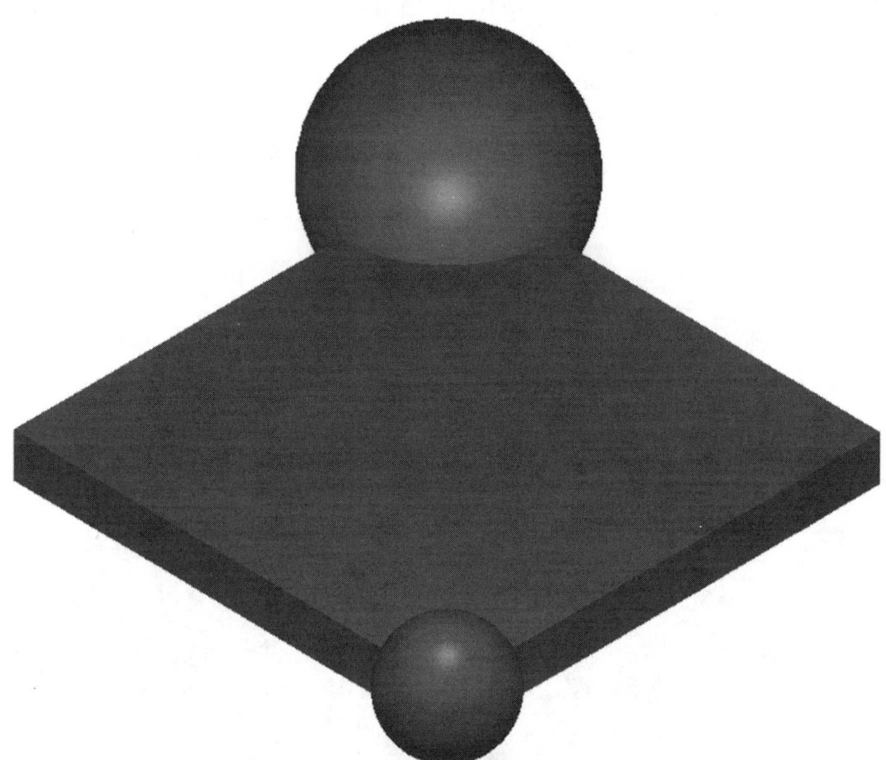

We now have two spheres at opposite ends of the slab.

This is much too simple for an adventurer like you. Tighten your boot laces and move out. I see some cylinders and cones ahead.

7. Set the active color to cyan (`co=7`).
8. Select the *Place Cylinder* command from the 3D Primitives palette. Turn the orthogonal setting off and move the active depth of the top view below the spheres. Activate the radius setting and input `2`. In the top view select a data point in the center of the slab to start the cylinder, then place a data point up and to the right of the first. This sets the axis for the radius. Move the cursor into the front view and define the height of the cylinder by placing a data point at the top of the view.
9. Set the active color to orange (`co=6`).

10. Activate the *Place Cone* tool. In the top view, place the first data point inside the slab, near the left side. Place the second data point just to the left, outside the slab, to define the primary radius. Move the cursor to the right view and place the third data point at the top. The fourth data point defines the secondary radius, and it should be fairly small.

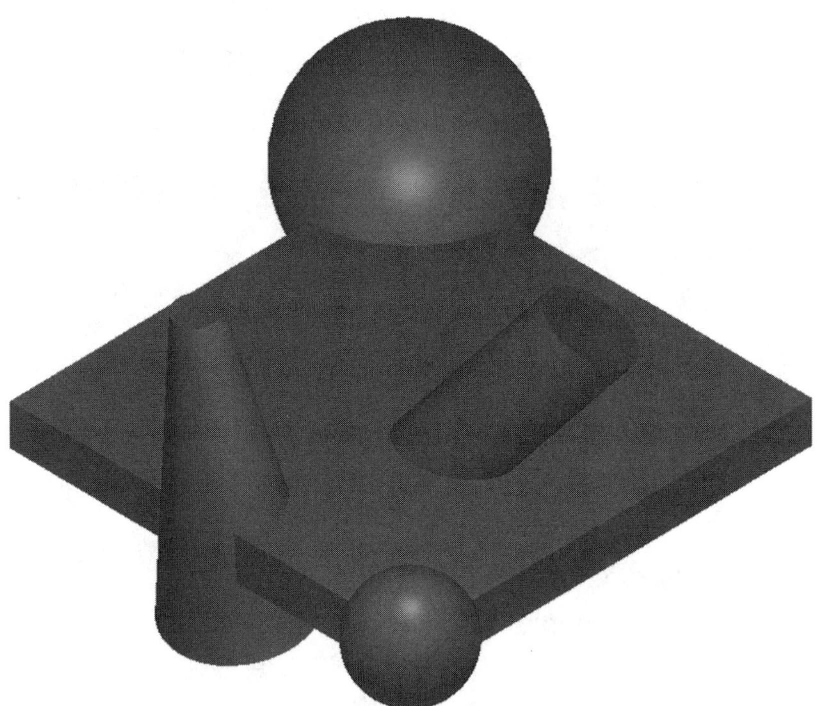

Iso view of slab with spheres, cylinder, and cone.

The Torus

The Torus icon.

By now you must be hungry; how about a big chocolate donut? Well, the chocolate might be difficult, but the donut is easy. In the world of 3D

MicroStation, a donut is referred to as a *torus*. It is defined by a primary radius, a secondary radius, and a projection angle.

The Place Torus tool settings dialog box.

The first data point determines the center of the torus. Once placed, a dashed line appears, following the cursor. The second data point determines the origin of the torus and the start angle. The third point determines the diameter of the torus and the final angle.

The Wedge

The Wedge icon.

The *Wedge* is the last of the 3D primitives. A wedge is a surface of revolution made with a rectangle. The first data point determines the outer edge of the wedge. The second point determines the center. The third point defines the angle, and the fourth defines the height.

Place Wedge
Type: Surface
Axis: Points
Radius: 0.0000
Angle: 0.0000
Height: 0.0000

The Place Wedge tool settings dialog box.

Exercise 2: The Torus and the Wedge

Why don't we spend a moment practicing placing a torus and a wedge?

1. Open the **Inside3D → Chapter 5 → Torus and Wedge** file.
2. Set the color to yellow (`co=4`).
3. Activate the 3D primitives palette **(Palettes → 3D → Primitives)**. Select the *Place Torus* tool. Make sure type is set to solid and the axis setting is on points.
4. Place the start point of the torus at the lower center of the top view.
5. Input `DL=,0.25`; this defines the center point of the torus.
6. As the cursor is moved around the screen, the diameter and angle change. Placing a data point will place the torus that is displayed; however, do not do that at this time.
7. In the tool settings palette, set the angle to 270 degrees.
8. The angle has now been constrained, but the diameter still follows the cursor.
9. Now set the secondary radius in the tool settings palette to 0.1 and press the Tab key.
10. The only effect the cursor has now is to mirror the torus.
11. Move the cursor to the right and place a data point.
12. Perform a fit in all four views.
13. Now it is time to have some fun with wedges.

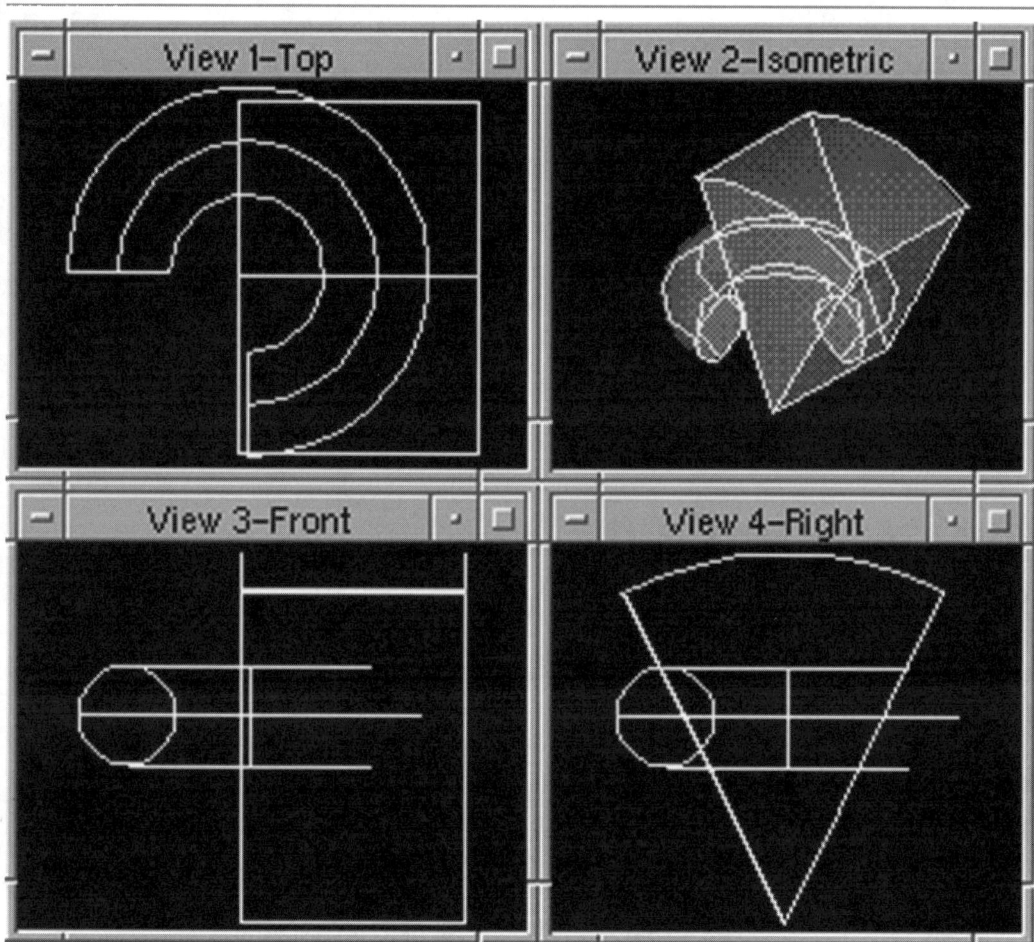

Top, Front, Right, and Iso views of torus and wedge.

14. Set the color to yellow (`co=4`).
15. Select the *Place Wedge* tool. Make sure type is set to solid and the axis setting is on points.
16. Set the Radius to 0.75 in the tool settings palette.
17. MicroStation now wants you to define the starting point of the wedge. Place a data point in the upper right corner of View 4. The

radius is constrained but the dashed line still follows the cursor, waiting for you to define the center of the wedge. Move the cursor to the lower center of the view and place a data point.

18. A wedge is now displayed. It updates as you move the cursor. Now move the cursor to the left until a roughly 30-degree wedge is displayed and place a data point.
19. To define the height, you must move to a different view. In the top view, move the cursor to the right and place a data point.
20. Now fit all views to view the torus and wedge. By doing a quick rendering of the view, you can really see the shapes well. Why don't you phong shade the Iso view **(View → Render → Phong)**?

Surface Creation

Well done; you have made it through the terrain of 3D primitives. Now things get exciting. Now we leave the stable, secure traveling behind us, and break out the ropes and climbing gear. Surface creation is where the fun and thrills are. If you can think of a surface, you can model it. But don't get overconfident; time and practice are required in this area. Stay close and don't fall. Off we go.

The 3D Free-Form Surfaces dialog box.

The Surface of Projection icon.

The first and most basic tool here is the *Surface of Projection* tool. This tool allows you to reuse elements from 2D drawings. This tool is also the next step for the elements discussed in Chapter 4. An element is projected or extruded, adding the third dimension. Boundary elements such as lines become plains, circles become cylinders, and plains become slabs. The original element remains and a new one is created.

Once again, the created element can be a surface (hollow) or solid (capped with a top and bottom). You will also notice your friend the orthogonal element is here. Some new options in the tool settings are Distance, Spin Angle, X-scale, and Y-scale. The Distance option is a dialog area to input the exact distance the boundary element is to be projected. Spin Angle allows the projected surface of the element to be rotated.

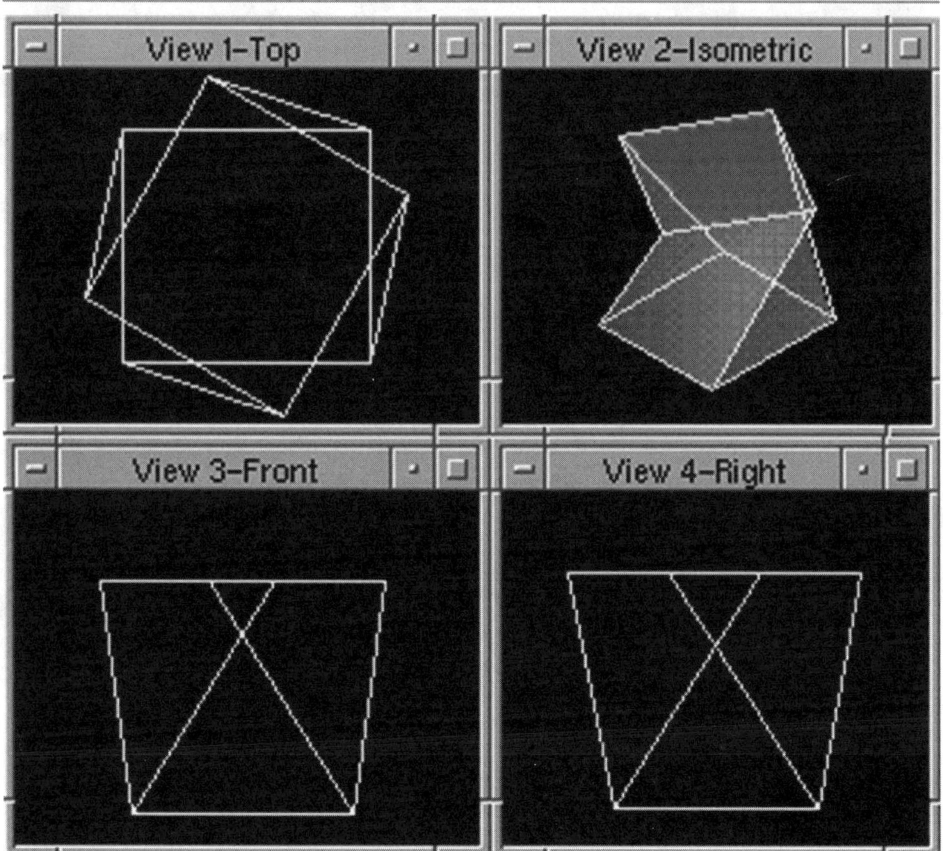

View of spin setting to surface of projection.

X-scale and Y-scale alter the scale of the projected surface.

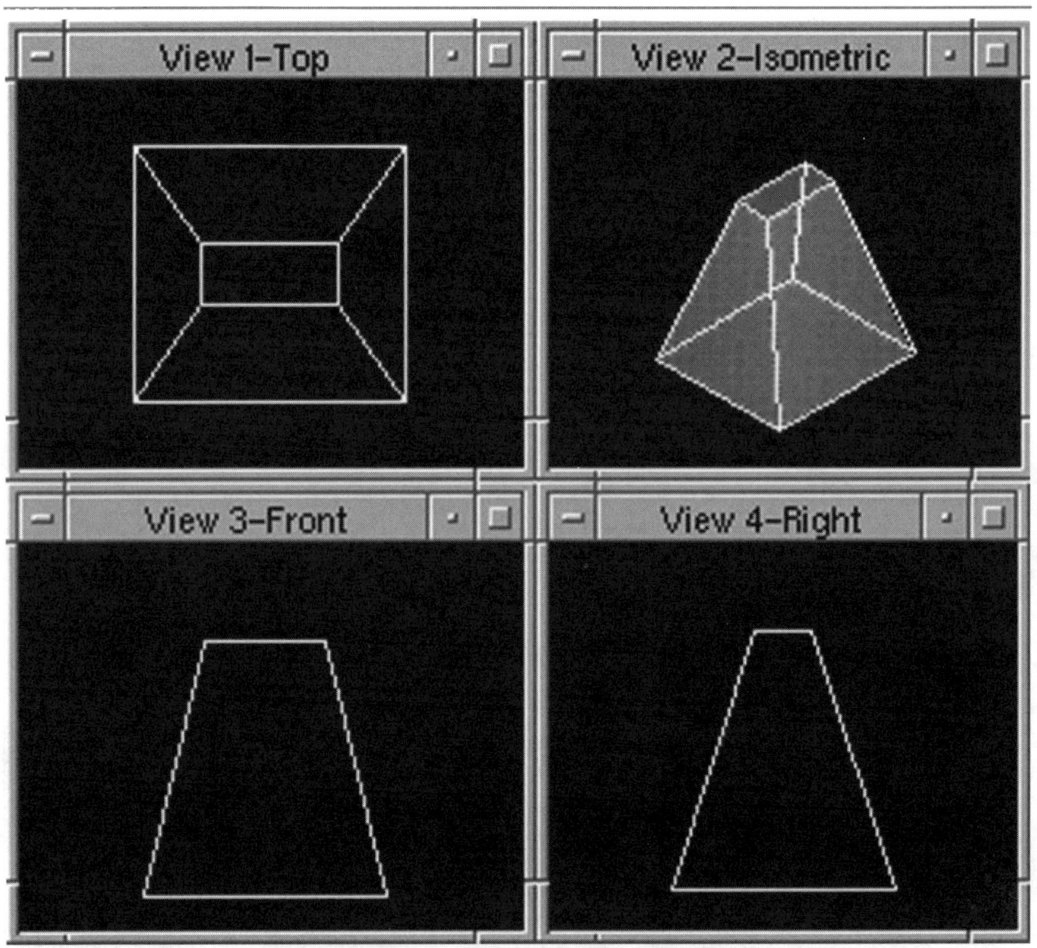

View of X,Y-scale setting for surface of projection.

These settings make this a powerful tool, but be careful—they can be tricky. Spend some extra time wandering about in this big green "settings" meadow.

The Surface of Revolution icon.

The next tool we encounter is *Surface of Revolution*. It is created by rotating a planer element about an axis of revolution. By using this tool on an arc or curved linestring, you can create a vase, wheel, or lamp stand.

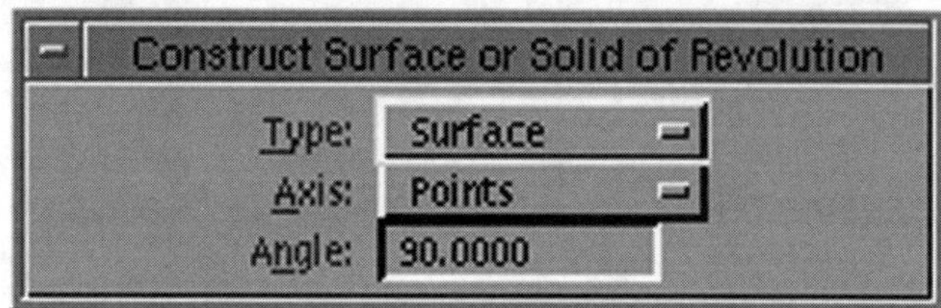

Surface of Revolution tool settings.

By now the tool settings should be like a comfortable broken-in pair of hiking boots. The only thing unique here is the angle option. It controls the angle the object is rotated. Ninety degrees rotates the element a quarter of a circle, 130 degrees rotates a third, 180 degrees rotates one half, and so on. Similar to *Surface of Projection*, the original element remains and a new one is created.

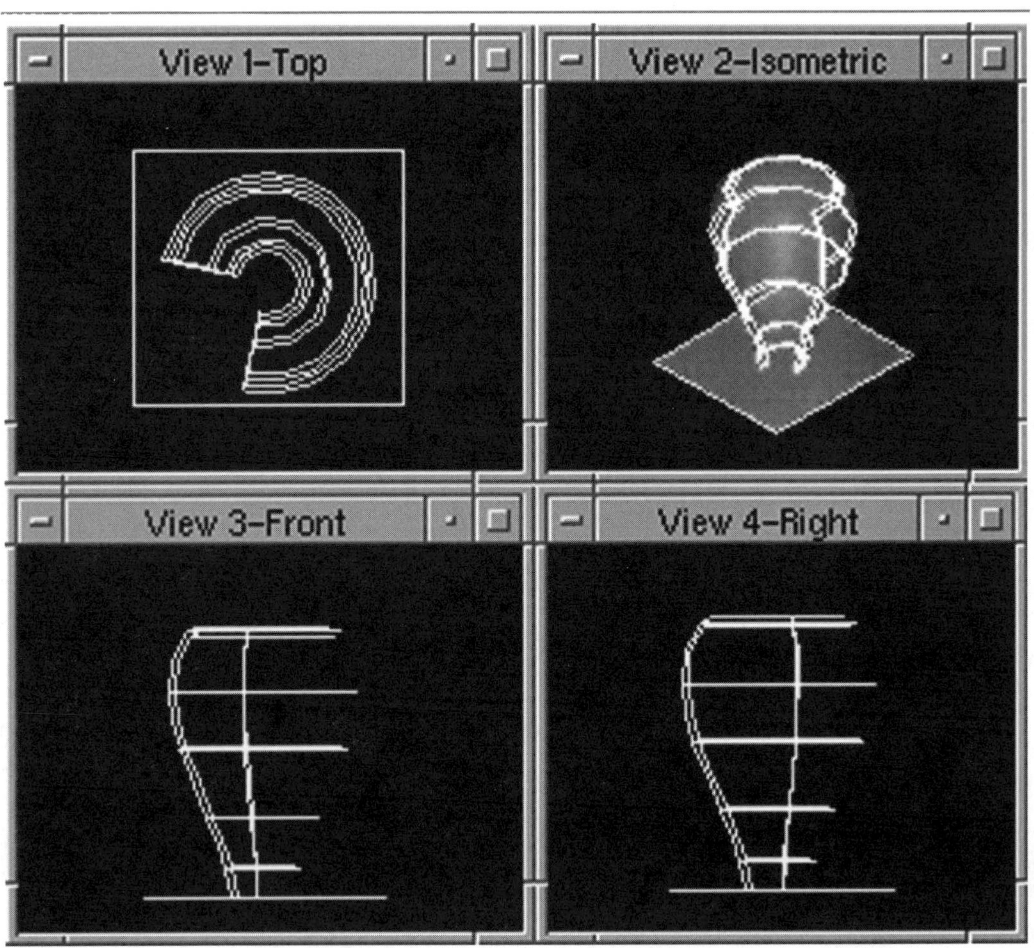

View of surface of revolution.

Exercise 3: Surface Creation

1. Open the **Inside3D → Chapter 5 → Surface Creation** file.
2. Set the color to yellow (`co=4`).
3. In the top view, place a circle that fills the view.
4. Activate the Free-Form Surfaces palette **(Palettes → 3D → Free-Form Surfaces)**. Select the *Surface of Projection* tool. Make sure type is set to solid and the axis setting is on points.
5. In the distance area input 0.1. The X- and Y-scales should be 0.3.
6. Define the circle as the element to be projected, and in the front view move the cursor so the object is projected upward. Then accept the new object with a data point.
7. Activate the *Place Circle by Diameter* tool. Now place a circle on the top of the cone. Tentative point to the top of the new cone and place a data point, then tentative point to the opposite side of the top and place another data point.
8. Now set the distance to 1.0 and the X- and Y-scales to 1.5 and project the new circle upward.
9. Change the active depth in the Top view to the top of the newest cone.
10. Place an orthogonal square that encloses the top of the new cone.
11. Activate the *Surface of Projection* tool and set distance to 0.1, the spin to 50, and the X- and Y-scales to 1.5.
12. Project the square upward to complete the candle holder.

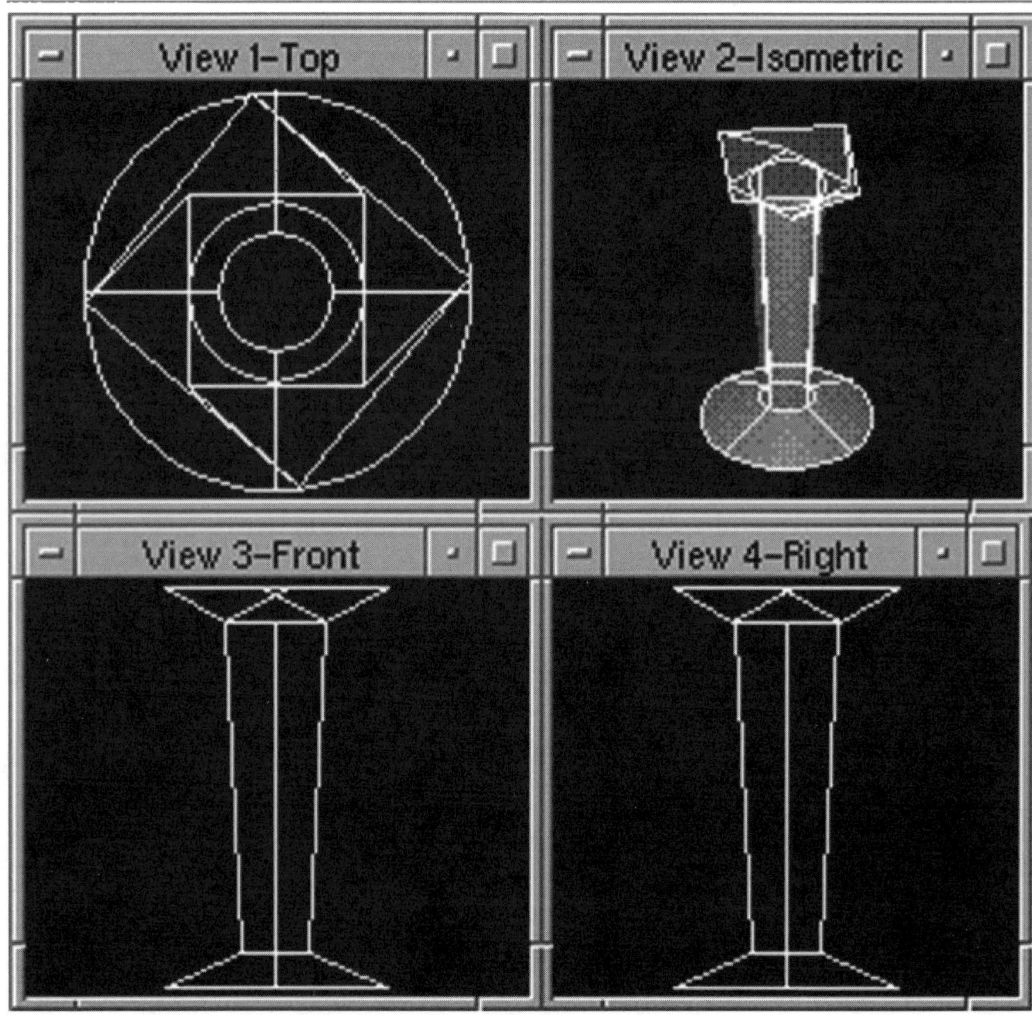

View of candle holder.

13. Open **Inside 3D → Chapter** 5 **→ Tire Creation.**
14. Attach the cell library *CHAP5.cel*. Make "tire" the active cell and place one copy in the front view.
15. Activate the Free-Form Surfaces palette **(Palettes → 3D → Free-Form Surfaces)**. Select the *Surface of Revolution* tool. Make sure type is set to solid and the axis setting is on points. In the angle area input 360.

16. Turn the axis lock on in the Locks palette.

17. In the front view, identify the *tire.cel* as the element to be rotated. In the same view, tentative point to the lower left end of the tire cell, then select a data point. With the axis lock set to On, the axis of rotation is constrained. With the axis of rotation in the horizontal position, place a data point.

18. The tire now takes shape; however, don't forget to accept the new object with a data point.

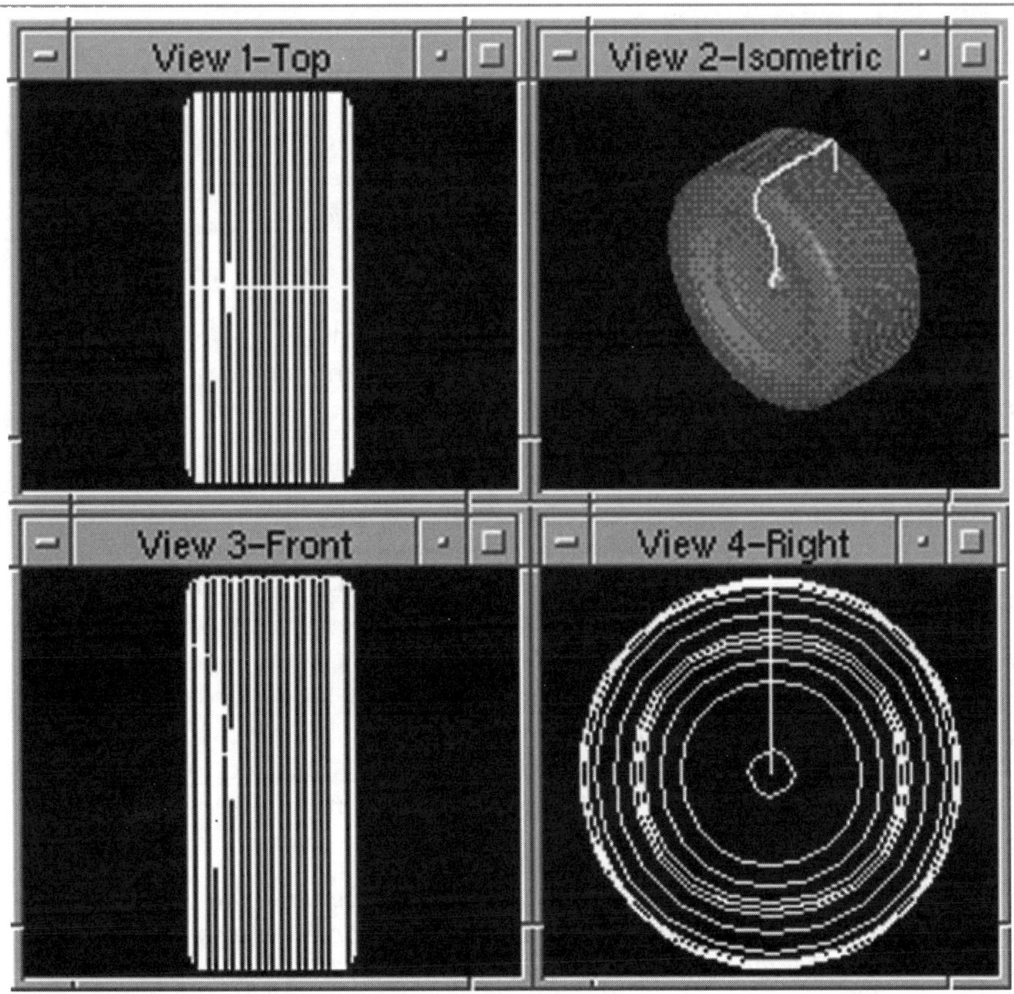

View of the tire.

✖ **CAUTION:** *Narrow paths with steep slopes ahead; safety equipment and ropes suggested. If you have them, smoke them! If you drink coffee (or better yet, expresso) get a cup! You are in for some real fun now.*

Free-form surface (NURBS) icon.

NURBS nonuniform rational B-spline surfaces are a thick forest that we will avoid for now. In Chapter 9, I will guide you through the mystical adventure of NURBS.

The Surface by Section or Network icon.

Surface by Section or Network is also a B-spline surface; however, it is uniform, unlike NURBS. This tool is useful in creating surfaces such as topography. Lines, linestrings, curved linestrings, shapes, and complex elements can be used to form these surfaces.

Construct Surface by Section or Network tool settings dialog box.

Cross-section and network are the only two tool settings for this tool. Cross-section allows the selection of elements in the order in which they are to be connected. When the last desired element is selected, placing a data point on no element allows the viewing of the surface. A second data point will accept the surface.

NOTE: *If the surface appears twisted or otherwise distorted, try reversing the direction of data points in the problem element. Use the Change Element Direction on the Modify Element palette.*

The network option on the tool settings palette allows a surface to be designed by interpolating a network of elements. When constructing the curves for the network, be sure that all curves in both directions (the U and V directions) of the network intersect.

The Construct Surface by Edges icon.

Construct Surface by Edge allows you to create an endless number of shapes. It basically uses the selected elements (lines, linestrings, curved linestrings, shapes, and complex elements) to define the perimeter of a B-spline surface. This surface can be 3D (the elements can have different Z-values from one another); however, all the edges must share the same end point as the next element in the edge.

The only time this does not hold true is when two elements are used. They do not have to share end points. MicroStation designs the missing two edges to form the surface. After selecting the two elements and placing a third data point, the surface is displayed. Sometimes the object appears to be twisted when it is displayed. By hitting reset once, the order of the points in the surface is readjusted. The object can then be accepted by placing another data point.

> **NOTE:** *It is simpler to use the Create Complex Shape command to form a surface by two elements with a common end point.*

Exercise 4: Topo and Chair

Why don't you stand up, stretch, and get ready to try these tools. First, let's try the *Surface by Section* or *Network* tool.

1. Open the **Inside 3D → Chapter 5 → Topography** file.
2. Attach the cell library *CHAP5.cel.* Make "topo" the active cell and place one copy in the top view. Now drop the cell.
3. Activate the Free-Form Surfaces palette **(Palettes → 3D → Free-Form Surfaces)**. Now activate the *Surface by Section* or *Network* tool. Set the tool settings palette to Cross-Section.
4. Select the shapes from the largest to smallest shape. After the smallest shape has been selected, place a data point off to the side.

This displays the new element. One more data point will accept the element. Do a quick rendering of the iso view to see the topo.

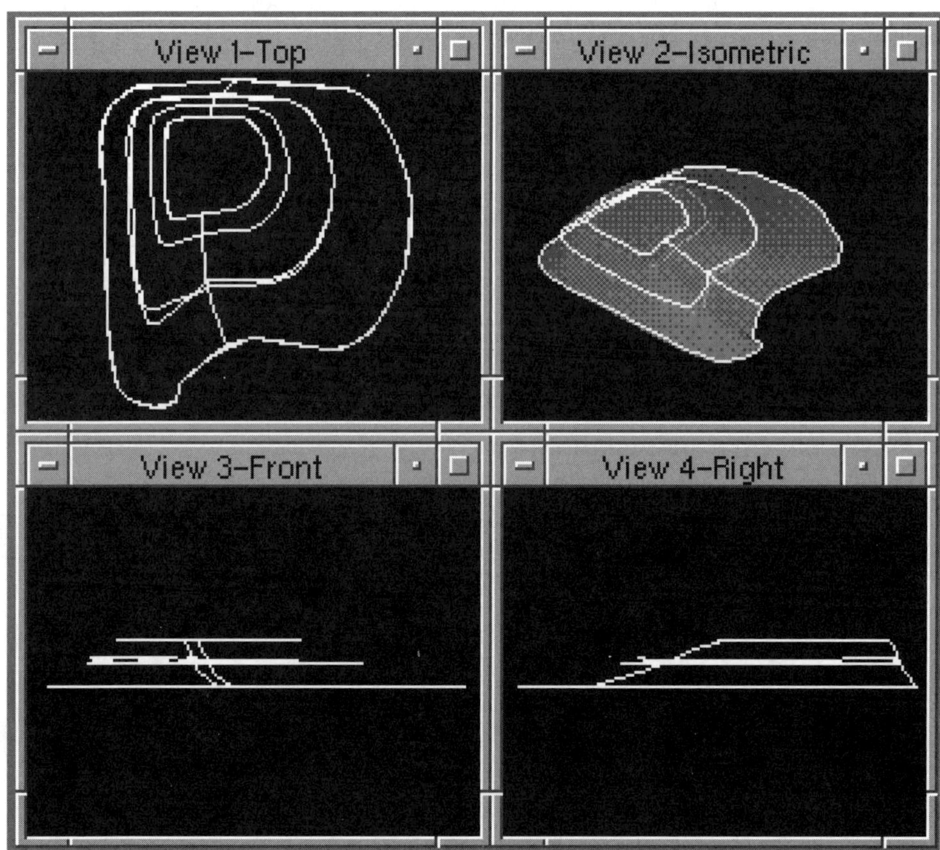

Rendering of the topo.

5. OK, that was easy. In this exercise you will create a chair. So when you finish, you'll have a place to sit and rest.
6. Open the **Inside 3D → Chapter 5 → Open Chair** file.
7. Attach the cell library *CHAP5.cel.* Make "chair" the active cell. Place one copy in the Top view and drop the cell.
8. What you see is the skeleton of the chair you will create. Three basic parts make up the chair, the actual seat, and the two legs. Each one is created separately.
9. Activate the Free-Form Surfaces palette **(Palettes → 3D → Free-Form Surfaces)**. Now activate the *Surface by Edge* tool.
10. First, you will create the seat. In the Iso view, select the linestring that forms the right side of the seat, then the top of the seat, then the left side of the seat, and then the front (or bottom). Enter a data point to view the element, and another to accept the element.
11. Now for the legs. Use the *Selection* tool to select the curved linestring. Hold the Control key down and select the line that forms the top of the leg. Now both should be selected. Now activate the *Surface by Edge* tool. The new element is now displayed; place a data point to accept the element.
12. Repeat the process to form the opposite leg.

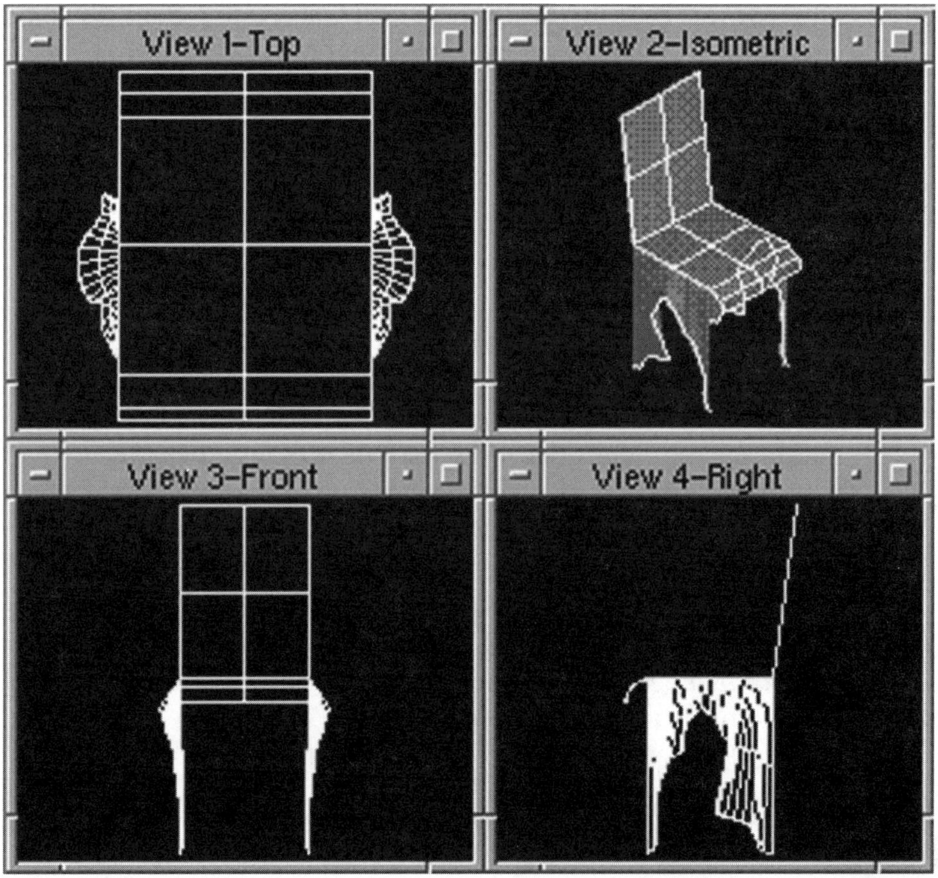

Creating a chair.

The Construct Tubular Surface icon

The *Tubular Surface* command is extremely useful for those of you adventuring in piping, facility management, or related fields. This command will convert a line, arc, linestring, curved linestring, or chained element into a surface. That surface could be a pipe or the section of a conveyor belt.

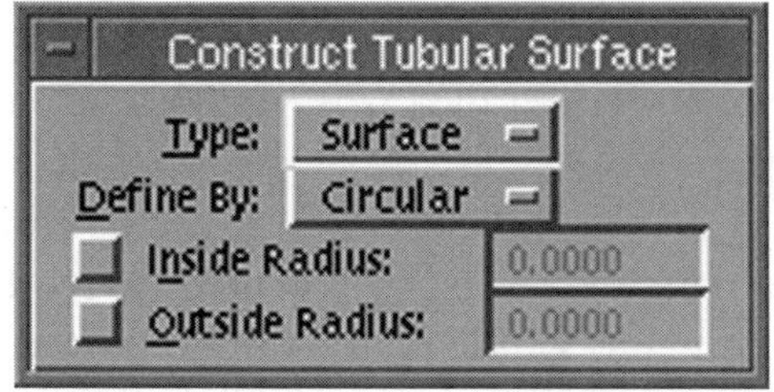

Construct Tubular Surface tool settings dialog box.

The type options function in the same way as the 3D Primitives: surface or solid. The Define By option controls the shape that is projected along the element. The circular selection projects two circles with an inner and outer diameter. Section allows you to determine the element that will be projected. The inner and outer dialog areas allow for exact values.

The Construct Skin Surface icon.

This tool can get many an adventurer out of a difficult situation. The *Skin Surface* tool creates a B-spline surface between two planar elements. A trace element is used to define the path of the surface. This element can be a simple line, a complicated curved linestring, or anything in between. The number of vertices in the trace element, as well as their location, determine the final shape of the surface.

The first data point determines the trace element. The second identifies the first planar element, and the third data point determines the second planar element. The fourth data point displays the surface and the fifth accepts the surface.

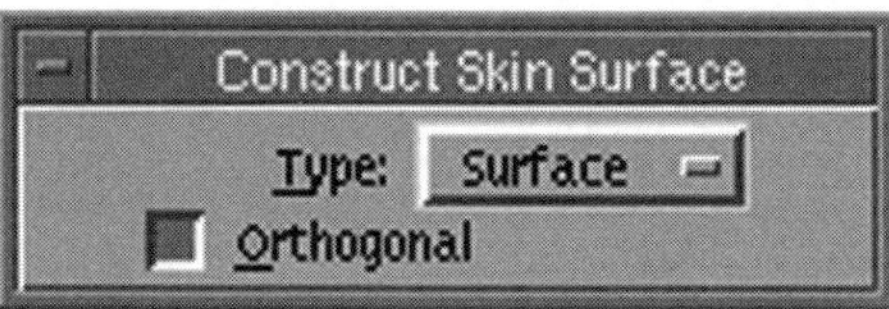

Construct Skin Surface tool settings dialog box.

The tool settings for the skin surface allow the created element to be a surface or a solid. Good old faithful orthogonal is here as well.

The Construct Offset Surface icon.

Construct Offset Surface is the last tool on the 3D Free-Form Surface palette. It allows the manipulation and repurposing of Surfaces of projection and revolution. The surface is offset scaled in all directions except the direction of the original projection of revolution.

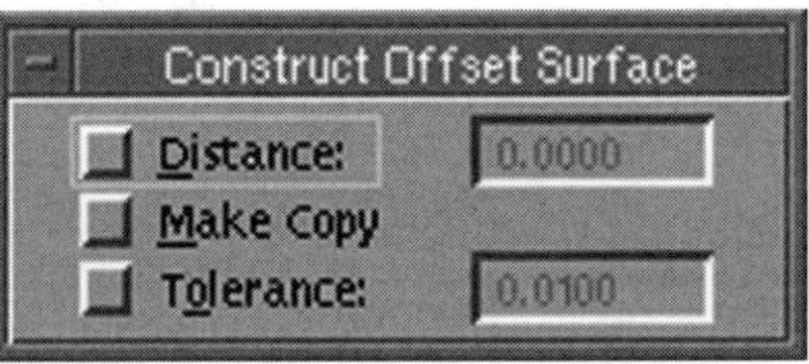

Construct Offset Surface tool settings dialog box.

An exact value can be input for the offset in the distance dialog area of the tool settings. The Make Copy option creates a new surface, leaving the original. Tolerance overrides the system tolerance.

Exercise 5: The Roller Coaster

Now let's go for a ride, a roller coaster ride. This exercise will use tubular and skin surfaces as well as the offset surface tool.

1. Open the **Inside3D → Chapter 5 → Roller Coaster** file.
2. Attach the cell library *CHAP5.cel.* Make "Coaster" the active cell and place one copy in the Top view. Now drop the cell and turn off levels 3 and 4 in all views.
3. Activate the Free-Form Surfaces palette **(Palettes→ 3D→ Free-Form Surfaces)**. Activate the *Tubular Surface* tool. Set the tool settings palette to Section.
4. Select the roller coaster track as the trace element. Now select the short line that crosses the track as the section. Enter a data point to view the surface and another data point to accept the surface. Now the track of the roller coaster is in place.

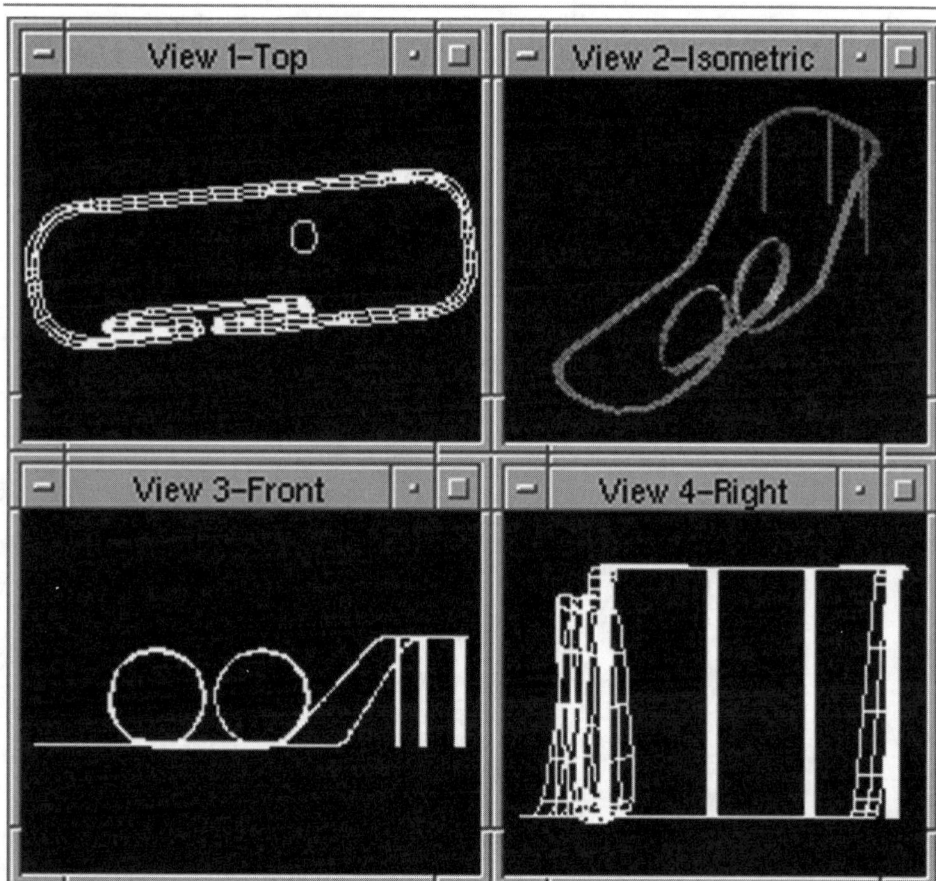

The track of the roller coaster.

5. Let's add some strength to the supports for the track. In tool settings change the setting to circular. Input `0.75` as the inside radius and `1.0` as the outside radius (make sure both radii settings are activated).
6. Data point on one of the columns. Even though the radii are set, you must still enter multiple data points to place the surface. When the column is displayed, enter a final data point to accept the surface. Repeat the process on the remaining three supports.
7. The roller coaster is almost ready to ride, but you must first add some lights and a loading station. Construct the skin surface and offset surface will help you to finish.

8. Turn level 3 on in all views. The required basic elements needed to create the lights are now visible.
9. Window in on one of the lights in the iso view. Activate the *Construct Skin Surface* tool. Select the curve linestring as the trace element. Now select the square base as the first section and the circle as the second section. Enter another data point to display the surface. Place a final data point to accept the surface. Repeat the process on the remaining three lights.

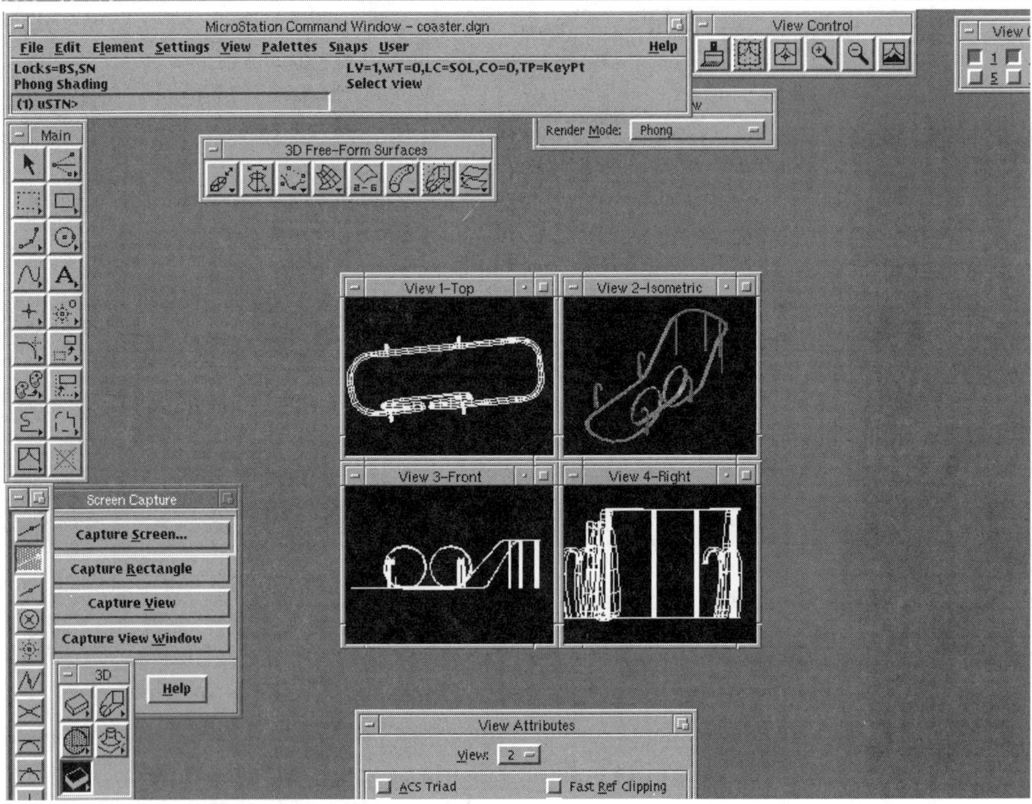

Adding lights to the roller coaster.

10. Turn level 4 on in all views and perform a fit in all views. Window in on the displayed arc and the two lights near it.
11. Activate the *Construct Surface of Projection* tool and set the distance to 35. Project the arc so that it remains between the two lights.

12. Now activate the *Construct Offset Surface* tool. Turn off the distance and tolerance settings and the turn on the copy setting. Select the projected arc and move the cursor away from the track until the second arc look appropriate, then place a data point.

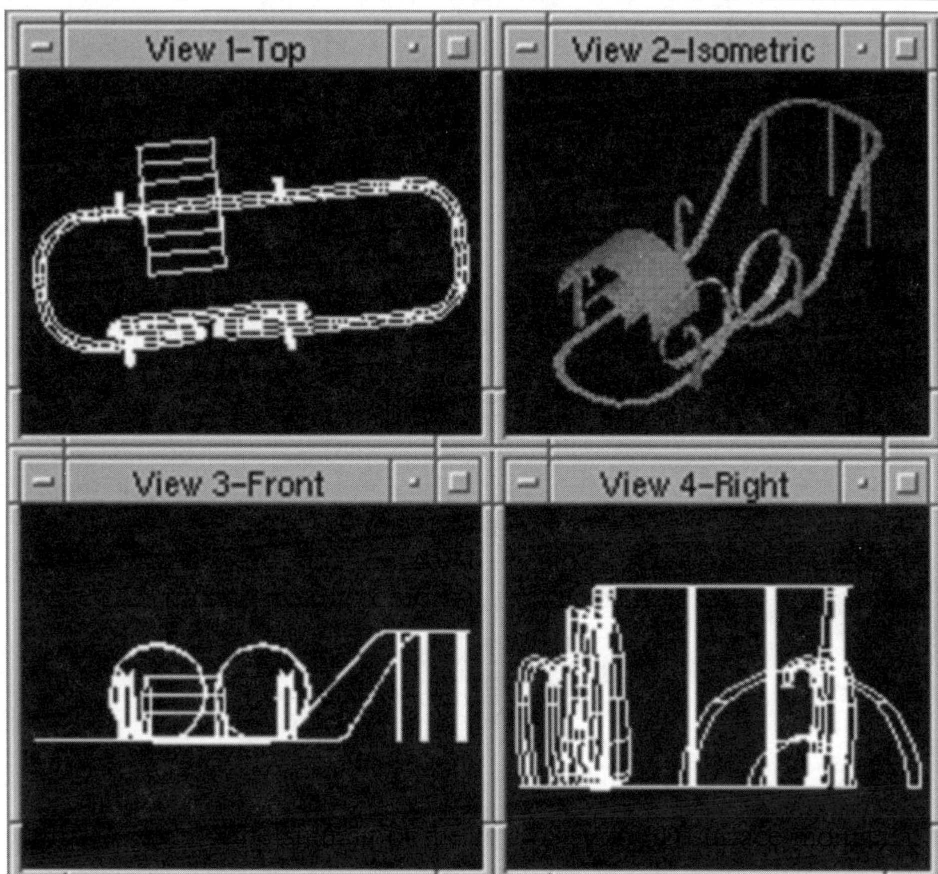

The finished roller coaster.

Summary

You have reached the summit of the MicroStation 3D surface mountain. I'm impressed; it's no small accomplishment. From this summit you can look out over the world and see all of the objects made of 3D free-form surfaces.

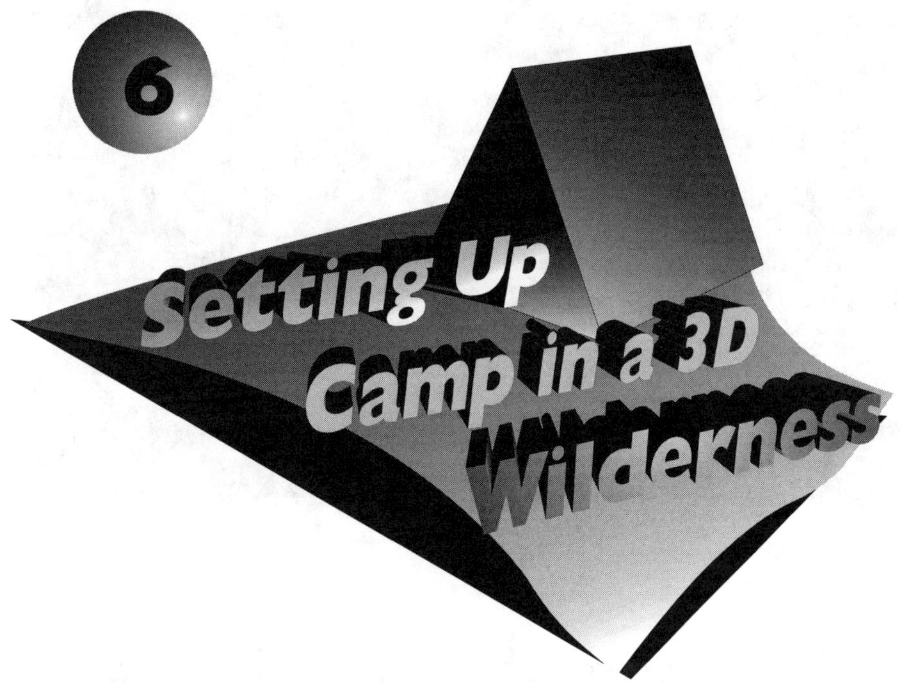

Introduction

In Chapter 1 you saw how the process of 3D construction took place. Now you will see how the museum from Chapter 1 was built. In this chapter you will focus on the techniques and processes used to construct the museum, using the tools you've learned about in the previous chapters.

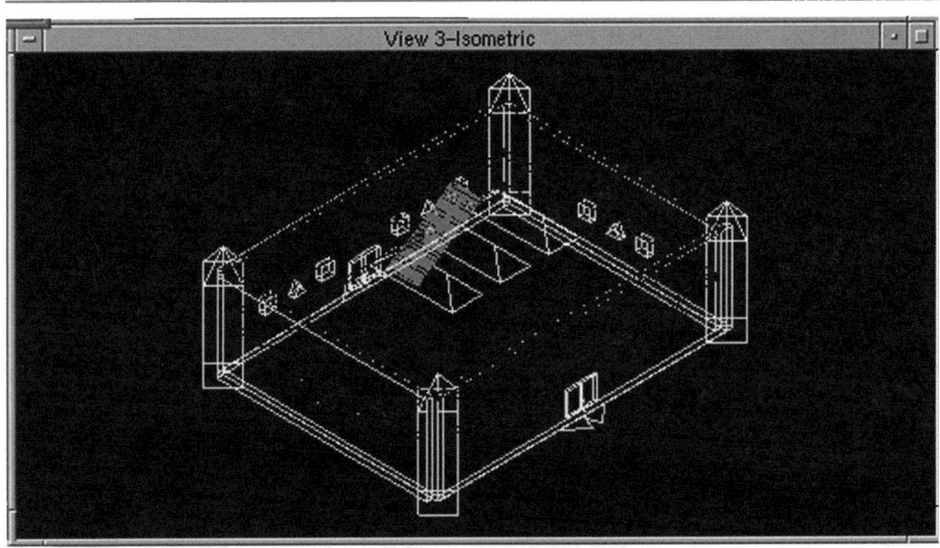

Iso view of the art museum.

Building Blocks of the Art Museum

Exercise 1: Building the Walls and Roof

To start the process, open the *Art Museum* file from the Inside 3D menu. This file has the completed building in it, so you can use it as a reference. You still have not learned all the 3D tools yet, so Safari Sam will stick to the topics we have covered.

You will start this project by opening the *Slab/Foundation* command from the Inside 3D pulldown menu, opening a file with the foundation of the art museum already placed. The foundation is a solid rectangular slab with dimensions of 100 feet x 75 feet x 2 feet and is on level 1.

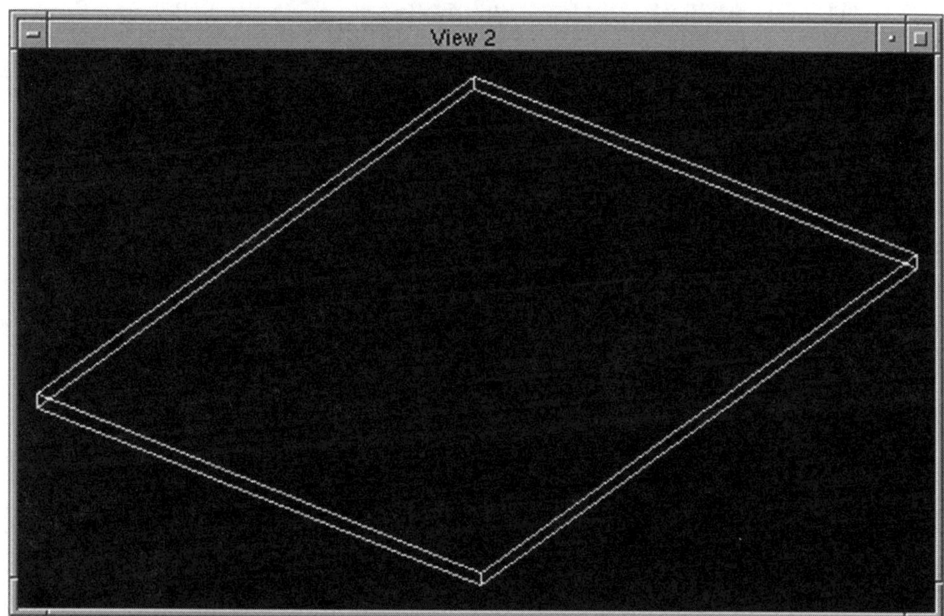

The foundation of the art museum.

The first thing you want to do is build the walls and roof. To do this you will use the *Place Slab* command. The height of all the slabs will be 30 feet, and they will be placed with their outside surface aligned with the outside surface of the foundation.

1. Change to level 2, select the *Place Slab* tool. The settings are Solid and Orthogonal. In the Top view snap to the lower left corner of foundation (point 1 in Iso view) to build the west wall and data point to accept. Make sure you select the top of the foundation. Snap to the upper left corner of the foundation (point 2 in Iso view) in the Top view and data point to accept. Key in the following for the walls' dimensions:

   ```
   DX=1
   DX=,,30
   ```

2. Copy the new wall 99 feet east in the Top view. Key in the following:

   ```
   DX=99
   ```

The two walls should look like the Iso views in the following figure.

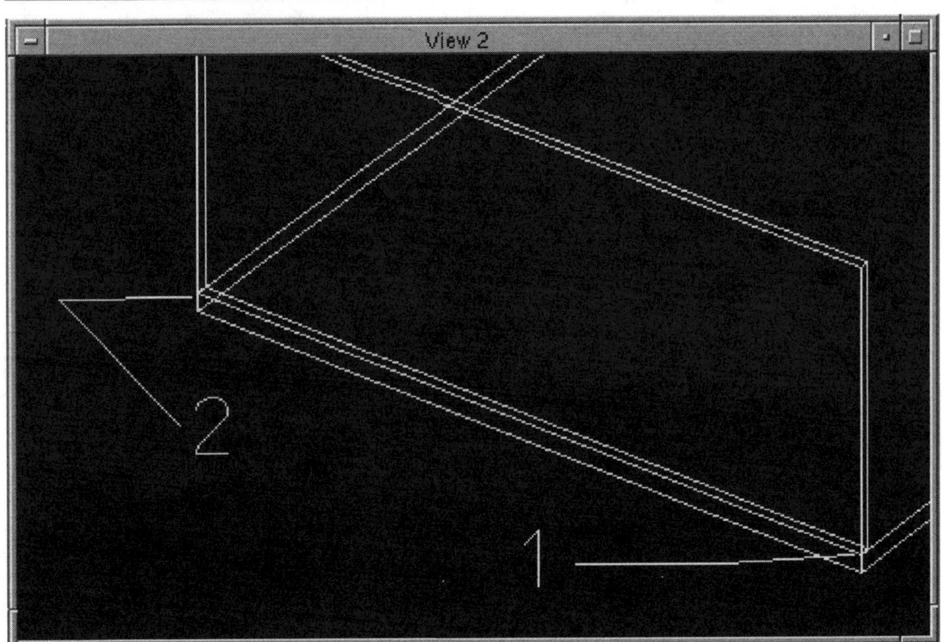

Placing the west and east walls and snap points.

3. Next, you will place the Front wall, also on level 2. Select the *Place Slab* tool and in the Front view snap to the same point that you started the west wall from (point 1, the front left corner of the upper surface of the foundation) and data point to accept. Next, again in the Front view, snap to the front right corner of the upper surface of the foundation (point 2) and data point to accept. Finally, snap to the outside upper right corner of the east wall (point 3), and data point to accept.

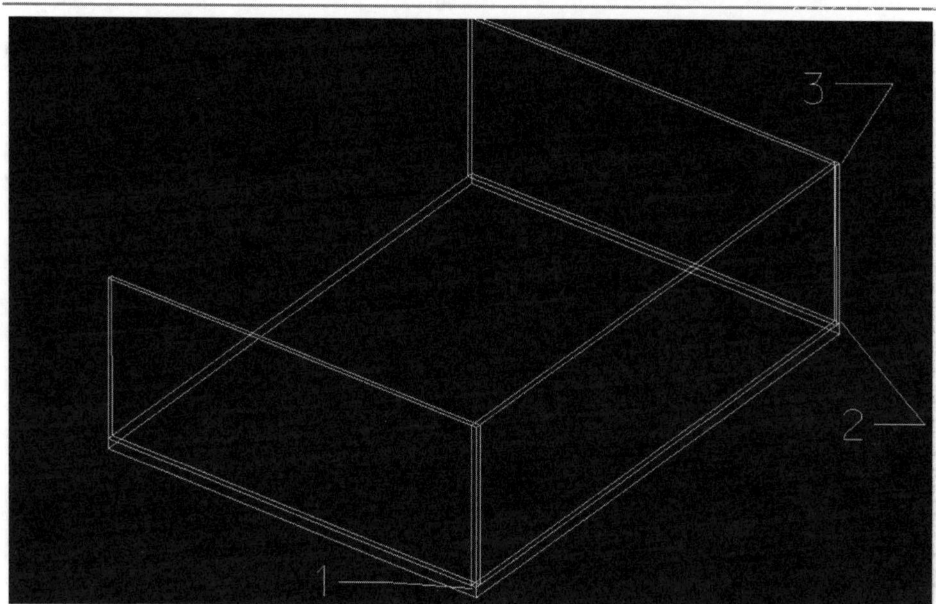

Placing the Front wall and snap points.

4. Copy the Front wall in the Top view, using the following keyin to create the back wall:

   ```
   DX=,74
   ```

5. Use the *Place Slab* command again to create the roof. In the Top view snap to the upper left corner of the west wall (point 1) and data point to accept. Next, snap to the upper right corner of the west wall (point 2) and data point to accept. Key in the following to complete the roof:

   ```
   DX=100
   DX=,,2
   ```

The building should now look like the following figure.

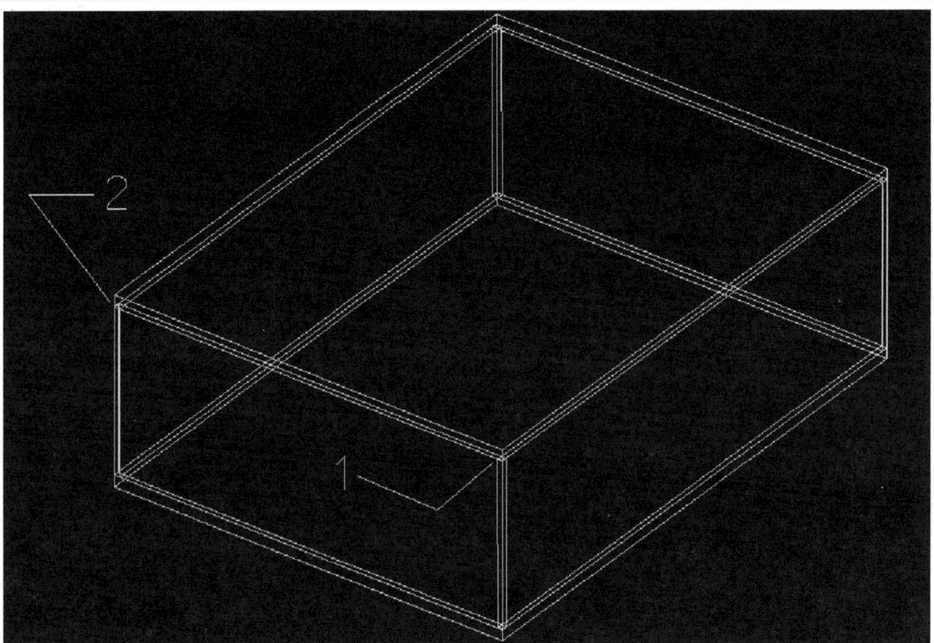

Art gallery with walls and roof placed.

Exercise 2: Adding More Detail to the Museum

The front elevation shows the front facade of the building. You will now create the doors, and later on you will learn how to remove the wall where the door is going to be placed. Right now they will just be placed.

1. Use the *Place Slab* command with the same settings as before. Set your Snap Divisor to 2 and use the following keyin:

   ```
   KY=2
   ```

2. In the Front view, snap to the middle of the front upper surface of the foundation (point 1) and data point to accept. Key in the following:

   ```
   DX=4
   DX=,7
   DX=,,1
   ```

NOTE: *When you do remove the walls using Boolean operations, make sure that the door is slightly thicker than the wall.*

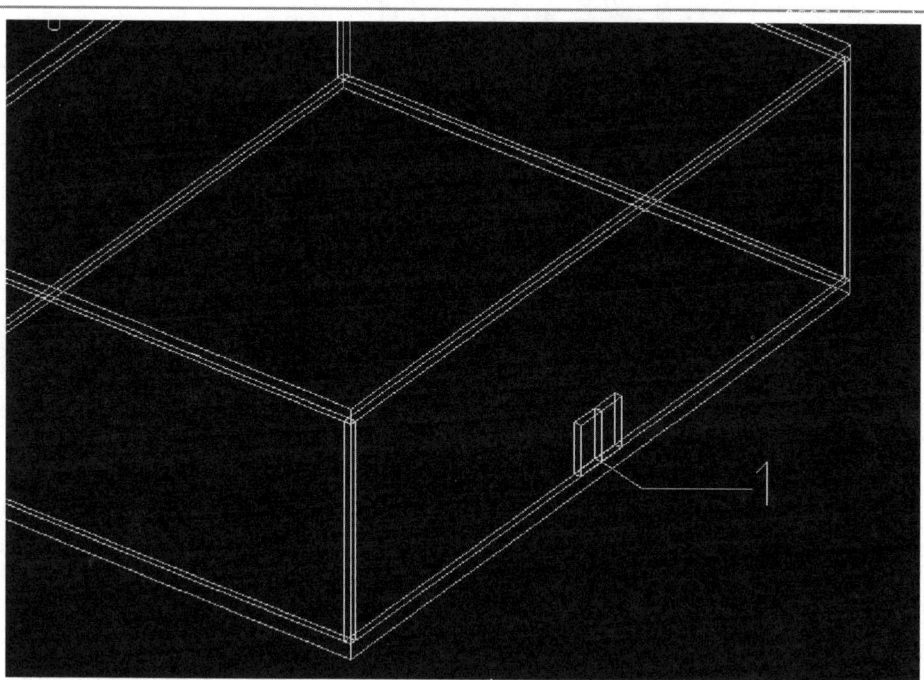

Placing a door using Place Slab.

3. In the Front view copy the door, using the following keyin:

   ```
   DX=-4
   ```

4. In the Top view copy the doors to create doors in the rear wall.
5. Next, you will place the second floor and the staircase. First, you will create the staircase. Stairs follow the 7-11 rule, that is, 7 inches high and 11 inches deep. We will be using 1 foot deep and 6 inch high stairs.
6. Select the *Inside3D → Chapter 6 → Staircase* file. The file is the same as the one you just finished, except a line has been placed to allow you to create the first stair. This line is 5 feet from the back door. You should be on level 4.
7. Use the *Place Slab* command in the Top view and snap to an endpoint of this new line, data point to accept, then snap to the other end, and data point to accept. Key in the following to create the first stair, which is 2 inches thick:

   ```
   DX=1
   ```

```
DX=,,.1666
```

8. Now in the Front view, *Move* the stair up 6 inches, using the following keyin:

   ```
   DX=,.5
   ```

9. Now copy the stairs in the Front view using the following keyin:

   ```
   DX=0.9166,0.5
   ```

Repeat the copying process until you have created 15 stairs. The top of the stairs should be at the midpoint of any wall.

1. Next you create the stair support. Use the *Place Block* command in the Top view and snap to the middle of the line used to create the first stair. Key in the following to create the block:

   ```
   DX=1,0.5
   ```

2. Now using the *Move* command in the Top view snap to the middle of the line on the block aligned with the construction line, and move to the middle of the construction line. The middle of the block should be the same as the middle of the construction line.
3. Copy the block in the Top view by snapping to the middle of the line on the block, opposite to the construction line, and snap to the middle of the bottom surface on the right side of the top stair. Finally, place a line from the corner of lower block to the corresponding corner of the upper block.
4. To create the support, use the *Construct Surface of Projection* command, using the lower block and line.
5. Use the *Construct Solid of Projection* command, with Solid and Orthogonal set, from the 3D Free Form Surfaces palette. Select the shape and the line you just created to create the floor.
6. Delete any construction elements used.

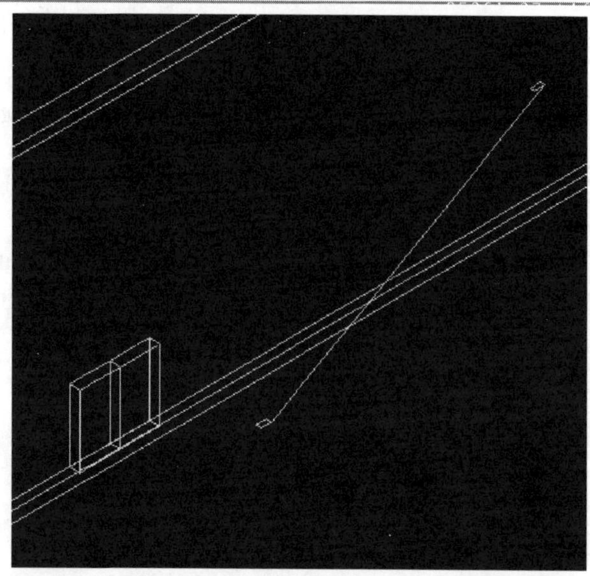

The blocks and line placed relative to the stairs in the Iso view.

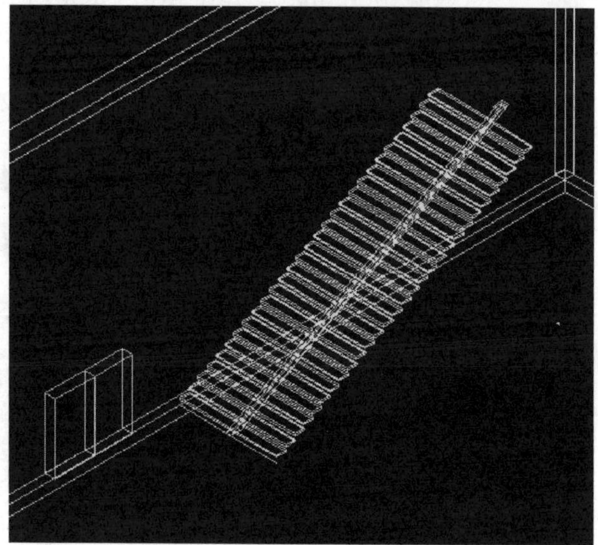

The stairs with their support rail.

Additional work can be done to create a handrail. This task will be left for you to complete.

The next step is to create the second floor. This will be done using both snap point and precision input.

1. Change to level 3, using the *Place Shape* command in the Top view. You start by snapping to the midpoint of the right side of the rear wall (point 1, this may take some effort, but just keep using the tentative point and check the Iso view to see if you snapped to the right point), continue snapping to points 2, 3, and 4, and data point to accept each snap point. Then key in the following:

```
DX=60
DX=,-10
DX=-50
DX=,-55
DX=80
DX=,55
```

2. Now snap to the end of the top front of the top stair, data point to accept, then snap to the top back of the top stair, and data point to accept. Then select *Close Element* from the Tool Settings palette.

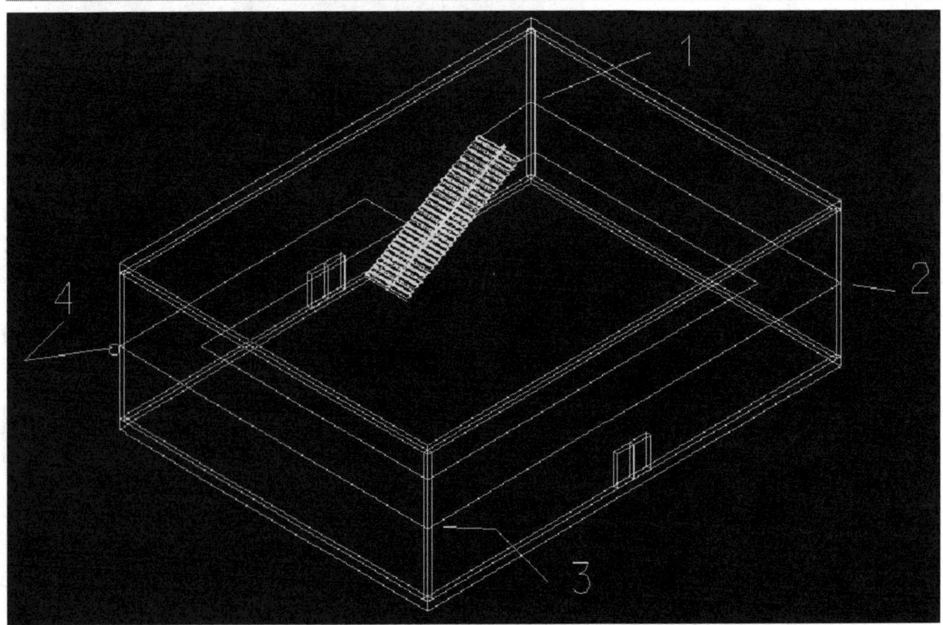

Snap point for placing the second floor with Place Shape.

3. In the Front view, use *Place Line* to draw a line from the top surface of the top stair to the bottom surface. This line will be used with the shape to create a solid of projection to create the second floor.

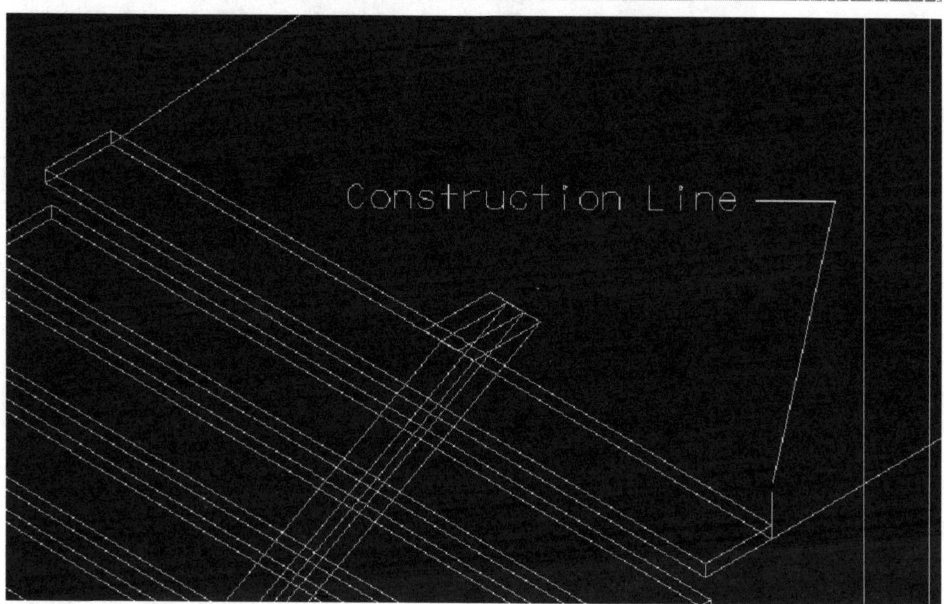

Place Line on top stair in Front view.

4. Use the *Construct Solid of Projection* command to create the support. Use the block on the bottom and the line between the top and bottom blocks to create the solid.

The next construction project is to create the four towers and their skylights.

1. Select the *Inside3D* → *Chapter 6* → *Towers* file. This file has all the construction elements you created previously, with the addition of a construction line in the lower left corner of the Top view. Go to level 5.
2. Select the *Place Shape* command and in the Top view snap to the end of this line and key in the following:

```
DX=7.0711,7.0711
DX=7.0711,-7.0711
```

```
DX=-7.0711,-7.0711
```

3. Then select *Close Element* to finish the command.

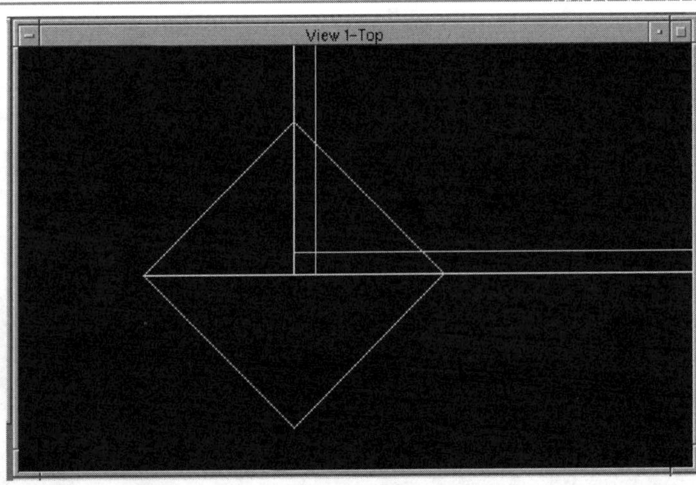

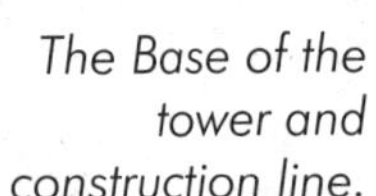
The Base of the tower and construction line.

4. Place a construction line to create a 30-foot tower, using *Construct Solid of Projection* from the newly placed shape.
5. Snap to the beginning of the construction line used to build the tower base. The beginning of this line is the center of the tower. Place a line going up through the center of the tower to a height of 35 feet.
6. Use the *Place Shape* command and in the Top view snap to a corner of the tower (point 1), then to the center point (point 2), and then back to an adjacent corner (point 3), producing an inclined triangle to close the shape.

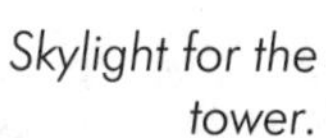
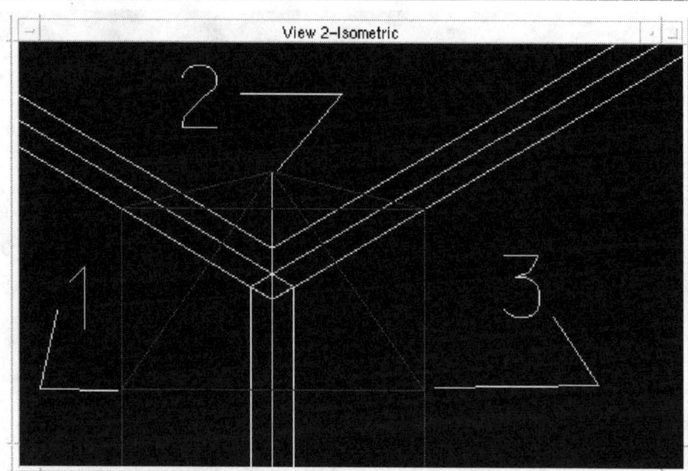

Skylight for the tower.

7. Repeat the process for the remaining three sides of this tower, and then do the other three towers.
8. Delete any construction elements used.
9. Select the *Inside3D → Chapter 6 → Skylights* file. This file contains construction elements to help build the skylights. Remember, you have not cut holes in the roof yet (you will learn about that in Chapter 10); right now you will place the 3D elements and later you can come back and finish the art gallery. Change to level 6.
10. Using the long construction element on the left (point 1 is the origin of the block), place a block with dimensions 30 feet x 10 feet. Move the block so that the center of the end of the block is aligned with the construction line.

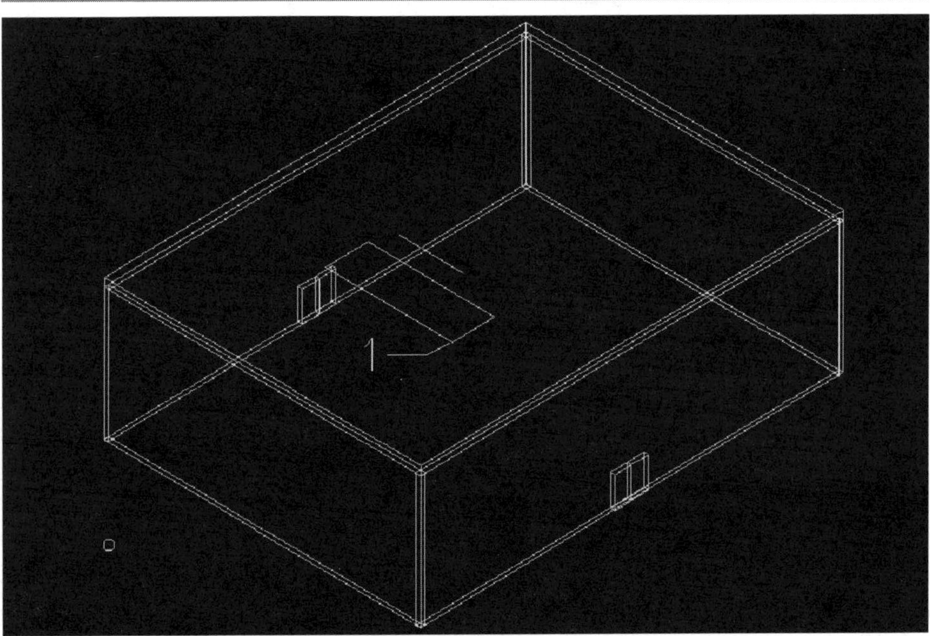

Block placed and moved with snap point for block origin.

11. Use the *Place Shape* command to draw the panes of the skylight. Use the snap points (1 through 4) indicated. Use the *Place Shape* command to make the end windows' panes. You need to draw four window panes. Delete the construction elements after you have finished.
12. In the Top view use the *Select Element* command, and holding the data button down, draw a window around the newly placed elements, to select them all. Then use the *Edit→ Group* command to group the elements.

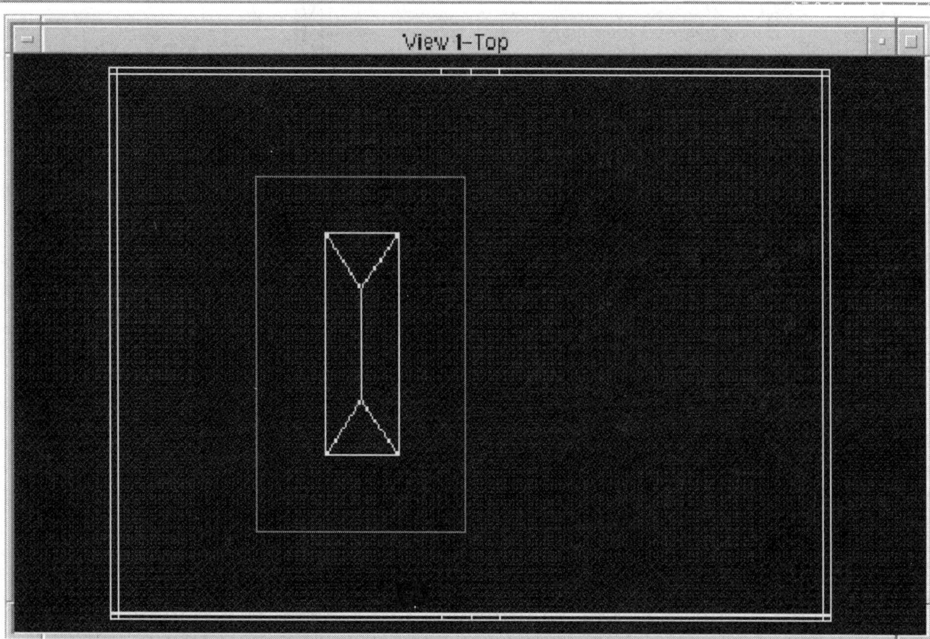

Grouped skylight elements.

13. In the Top view, copy the grouped elements over 15 feet. Use the following keyin:

    `DX=15`

14. Copy these elements again by 15 feet, creating three skylights.

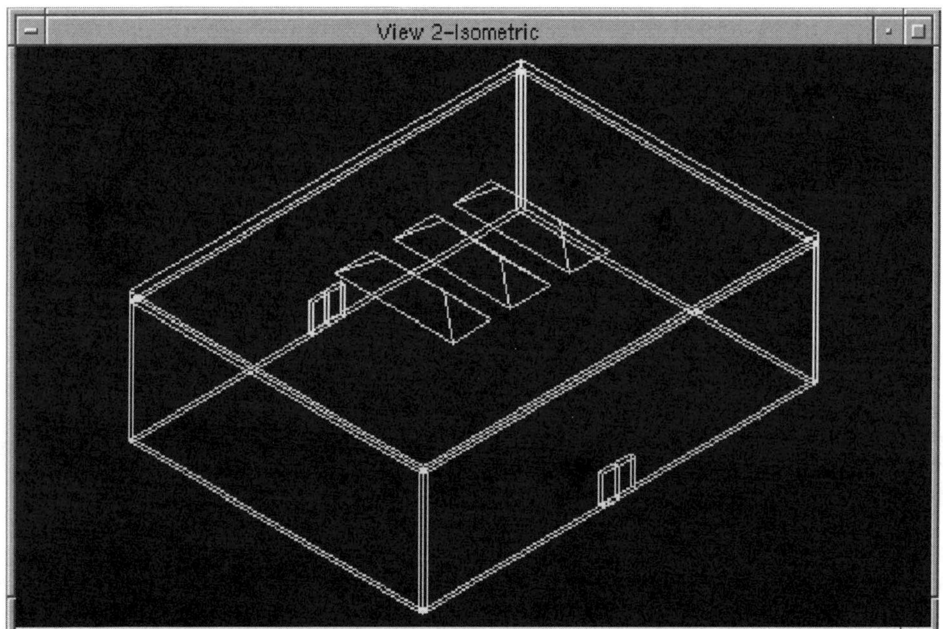

Skylights for the roof.

Adding windows is easy enough, using *Place Shape* and a construction line (the construction line is perpendicular to the shape and is 1 foot long, since the walls are 1 foot thick) and then using the *Construct Solid of Projection* command. Create any shapes you want. Safari Sam used squares and triangles.

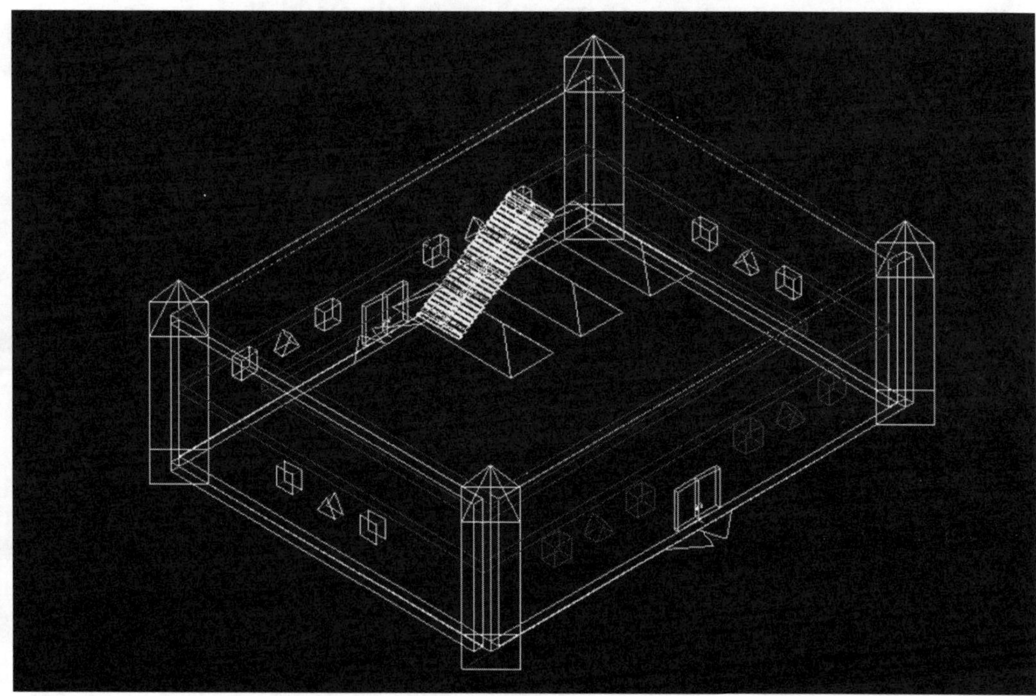

Iso view of the 3D Art Gallery with windows placed.

Exercise 3: Adding Art to the Museum

The last step is to add art to the gallery. First, you will set your active depth to the level of the first floor (the top of the foundation).

1. Select the *Set Active Depth* command. Select the Top view, then in the Front view snap to the top of the foundation, and data point to accept.
2. Use the *Place Slab* command to create a pedestal. Enter a data point anywhere in the Top view to begin the slab. Use the following keyins to create a slab 2 feet x 2 feet x 4 feet.

```
DX=2
DX=,2
DX=,,4
```

You can copy this pedestal wherever you like and then add 3D sculptures on top of it. Some examples might be spheres, cubes, and any mixture of the 3D primitives you know. Later on you will learn how to map these objects with a texture map of any surface imaginable, such as marble, metal, wood, or anything else.

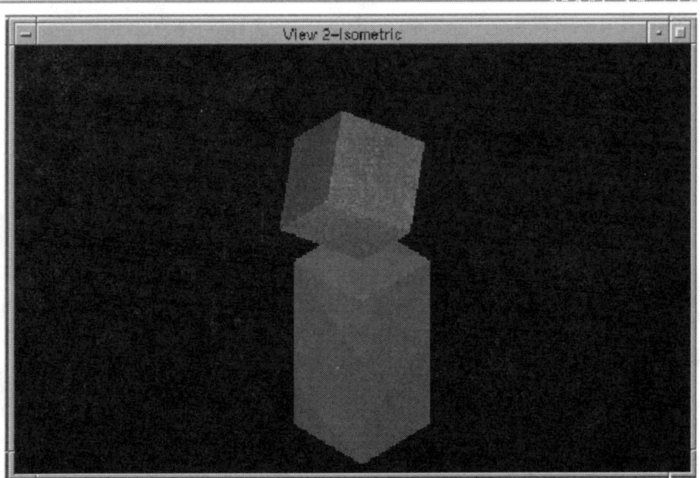

A sculpture of the 3D design cube.

Later you will also learn about placing cameras in the gallery to view the art in different ways.

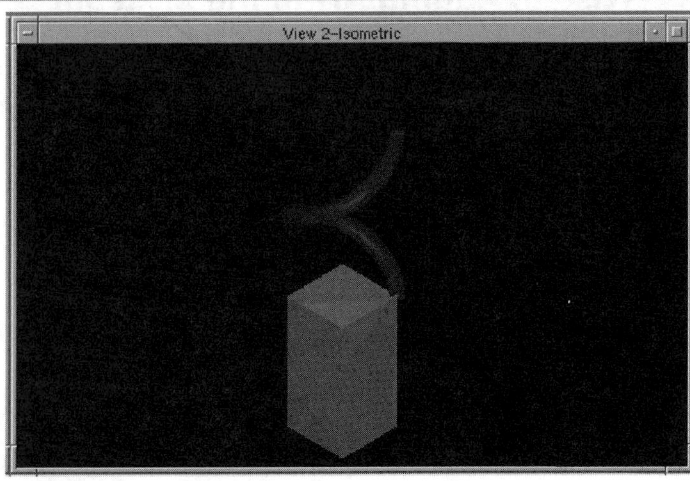

A 3D spiral sculpture.

Using other 3D tools, sculptures can be created.

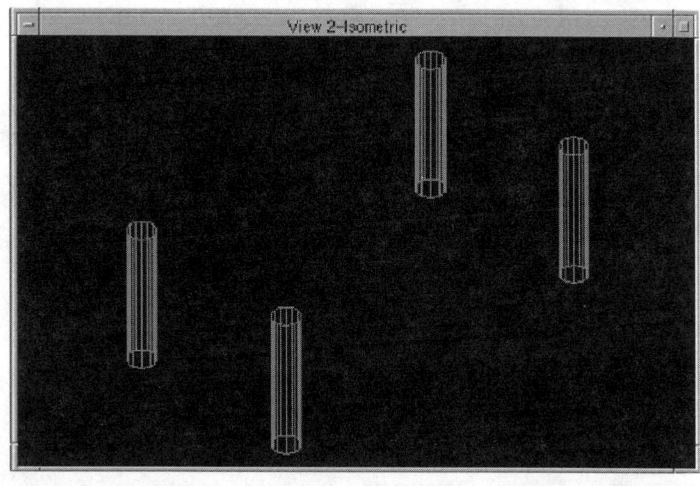

Creating a bench by first creating cylinders for feet.

Create a bench using *Place Cylinder* and *Place Slab* commands.

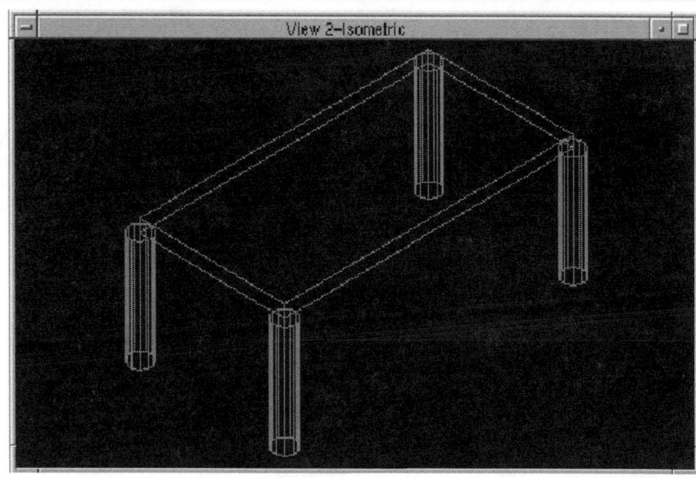

Adding a slab to create a bench seat.

Use *Place Cone* to create plant holders. You should have a good enough grasp of the process to create these on your own.

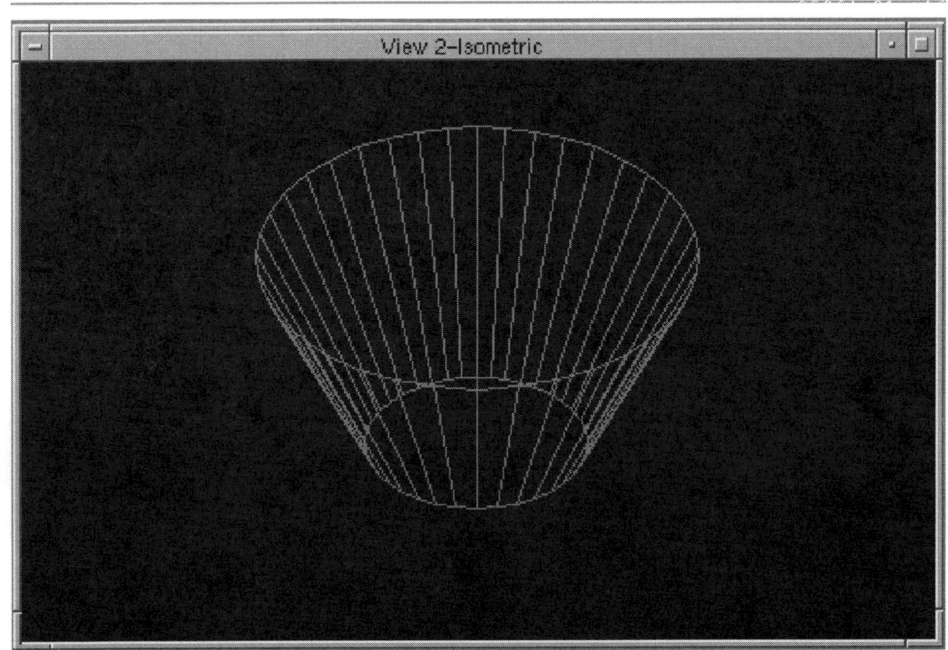

Plant holder placed using Place Cone.

Finally, use the *Render View* command and the Phong setting to render the gallery. Using cameras you can go inside the gallery to view the inside.

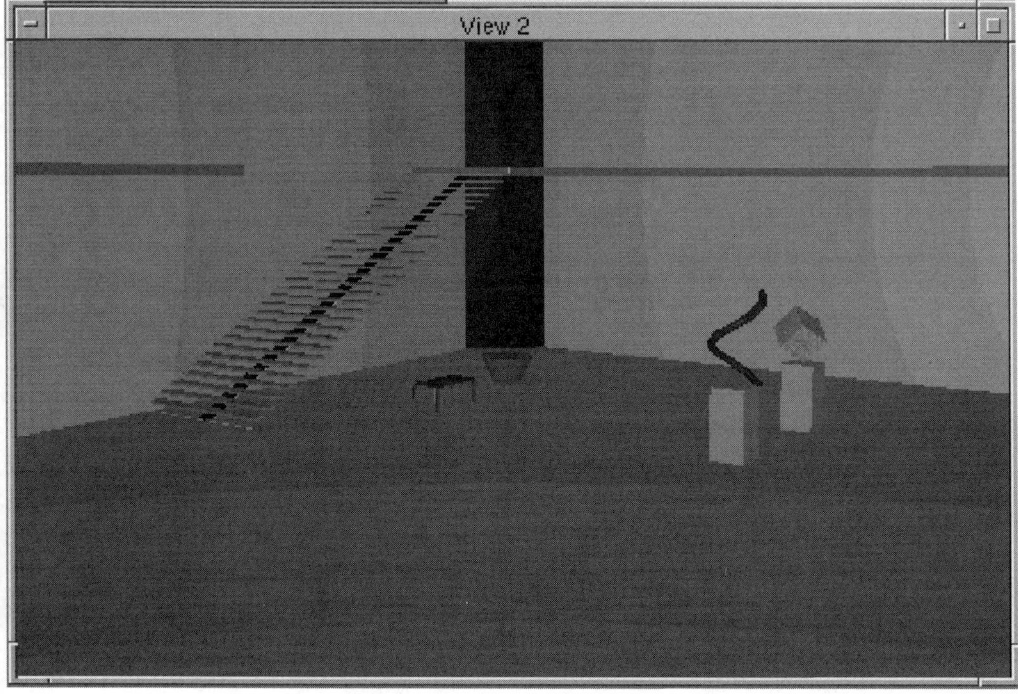

A view of the lower hall.

A rendered view of the lower part of the gallery.

Summary

The 3D process is based on simple 2D elements that help us create more complex 3D elements. Using previously created 2D files, you can create the 3D part using those 2D monument points. Using the information you already have or using information that is logical is important to the 3D design process. You must know where you are going in the 3D world before you can get there; therefore, using construction elements and precision input is the *best* way to proceed. Learn these lessons well and you will be able to model even more complex shapes, like B-splines, where construction elements and precision input are the only way that B-splines can be drawn.

7

Using MicroStation's Camera

Introduction

Up to this point, you have had a few glimpses of the many uses of MicroStation's elements, surfaces, and objects. Now, let Safari Sam guide you through the different methods of viewing them. You can have a great model, but if you can't display it in a way that people understand and appreciate, it isn't worth much.

MicroStation gives you an infinite number of ways to view files. You have been exposed to many of them already, such as the orthographic views (top, right, front, etc.) and other standard views. You have also been exposed to the method of rotating views yourself. Clipping planes, display volume, and saved views should also be "traveled ground" at this point in your journey.

We now move on to an adventure that more closely represents the "real world." For the most part, you have been viewing models as wireframes, which are simple line representations of real objects. This chapter will put some skin on those objects. It will also allow you to adjust the characteristics of the views, or camera shots, used to view your model in MicroStation. We will leave the really professional trailblazing stuff, creating extremely realistic photos, for Chapter 12. We have a lot of ground to cover, so let's get started.

The Camera

One of the most basic and useful tool in creating snapshots in MicroStation is the camera. You have used many other helpful methods of view manipulation on other parts of our adventure. However, the camera adds a whole new level of control. It allows you to position the point from which you view your creation. It allows you to determine the *target* or portion of the creation upon which you wish to focus. It even allows you to control the type of lens you want, a wide fisheye, a narrow telescopic, or anything in between. It is an extremely valuable tool for any adventurer who wants to take snapshots for the photo album.

Let us take a moment and review the basics of photography. The MicroStation camera is based on the concepts of the real camera. Target (or focal point), angle (or field of view), lenses, and lighting are all similar concepts.

Shutter speed and aperture are not found in MicroStation's camera. There is no limitation on the amount of time the image is exposed. As for the aperture, MicroStation's entire image is in focus. Unlike real cameras, you don't have to decide which portion of the image you want in focus. In fact, you don't have to focus at all; MicroStation takes care of that. Don't you wish all cameras were this easy? Each of the eight views in MicroStation can have a different camera setup, and they can be turned on and off at will.

Below is a list of the basic camera tools and terms and the MicroStation equivalents.

Standard Camera	MicroStation Camera
Focal point	Target
Field of view	Angle
Lenses	Lens
Lighting	Lighting (6 types)
Filters	Adjustment of the colors of lights
Shutter speed	No limitation
Aperture	No limitation

All of the settings for the camera can be controlled on the Camera subpalette *(View → Camera)*.

This subpalette contains the settings for the camera. Let's go through them one at a time.

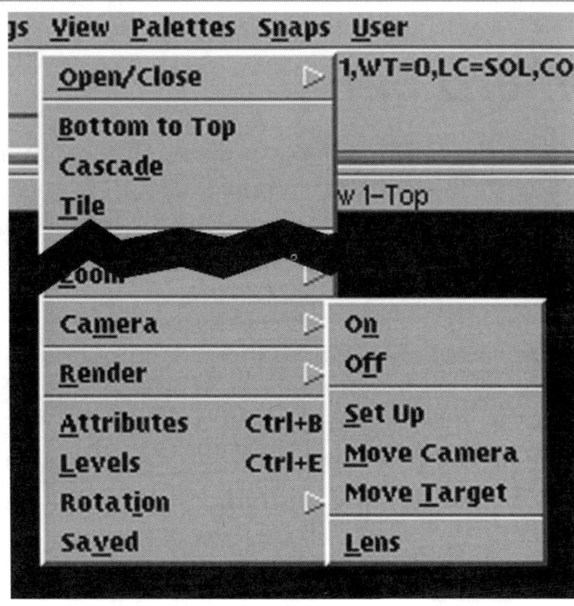

Camera subpalette.

The first section of the subpalette allows you to turn a view's camera attribute on or off. This is an extremely valuable tool. The camera allows wonderful views and images; however, they are difficult to work in. It is like trying to paint a masterpiece while wearing someone else's thick prescription glasses.

Working in some camera views is like a painting with thick glasses.

Camera Setup

The "View Pyramid" is used to display the portion of the model that will be visible in the camera. When positioning the camera and the camera target, the view pyramid is displayed in dashed lines. Its size and shape are determined by the lenses selected and the location of the camera and its target.

When the camera is defined, the clipping planes must be positioned. The front and back clipping planes limit the amount of the model that is displayed in the camera view. The front clipping plane is defined first. Clipping planes can be used to eliminate objects in the foreground or background that are not desired in the camera view.

Image of View Pyramid and camera setup.

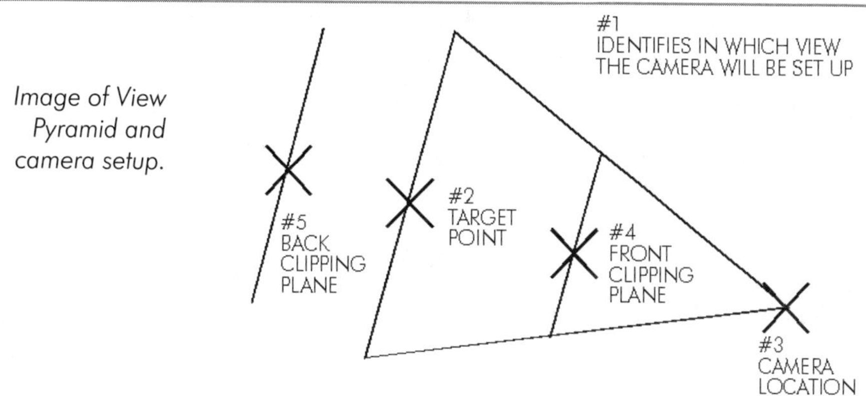

The *Set Up* tool *(View → Camera → Set Up)* is next on the subpalette. It allows you to designate the camera view and control the other aspects of the camera. This is the most important tool on this palette. We will be spending some time here, so fill your canteen and grab a freeze-dried snack.

After selecting the *Set Up* tool, MicroStation begins to prompt you for information. The first thing to define is the view you want to establish as a camera view (Step 1). Placing a data point in the desired view defines it as the camera view.

The next piece of information is the camera target (Step 2). This can be defined in any open view. The Top view is usually used for this purpose. You can place a data point over any portion of the view to define the target.

Once the target is defined, a pyramid is displayed and follows the cursor. The pyramid represents the field of view of the camera. The actual location of the camera must now be established (Step 3). This is usually done in the same view as the target but can be placed in any open view. When the desired location is found, enter a data point to place the camera.

When the camera location is defined, the view pyramid is locked in position. MicroStation now needs you to define the front and back clipping planes (Steps 4 and 5). As the cursor moves around the screen, the visible portion of the pyramid changes. The front plane is defined first with a data point, and then the back plane is defined with another data point.

After the back clipping is placed, the camera is totally defined and it is turned on in the designated view.

➾ **NOTE:** *At any point before placing the back clipping plane, hitting reset will undo the last camera setting, going back to the designation of the camera view. So you can try different options before actually accepting the camera setting.*

Once placed the camera can be redefined or modified, so don't panic if you don't like the settings you have. *Move Camera* and *Move Target* allow you to manipulate a camera after it has been located. Both tools work just as they do in the setup camera. First, select the view with the camera to be modified, and then move the camera or target in any of the open views. The front and back clipping planes must also be redefined.

➾ **NOTE:** *When you create a camera view that you like, I suggest you create a saved view, so you can always come back to it.*

Lenses

Now for the real fun—lenses. Just like a real camera, you can put many different lenses on your camera. From fisheye to telephoto, MicroStation has the lenses you need. The lenses can be selected and adjusted on the Lens palette *(View → Camera → Lens).*

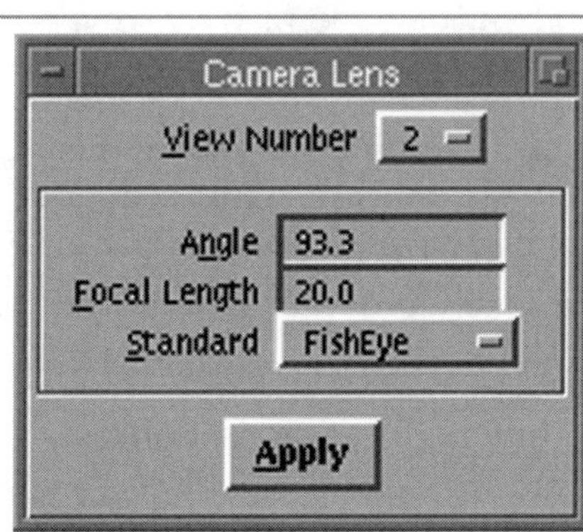

Camera Lens palette.

A combination of angle and focal length defines the MicroStation lenses, just as in real cameras. As the angle is increased, the focal length decreases.

Below is a list of the standard lenses. It's a good idea to start with the standard lenses, and then modify them as necessary.

Lens	Angle	Focal Length (mm)
Telescopic	2.4	1000
Telephoto	12.1	200
Portrait	28.0	85
Normal	46.0	50
Wide	62.4	35
Extra-wide	74.3	28
Fisheye	93.3	20

Once your camera is defined, adjusting the lens can be an effective way to learn more. There is no better way of understanding the camera and its lenses than to actually use them. If you are up to it, I have planned a short excursion. Grab your camera and stay close.

1 Telescopic
2 Telephoto
3 Portrait
4 Normal
5 Wide
6 Extra-wide
7 Fisheye

Standard lenses.

Exercise 1: The Camera and Lenses

1. Open **Inside 3D → Chapter 7 → Camera.**

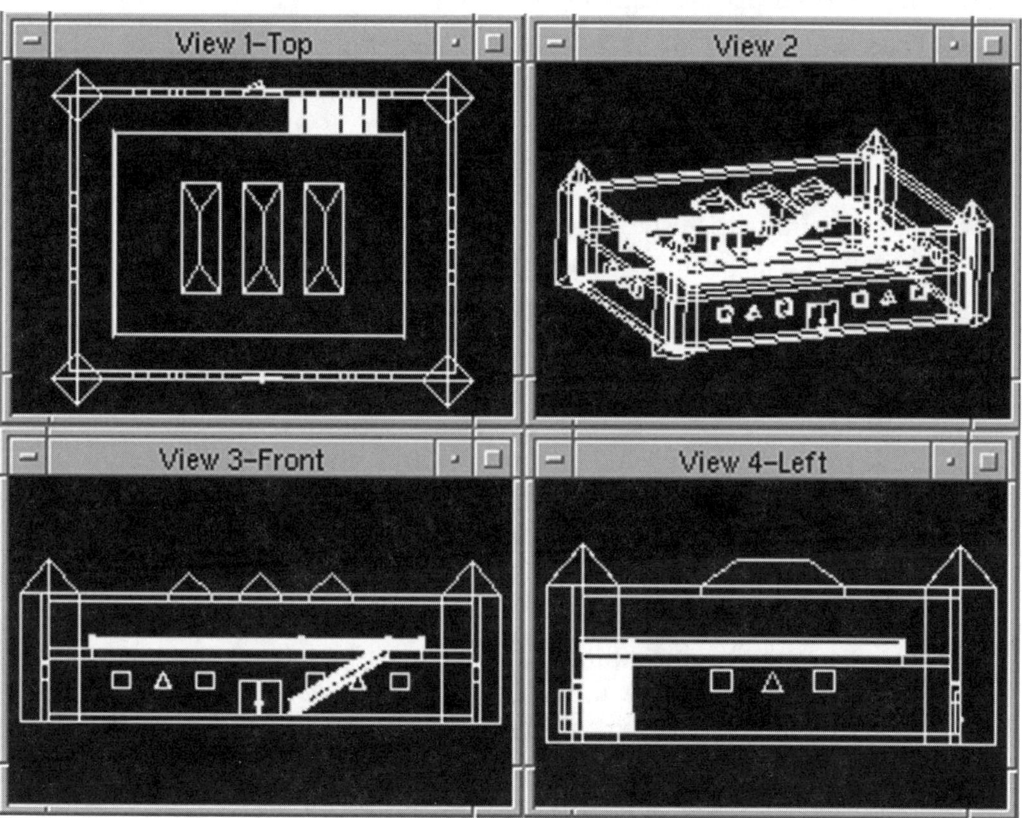

The Art Gallery.

This is the art gallery that I introduced you to in Chapter 1. We will spend some time here, studying the MicroStation camera and its operation. You will learn to set up the camera, its target, its lenses, and learn how to adjust them. Let's start.

1. Open the Lens palette **(View → Camera → Lens)**. Set the lens for the camera in View 2 to a normal lens.
2. Activate the camera setup from the Command window View subpalette **(View → Camera → Setup)** or type

```
Set Camera Definition
```

in the Command window dialog area.

The first thing MicroStation asks is which of the open views you desire to set up as a camera view.

3. Place a data point in View 2.

   MicroStation now wants the target of the camera to be defined. This is most often done in the Top view.

4. Place a data point in the Top view (View 1) near the staircase.

   Once the target is established, the view pyramid of the camera is displayed. The location of the camera now follows the cursor about the screen. As the camera moves, its view pyramid is automatically adjusted.

5. Place a data point in View 1 (Top view), near the lower left corner of the gallery.

   Now you must establish the front and back clipping planes.

6. Establish the front clipping plane by placing a data point in View 1, just inside the second story walkway. Now, establish the back clipping plane outside the gallery by placing a data point in the upper right corner of View 1.

   As soon as the back clipping plane is established, the camera view is updated. You are now looking through the MicroStation camera. This tool will allow you to visualize and communicate many concepts and ideas. But we are not finished. Once the camera is established, it can be adjusted. I will guide you through the process of adjusting the camera, its target, its clipping planes, and its lenses.

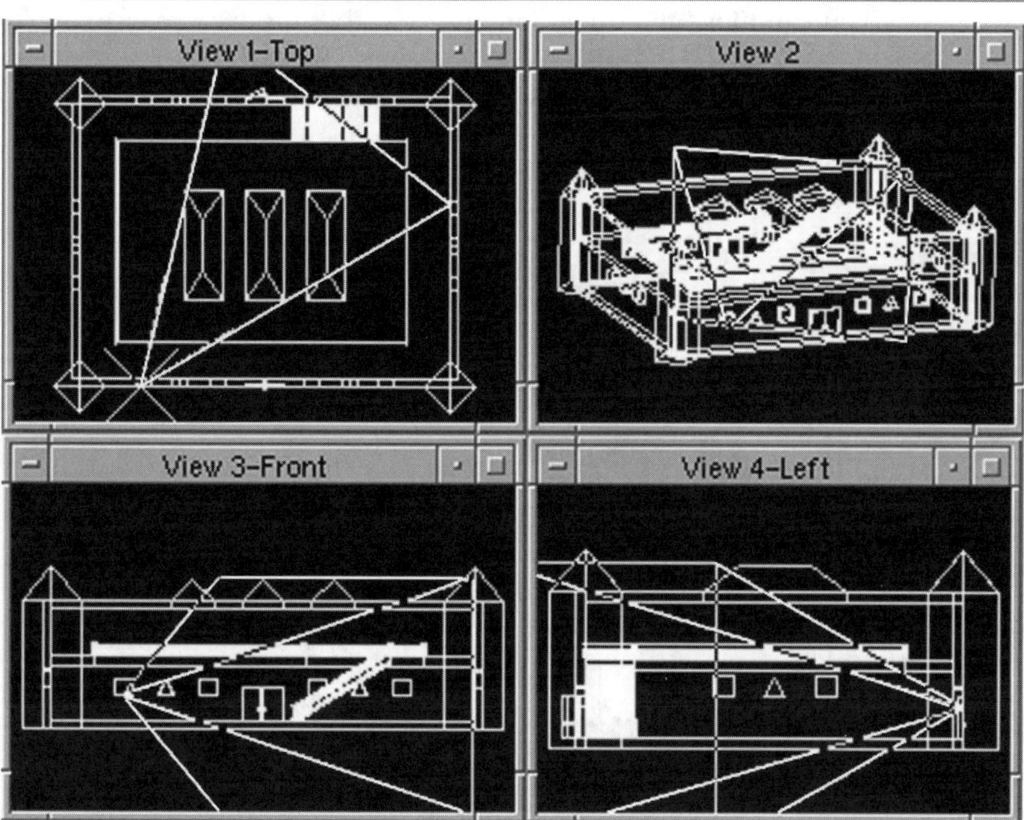

The camera view pyramid.

7. Activate the *Move Camera* tool **(View → Camera → Move Camera)** or key in `Set Camera Target` in the Command window.

   The first step is to define which of the views of the camera you want to modify.

8. Place a data point in View 2.

Now you can move the camera in any view, but for this exercise use the front view.

9. Move the camera in the Front view, so that it appears perpendicular to the stair in the Top view. Place a data point in the Front view. Then move to the Top view and establish the front and back clipping planes.

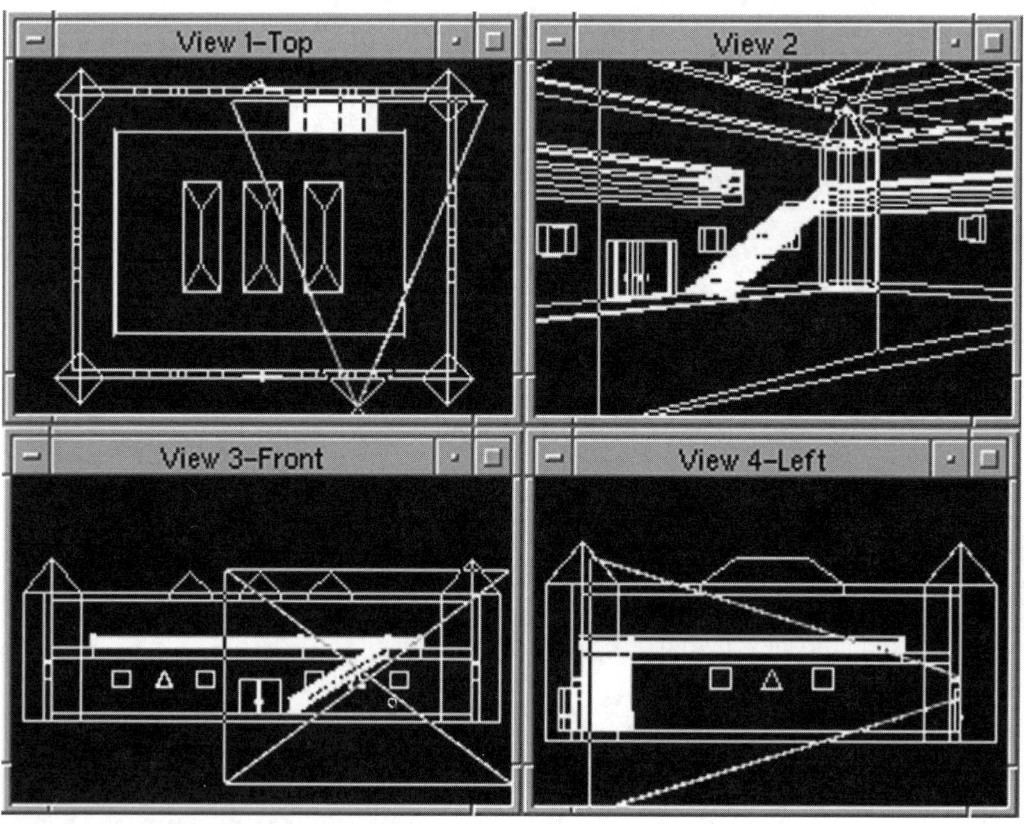

Move camera diagram.

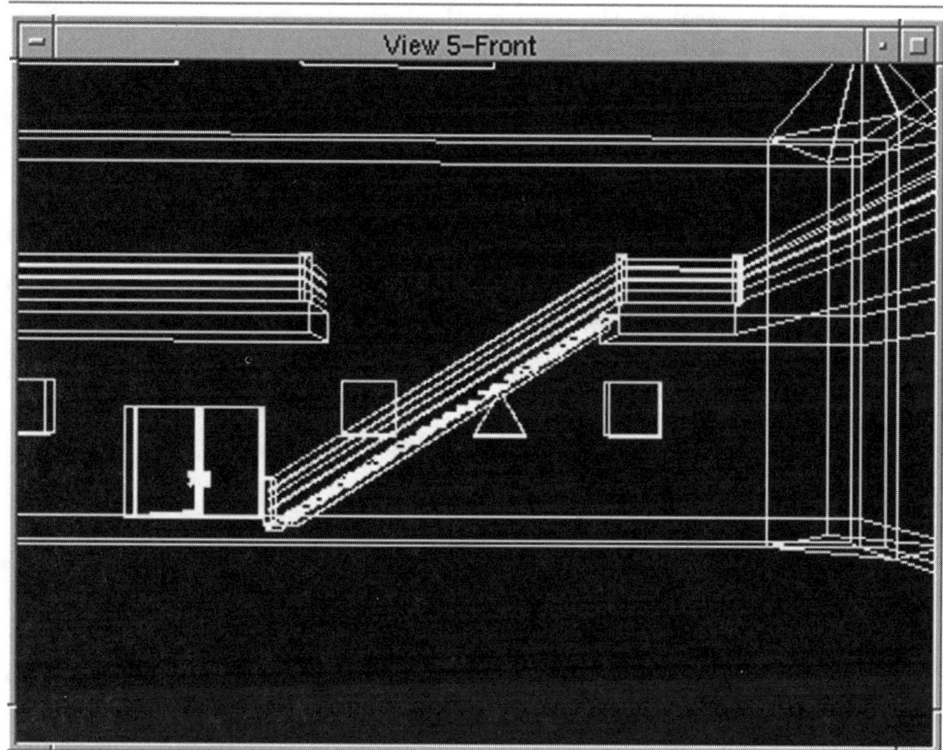

View of camera perpendicular to stair.

Perspective

Perspective is what everyone wants and few have. Safari Sam is here to give you some perspective on perspective. Let's start with *Parallel Projection*. The standard MicroStation perspective is a parallel projection. Three-dimensional objects are displayed on a 2D surface (your screen). Objects in the distance are in the same scale as those in front. This is great for creating models, but lacks realism.

For that reason, MicroStation has *perspective projection*. Elements at a distance are displayed scaled down appropriately to those in front. Perspective projection is basically the camera view. MicroStation supports two types of perspective projection, two point and three point. In a two-point perspective all vertical lines (lines in the Z-direction) are displayed parallel to the screen. Only elements in the X- and Y-directions are projected in perspective. In a three-point perspective, lines in all three directions are projected in perspective.

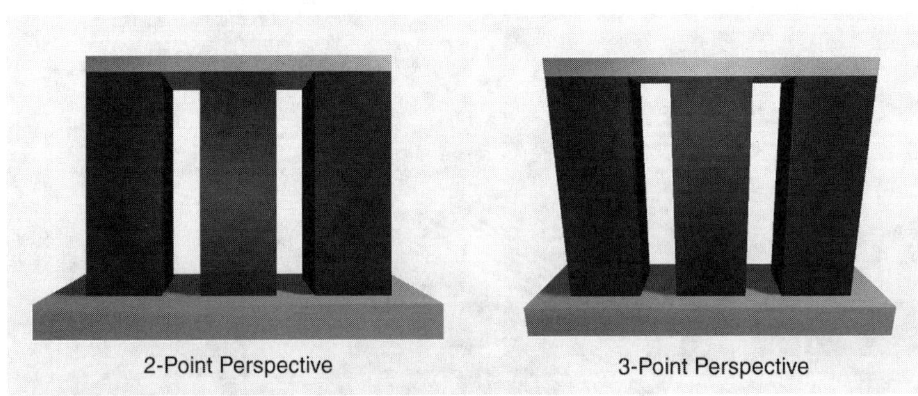

Two-point and three-point perspective.

A two-point perspective is achieved by defining the camera and target in the same view at the same Z-depth. A three-point perspective can be achieved by placing the camera and target in the same view but at different Z-depths or by placing the camera and target in different views. There are a few basic concept used to establish the desired perspective. These concepts are covered in Exercise 2.

Exercise 2: Perspective

1. Open **Inside → 3D Chapter 7 → Perspective.**

   This sculpture will help to explain the concept of two-point and three-point perspectives.
2. Set your camera lens to Extra-wide.
3. Activate the *Camera Setup* tool **(View → Camera → Setup)**.
4. Set the camera view by placing a data point in View 2.
5. In View 1 place the camera target at the center of the sculpture.
6. Now locate the camera at the lower center of the same view.

7. Include the entire sculpture in the clipping planes.

You now have a two-point perspective of the sculpture in View 2. Both the target and the camera were placed at the same Z-depth in the same view.

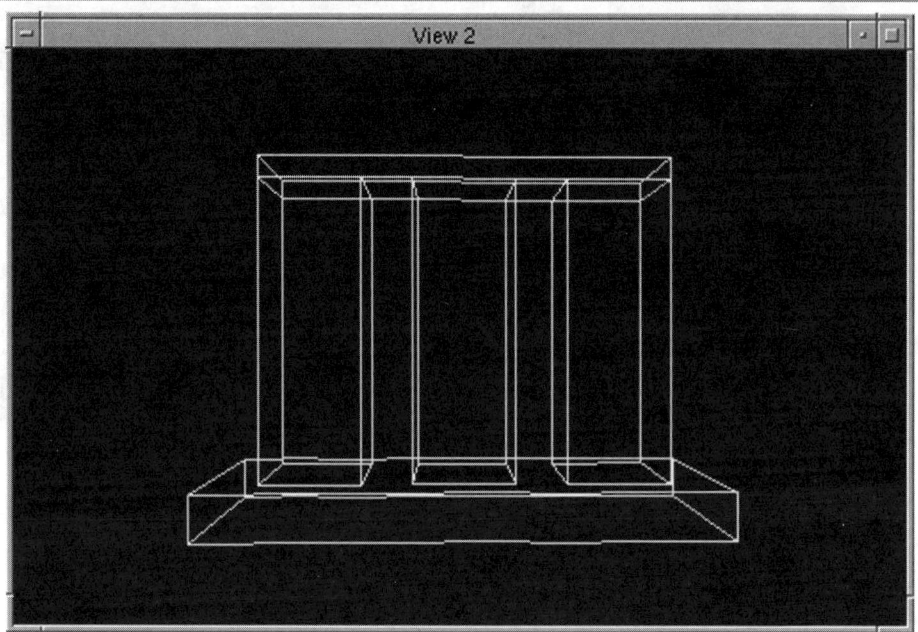

Two-point perspective of sculpture.

Now let's turn the sculpture into a three-point perspective.

1. Activate the *Move Camera* tool.
2. Identify View 2 as the camera view to be altered.
3. Relocate the camera by moving the cursor into View 4. Place a data point toward the bottom of the sculpture, far enough to the right that the pyramid includes the entire sculpture.
4. Redefine the clipping planes to include the entire sculpture.

Look at that, a three-point perspective. Notice how the vertical lines now have perspective, as well as the other lines.

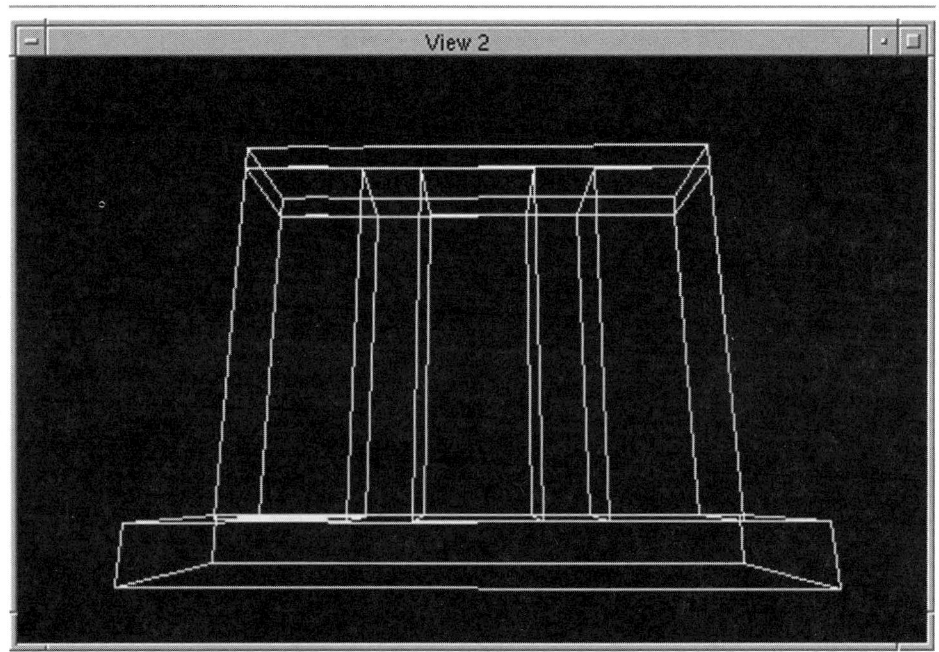

Three-point perspective of the sculpture.

Basic Rendering Routines

So, now you have the camera set up and adjusted. The next obvious step is to take a picture. There are just a few more settings to be considered before the picture is taken. What type of rendering technique (what type of film, black and white or color, etc.) are you going to use? How much of the view should be rendered, the entire view or just a section (a fence)? Lighting must also be considered. We will cover how much of the view should be rendered here; lighting will be covered later.

Let's tackle the rendering of just a section of the view option first. A fence can be used to designate the portion of the view to be rendered. This can be useful in checking your setup. You can view a portion of the view fully rendered, without having to wait for the entire view to be calculated. If this option is used, you must tell MicroStation not to render the view, but just the fence. This is done by keying in `RENDER FENCE "TYPE OF RENDERING"` ("type of rendering," i.e., Hidden Line or Phong).

The other option is the rendering technique to be used. The following list presents the techniques in increasing order of their realism.

Wiremesh
Hidden Line
Filled Hidden Line
Constant Shading
Smooth (Gouraud) Shading
Phong Shading
Phong Stereo Shading
Phong Anti-alias Shading

These techniques are located on the View pulldown menu of the Command window *(View → Render)*.

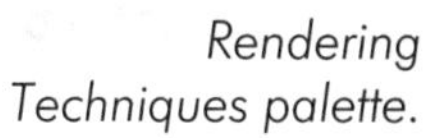

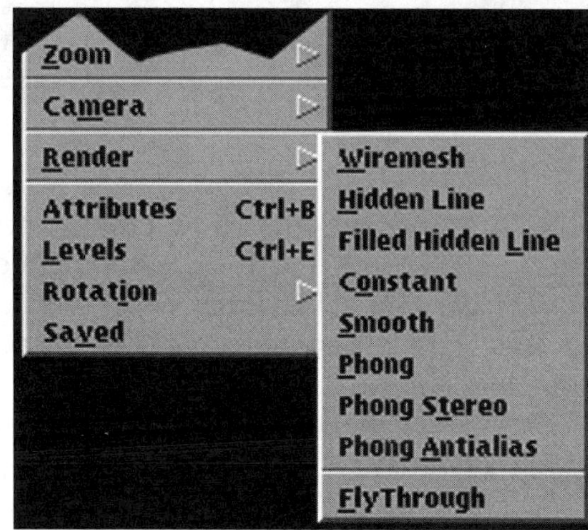

Rendering Techniques palette.

The basic display of the model is the wireframe. It is a simple display of MicroStation's elements. Every surface of every element is displayed. The wireframe is the quickest technique to display and work with.

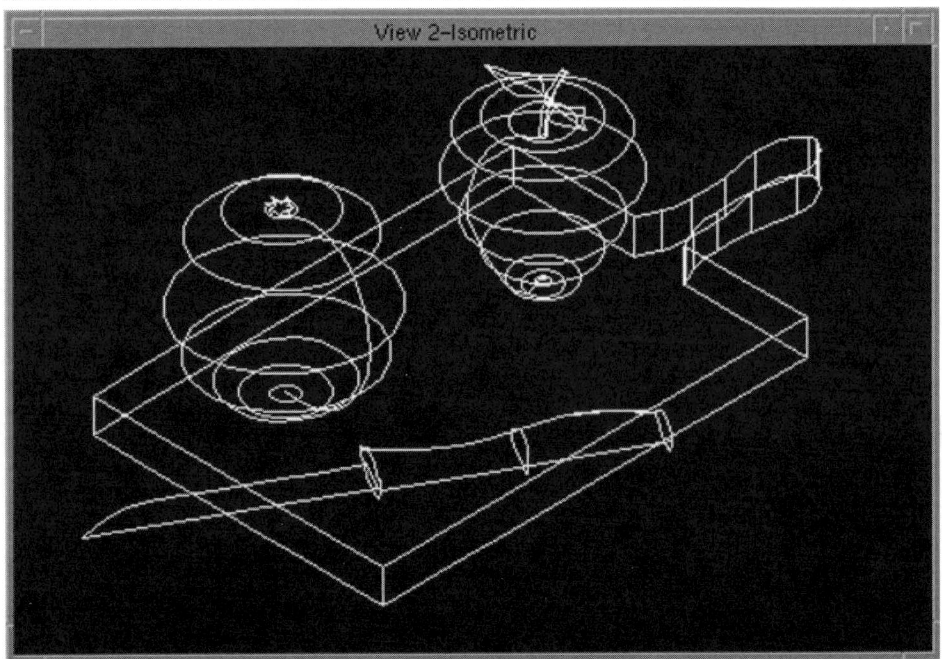

Wireframe view of still life.

However, it can become difficult to understand if the model becomes complicated. Because of this, there are other rendering techniques. So don't fret, we have special binoculars. You can view the model in different ways with these binoculars. As your guide, it is my duty to show you how to use these binoculars.

Wiremesh

The first setting is *wiremesh*. It is the most basic of the rendering techniques. It is similar to wireframe; however, it adds a mesh to complicated elements. This makes them more recognizable and understandable. All surfaces are displayed (no lines are hidden) because all elements are treated as transparent elements.

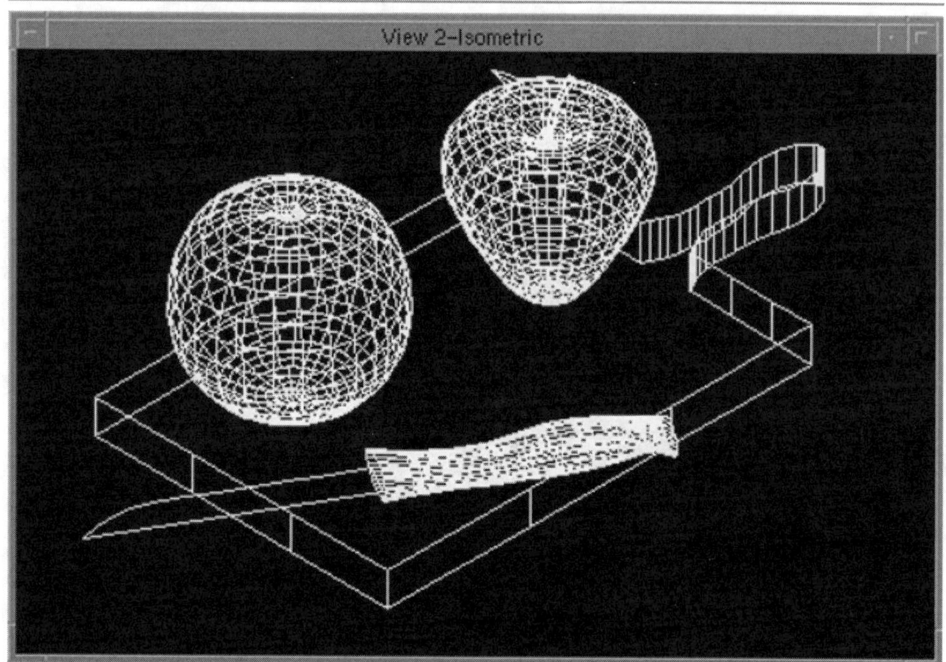

Wiremesh view of still life.

Hidden Line

Hidden Line is the next rendering technique. No surprise here, adventurer, this technique removes the lines that would not be visible in the view. This process takes as long as some of the more realistic techniques; however, it is one of the most informative, especially if color output or display are not options.

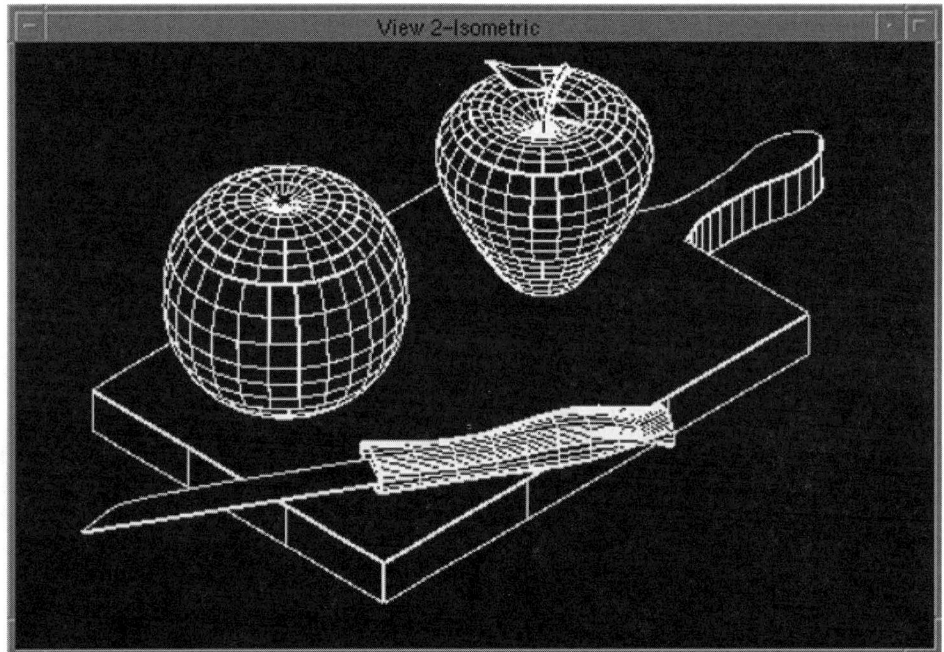

Hidden Line technique.

NOTE: *A hidden line 2D file can be created for documenting the model. This is done from the Command window (File → Export → Visible Edges).*

Filled Hidden Line

The next rendering technique is *Filled Hidden Line.* It is similar to the hidden line itself. The difference is that surfaces are filled with solid colors. The color of the element determines the color that fills the surface. The Filled Hidden Line technique is useful when a limited number of colors are available.

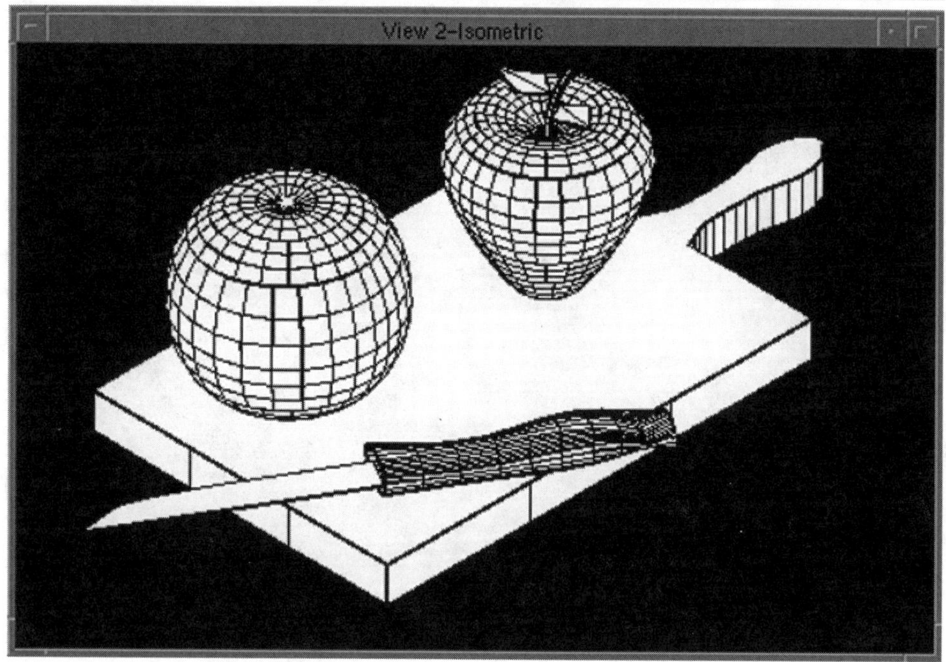

Filled Hidden Line technique.

Constant Shading

Constant Shading fills each surface with a single constant color. Contoured or curved surfaces are displayed as a mesh of polygons. The color of the surface is determined by the element color, material definitions, and lighting.

Constant Shading technique.

Smooth (Gouraud) Shading

The next shading style is *Smooth (Gouraud) shading.* In 1971 Henri Gouraud introduced this type of shading. Like smooth shading, contoured or curved surfaces are displayed as a mesh of polygons. The major difference is in the shading algorithm. Colors are calculated around the edges of each polygon, and then blended across the polygons' interiors. This adds a smooth appearance and eliminates the tile effect seen in constant shading.

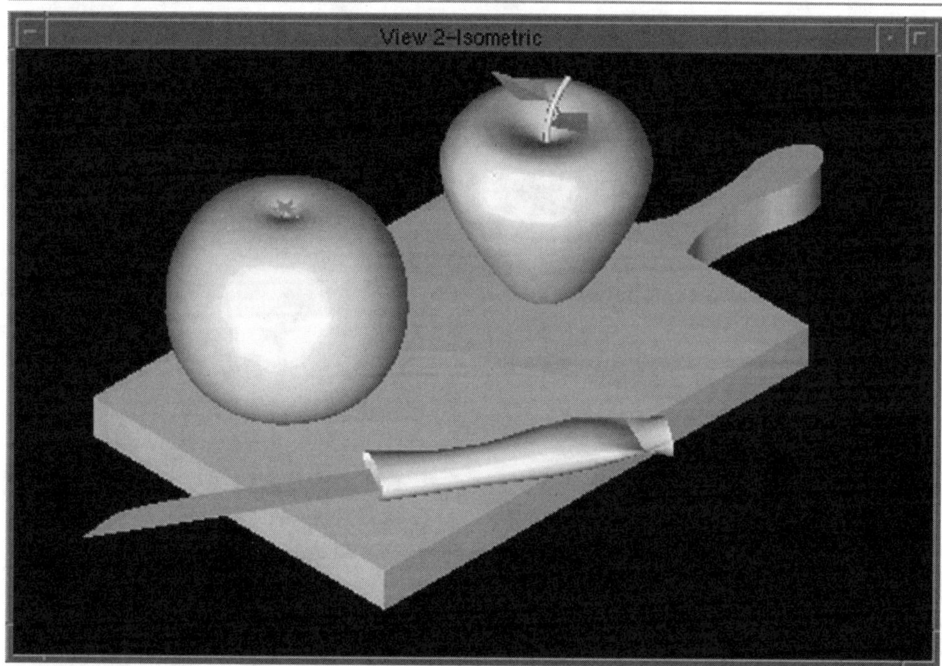

Smooth (Gouraud) shading.

Phong Shading

Bui-Toung Phong developed a new shading method in 1971, *Phong shading.* Phong shading is the most realistic shading method that MicroStation supports. Unlike the other shading techniques, phong shading calculates each pixel individually across a polygon. There is no tiling or averaging, but it does require a considerable number of computations. If your computer is not of the fastest breed, use constant and smooth shading to set up a view before using phong shading.

NOTE: *This is also a good place to use the Render Fence tool.*

Phong is also the only shading technique that renders shadows, bump maps, and atmospheric distance cueing. It is also required to show a pattern map in the exact desired position.

Phong shading.

Phong Stereo Shading

If you still have the sunglasses that are shipped as standard equipment with MicroStation, get them out. *Phong Stereo shading* is the reason they are supplied. This shading technique renders the view twice, first for the right eye and then for the left. The two images are then "color coded," one red and the other blue. A center image is created where the two intersect. With the glasses the image takes on a 3D effect.

Phong Stereo shading.

Phong Anti-alias Shading

I have saved the best for last. In *Phong anti-alias shading,* images are rendered using the Phong shading method and then the image is anti-aliased. Anti-aliasing is based on the theory of *fractal sets,* developed by Dr. Benoit Mandelbrot, a French mathematician.

The image is rendered using the phong shading technique. It is then rendered again in a slightly shifted position and an average is taken of the two. If the settings are at the defaults, the image is actually rendered four times, at four different positions and averaged. The quality of the anti-aliasing is controlled by the Anti-aliasing Grid Size in the rendering settings box *(Settings→ Rendering→ General).* The default of 2 produces excellent results. Increasing the grid size to 3 produces slightly better images, but requires additional passes. Phong anti-alias shading is by far the most time-consuming technique. However, the results are just as consuming—they are beautiful.

Phong anti-alias shading.

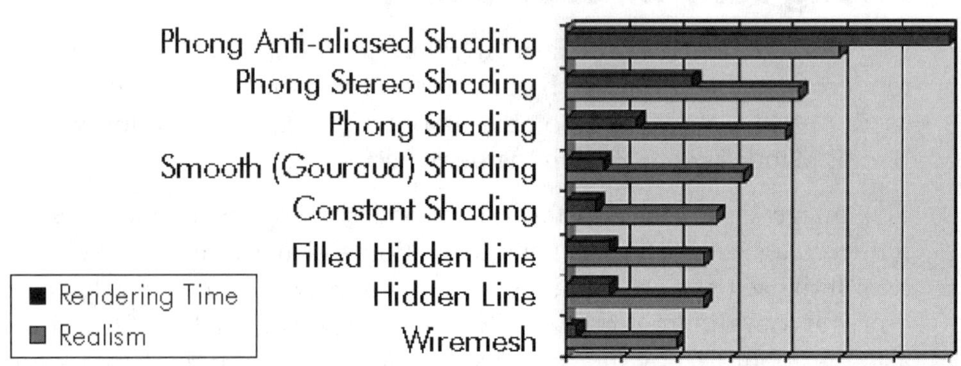

Rendering time versus realism.

Shading Techniques

What's that you say? The sun is going down, the shadows are growing, and perhaps we should set up camp? Don't worry, here in the MicroStation wilderness we have complete control over lights and shadows. The sun sets when we tell it to set, we turn the lights on and off, and we tell the shadows where they should fall. This section of Chapter 7 is an introduction to lights and shadows. You real trailblazers will need to peek at Chapter 12 for the advanced stuff.

Before you take control of the lighting, MicroStation simply provides you with a flashlight attached to your safari helmet, just like a miners' helmet, you are not able to edit this light source. When a view is shaded, this light is used to illuminate the model. Once you take control of the lighting, MicroStation supports two main types of lights, global and source. Global is just that, it affects the entire design cube. Source light is used to simulate the headlights of a car, a spotlight, and other similar light sources.

Global lighting (Settings → Rendering → Global Lighting) has three options: ambient, flashbulb, and solar. *Ambient lighting* increases or decreases the lighting of every surface of the model. It becomes the lighting coefficient of areas of the model that receive no other lighting whatsoever. The deepest, darkest part of your model will be as light as the ambient light. The value can be defined between 0 (none) and 1 (full). A value of 0.10 to 0.40 is most effective.

Flashbulb lighting is similar to the default light on your safari helmet. It is based on the concept of a flash for a camera. It is also defined between 0 (none) and 1 (full).

Solar lighting is intended to simulate the sun. A high degree of realism is possible with this tool. To truly understand this tool, see Chapter 12.

Now we move to *Source Lighting (Settings → Rendering → Source Lighting)*. MicroStation has three options here as well: Point, Distant, and Spot. Source lighting is created by placing a lighting cell. It contains the characteristics of the light and can be moved and modified. A *Point light* radiates in all directions from the cell. It is useful in adding general light to a dark portion of a model. *Distant lighting* is a directional ambient light. It can be used to increase the ambient lighting in just a portion of a model.

When the distant light cell is placed, it is given a direction. *Spot lighting* acts like a real spotlight. When the spotlight cell is placed, it needs a direction. However, it also requires a Cone Angle (the hot spot) and a Delta Angle (the falloff or fade).

Now that you have a general understanding of lighting, let's deal with shadows. Shadows are a subtle but important part of shading. However, they do require more time to render, so use them appropriately. Only certain MicroStation lights cast shadows. The following table lists the lights that do and do not cast shadows.

Lighting Technique	Casts Shadow
Ambient	No
Flashbulb	No
Solar	Yes
Point	No
Distant	Yes
Spot	Yes

Even though a light that can cast shadows is present in the model, it will not cast them if the Shadows option is turned off in the Rendering View Attributes settings box *(Settings→ Rendering → View Attributes)*. See Chapter 12 for a complete discussion of shadow maps and techniques.

Below are two images of the still life. Both have been rendered with phong anti-alias shading. One has lighting and shadows added and the other does not. The value of lighting and shadows is clear. Don't underestimate their effectiveness in communicating an idea.

Still life without lighting and shadows.

Still life with lighting and shadows.

Summary

In this chapter you learned how to work with MicroStation's camera and to change your lenses and perspective. You also learned how lighting and shadows can affect the presentation of your model, making it much more realistic. For more advanced shadow techniques, see Chapter 12.

Now you are ready to tackle auxiliary coordinate systems. Follow Safari Sam into the next chapter.

Introduction

Many MicroStation adventurers have heard of the realm of auxiliary coordinate systems (ACSs), and very few have ventured there. The fork in the road could lead to trouble; which way do you go? Many users are intimidated by the concept of using different coordinate systems, but Safari Sam will show you the way to use a powerful new 3D tool.

The coordinate system you have used so far is the Cartesian coordinate system. The Cartesian system is named after the famous French mathematician and philosopher, Renee Descartes, and uses X, Y, and Z coordinates to specify points in space. The other coordinate systems you will learn about are the spherical and cylindrical coordinate systems. Furthermore, you will learn how to manipulate the Cartesian coordinate system (hereafter referred to as the rectangular coordinate system) to create additional

Cartesian coordinate systems to facilitate the placement of elements on any plane.

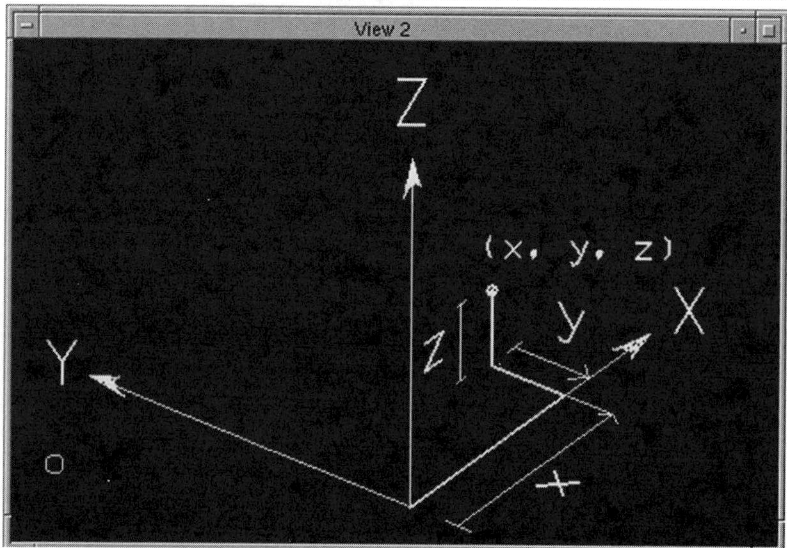

Rectangular coordinate system and its coordinates: X, Y, and Z.

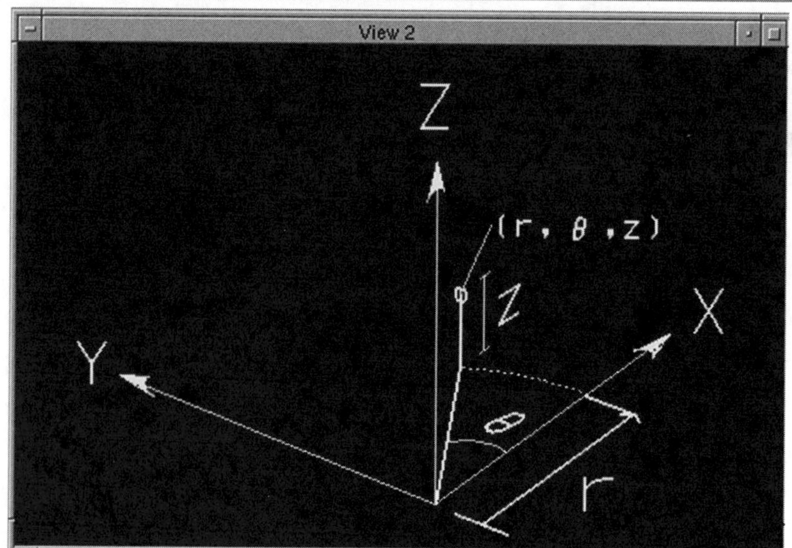

Cylindrical coordinate system and its coordinates: radius, angle, and height.

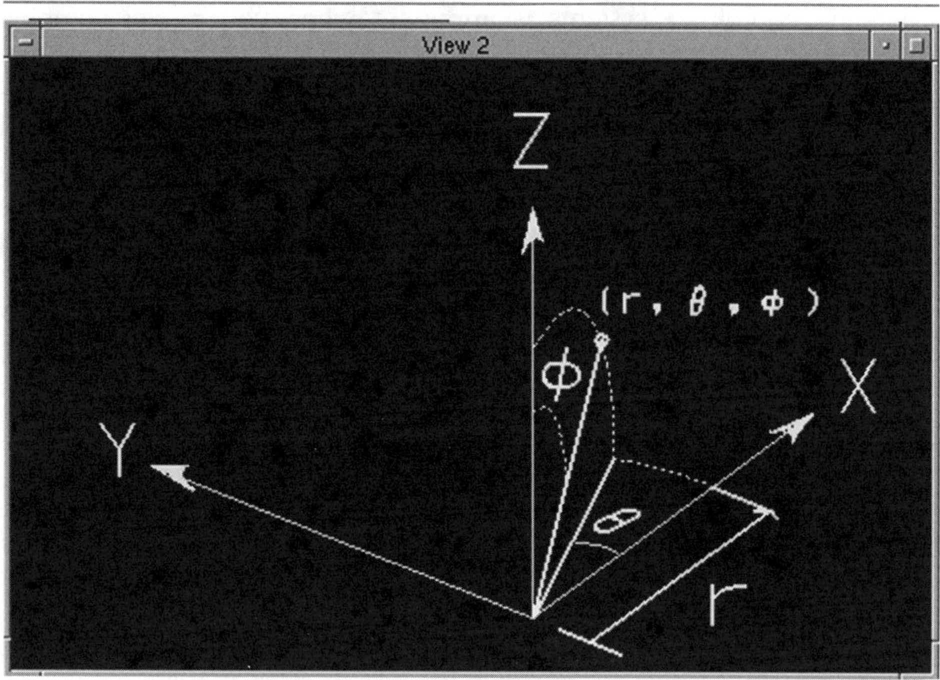

Spherical coordinate system and its coordinates: radius, angle 1 (θ), and angle 2 (φ).

Rectangular Coordinate Systems

The most commonly used coordinate system is the rectangular coordinate system (RCS). You have been using this system for 2D work. The 3D rectangular coordinates are defined by the X, Y, and Z-axes (both in the positive and negative directions). By creating a new RCS, we are merely changing the location and possibly the orientation of the auxilliary origin, from the global origin. The Auxiliary Coordinate System palette is found under the Palettes menu.

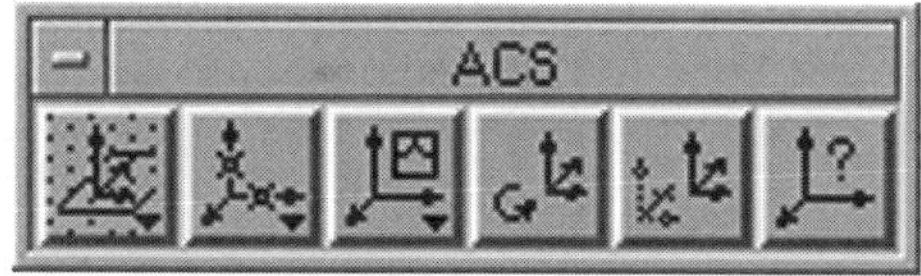

The Auxiliary Coordinate System palette.

Changing coordinate systems is important when you model complex objects, such as buildings and highways. In the case of a house, you have many different objects in different planes, such as the roof tile, which is in the plane of the roof. Drawing the roof tile on a standard pitched (inclined) roof is difficult, because the tile is inclined in two planes and is placed in a third. To alleviate the problem of calculating all those angles, MicroStation allows you to attach a coordinate system so that placing objects is easy. In the case of your roof, you attach a coordinate system so that the X,Y plane is parallel to the roof and the positive Z-axis sticks out orthogonally to the roof. By using this system, roof tiles can be placed in the new X,Y plane at z=0. Safari Sam will show you how.

Exercise 1: Defining a Rectangular Coordinate System

1. From the Inside 3D pulldown menu, select the *RCS Definition* command under the *Chapter 8* heading. Working Units are architectural.

   The file that opens is a 3D file with a slab placed in it. The slab was placed with its lower left corner (slab origin point) at the XY=0,0,0 point and is 15 feet by 15 feet and 1 foot thick. Furthermore, the slab was rotated in the Right view by 45 degrees, about the lower left corner in the Right view.

   Next, you will learn how to define an RCS and attach it to the roof.
2. Open the ACS palette from the Palette menu.
3. Select the *Define ACS (By Points)* tool from the ACS palette, and make sure that you select Rectangular from your Tool Settings palette.

The Define ACS (By Points) icon.

4. The first point you define will be the origin of the new ACS. Snap to the lower corner in the lower left of the Top view.

Note that there are two corners in the lower left of the Top view: the upper corner is at XY=0:0.0:0,0:0 (absolute coordinates), and the lower corner is at XY=0:0,-0:8 1/2,-0:3 1/2. Do you know why this lower corner is at a negative Y and Z? Think about the rotation operation and look at the Right view.

You wish to attach your RCS to the xy=0,0,0 point. Hence, you are not changing the global origin. You are, however, going to rotate the axis.

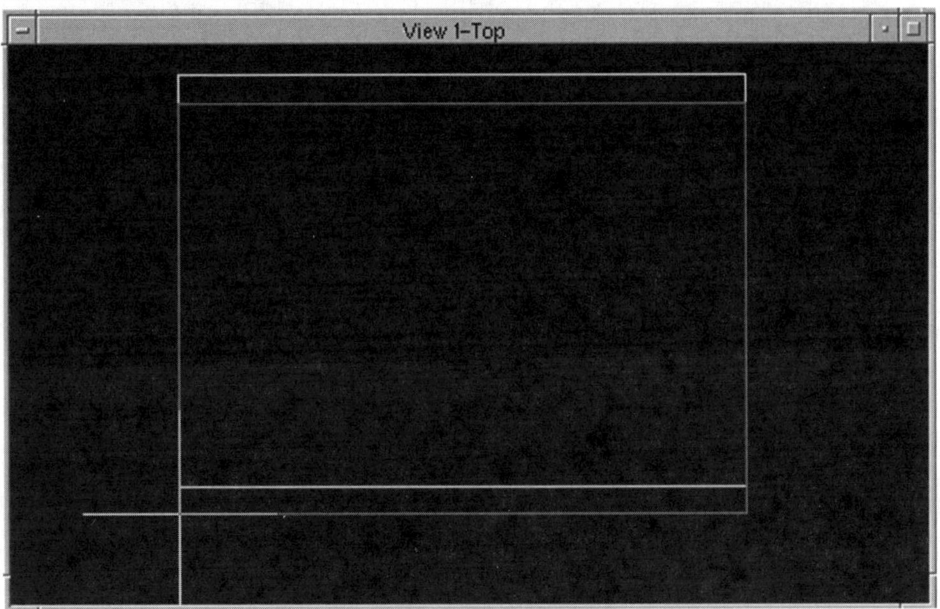

Definition of the RCS Origin by snapping and then accepting to existing elements in the Top view.

✔ **TIP:** *To get the coordinates of a point in space, use the Move Element tool from the Modify Element palette to snap to the location in question. Hence, snapping to the slab element in the lower left corner will give a coordinate readout. The display of the coordinates is in the Command window in the Status Field.*

MicroStation Command Window - ch8rcs2.dgn
File Edit Element Settings View Palettes Snaps User Help
0:0 , -0:8 1/2, -0:3 1/2 LV=1,WT=0,LC=SOL,CO=86,TP=KeyPt
Define ACS (By Points) Enter first point @x axis origin
(1) TYPE (R|C|S) :

Location of coordinates in Status Field of Command window after using the Move tool to successfully snap to an existing element.

5. You now need to select the positive X-direction. Snap to the upper corner in the lower right of the Top view. Once again, there are two corners in the lower right of the Top view. The upper corner is located at XY=10:0,-0:8 1/2,-0:3 1/2. Accepting the snap point sets the direction of the positive X-axis.

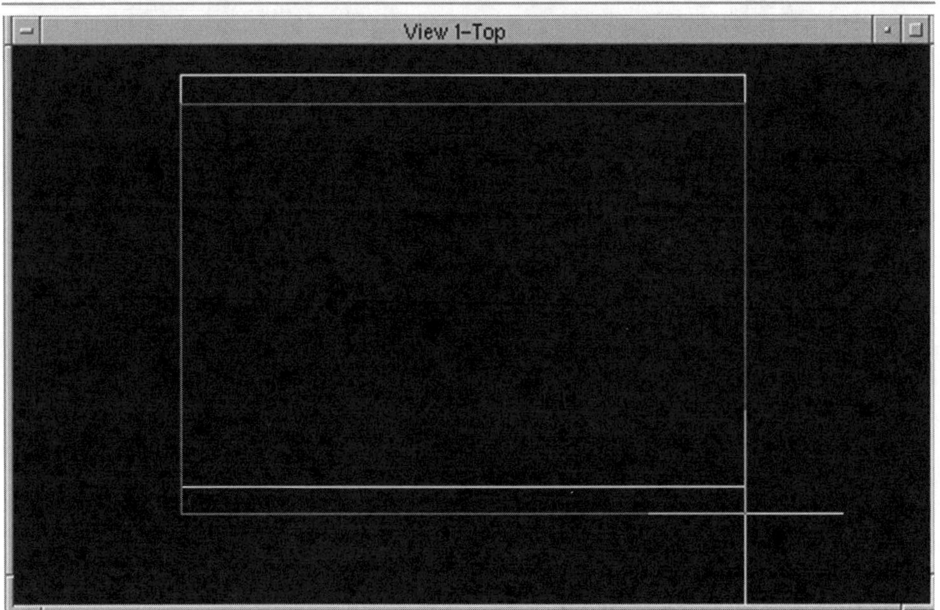

Definition of the positive X-axis direction by snapping and then selecting a data point to accept in the Top view.

6. Finally, you need to specify the direction of the positive Y-axis. Snap to the upper corner in the upper left of the Top view. Once again there are two corners in the upper left of the Top view. The corner of interest is located at XY=0:0,6:4 3/8,6:9 3/8 (absolute coordinates). After you have selected the correct corner, select a

data point to accept. If done correctly, the message "New Coordinate System Defined" appears in the message area of the Command window.

➾ ***NOTE:*** *The Z-axis direction is automatically defined by the positive X and Y axes, since it must be orthogonal to both.*

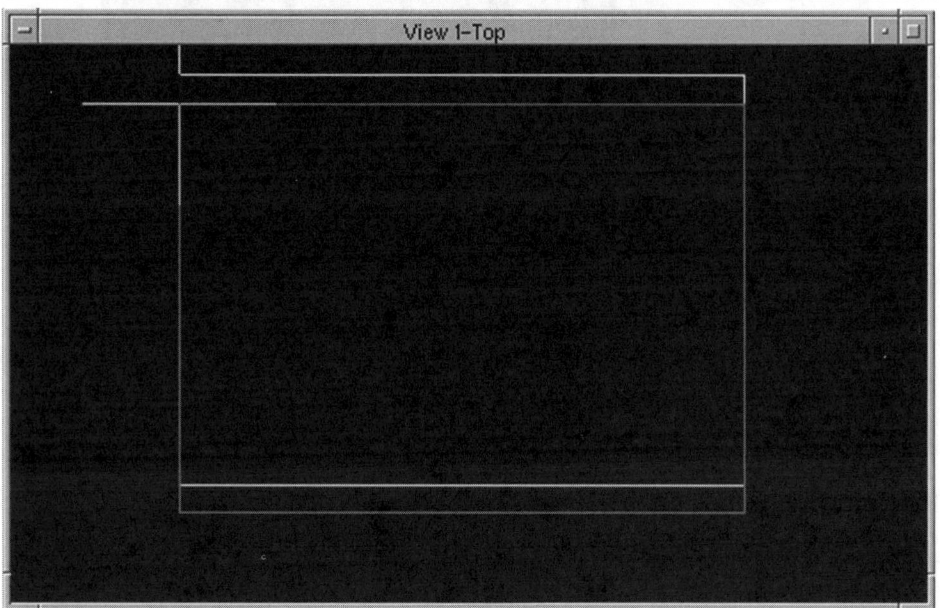

Definition of the positive Y-axis in the Top view by snapping to existing elements.

7. To see the new coordinate system, go to the **View → Attribute** dialog box and turn on ACS Triad by selecting All. This command shows the location and direction of the three axes you just defined.

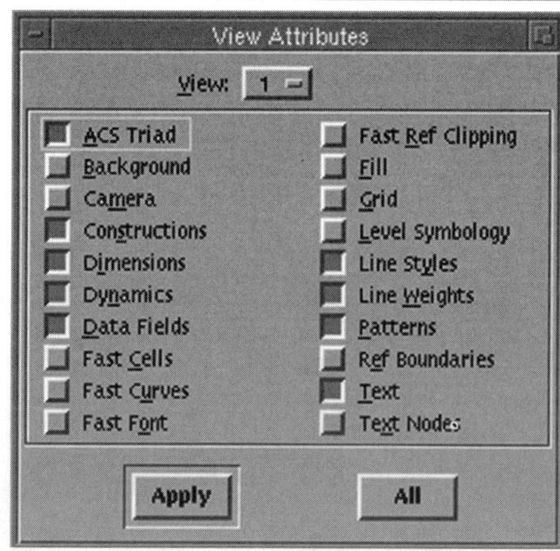

View Attributes dialog box with ACS Triad activated.

Standard MicroStation views with ACS Triad turned on for all views.

Congratulations, you have just defined a new RCS. You can now use it to place elements. The RCS has its X,Y plane aligned with the roof surface, since that is what we used to define it. Remember, the default coordinate systems (absolute and view) are still available to us, so XY= and DI= still work.

Precision Input Using Auxiliary Coordinate Systems

You now have an additional way of entering data for your design file, by ACS precision input. Let's review what the other precision inputs are.

Keyin	Description
XY=	Absolute point
DI=	Relative to view axes point
DL=	Relative to design plane
DX=	Relative to absolute axes point

Now we will see how we can use keyins for ACSs.

AD=	Relative to ACS axes
AX=	Relative to ACS view axes

These keyins are helpful when placing elements in the ACS plane. The precision keyins are the best way of creating elements in the ACS plane. However, other methods are also available to the curious adventurer. The way to create elements in the ACS plane without using precision input is to use the ACS Plane Lock from the *View → Locks → Full/Toggles* dialog box. These locks can also be activated from the Tool Settings palette when you select the D*efine ACS (By Points)* tool.

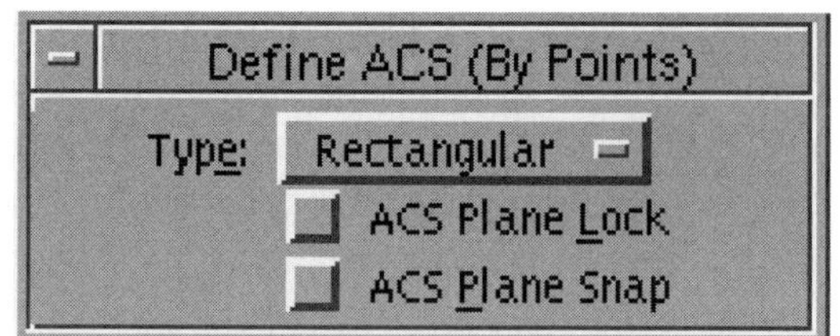

Tool Settings palette when Define ACS (By Points) is selected.

By activating this palette you set the ACS Plane Lock and ACS Plane Snap. The *ACS Plane Lock* command will only allow data points to be entered on the ACS X,Y plane (in our case, the X,Y plane of the ACS is aligned with the roof), and no data points can be entered in the design cube anywhere else. If ACS Plane Lock was off, a data point would be entered at the active depth parallel to the screen view in which it was drawn.

Similarly, the *ACS Plane Snap* command will allow snapping only to the ACS X,Y plane and nowhere else in the design cube.

Now that you have created an ACS of your own, it would be helpful to save it. From the Settings menu choose Auxiliary Coordinates *(Settings → Auxilliary Coordinates)*. The dialog box for Auxiliary Coordinate settings is presented. The dialog box is divided into two halves, Active ACS and Saved ACS.

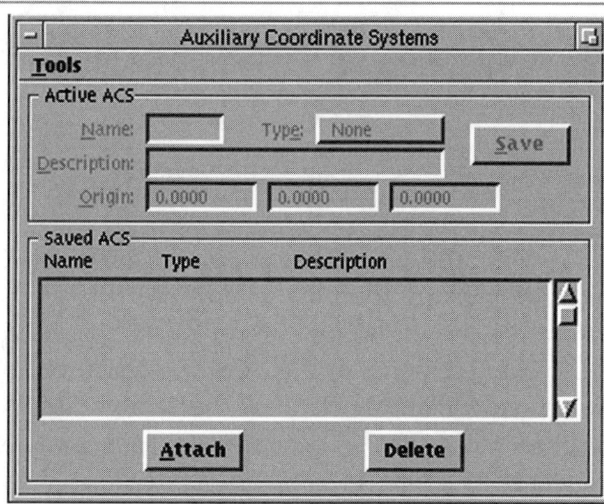

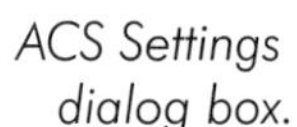
ACS Settings dialog box.

All fields in the dialog box are grayed out if you have not defined an ACS. If you have defined an ACS, the origin of it is displayed, and you can fill in the fields. Name your ACS in a logical fashion, for instance, rectangular ACSs could be named RECT1, and spherical systems could be named SPHER1.

You may also enter a description of the ACS. In your case the description could read "Roof of house." Next, make sure the type is set correctly (either Rectangular, Cylindrical, or Spherical), and then click on the Save button. At this point the Name, Description, and Type appear in the Saved ACS list. Now let's use the new RCS you just defined. If you have not saved it yet, do so now.

Exercise 2: Using a Rectangular Coordinate System To Create Elements

Using the RCS you previously created, you will now place tiles on the roof. Please note that the tiles you will create are not actually based on real tiles, but are just a way of demonstrating the coordinate system.

1. Open the **Settings → Locks → Toggles** dialog box, and turn on the ACS Plane lock. Now all data points you will place will be relative to the RCS you defined.
2. Using precision input you will create the first tile. Select the *Place Line* tool and enter the following keyin in the Command window:

   `AX=0,0,0`

   This will start the line at the origin of the RCS.
3. Next, enter the following keyin:

   `AD=1:6`

   which creates a line 1.5 feet long, aligned along the positive X-axis of the RCS. Reset to finish the command. Since you are working with precision input, the view is not important, but most of your work will occur in the Front view.
4. Select the *Place Line* tool again and snap to the middle of the newly created line (set your snap divisor to 2, or use the `KY=2` keyin) in the Front view, and key in the following:

   `AD=,,0:6`

This creates a 6-inch line perpendicular to the RCS X,Y plane, in the positive Z-direction. Reset to complete the command.

5. Select the *Place Arc By Edge* command and snap to the RCS origin (point 1, the beginning of your first line) in the Front view. Data point to accept when you have selected the right point, then snap to the top of second line you drew (point 2), and data point to accept.

Place Arc By Edge command.

✖ **WARNING:** *Make sure that the ACS Plane Snap lock is not turned on.*

6. Finally, snap to the end of the first line you drew (point 3), again in the Front view, and data point to accept. You should now have an arc perpendicular to the surface of the roof.

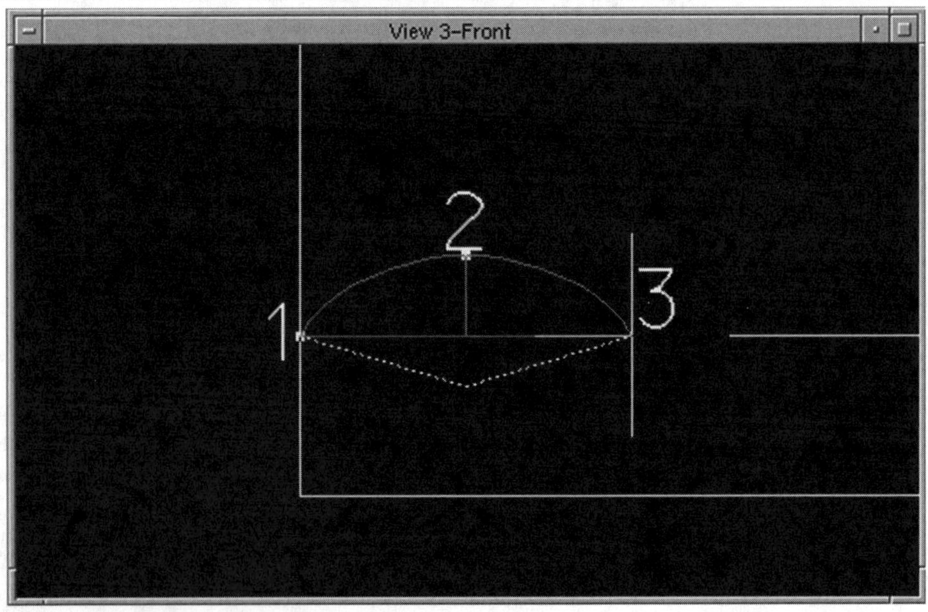

Place Arc By Edge command with placement points.

7. Select the *Copy Element* command and select the arc you just drew in the Front view. Then key in the following, creating a second arc 3 inches above the first.

   ```
   AD=,,0:3
   ```

8. Select the *Place Line* tool and snap to the end of the first arc (point 1) in the Front view and data point to accept. Next, select the end point of the arc above it (point 2), data point to accept, and reset to complete the command. Place a similar line from point 3 to point 4.

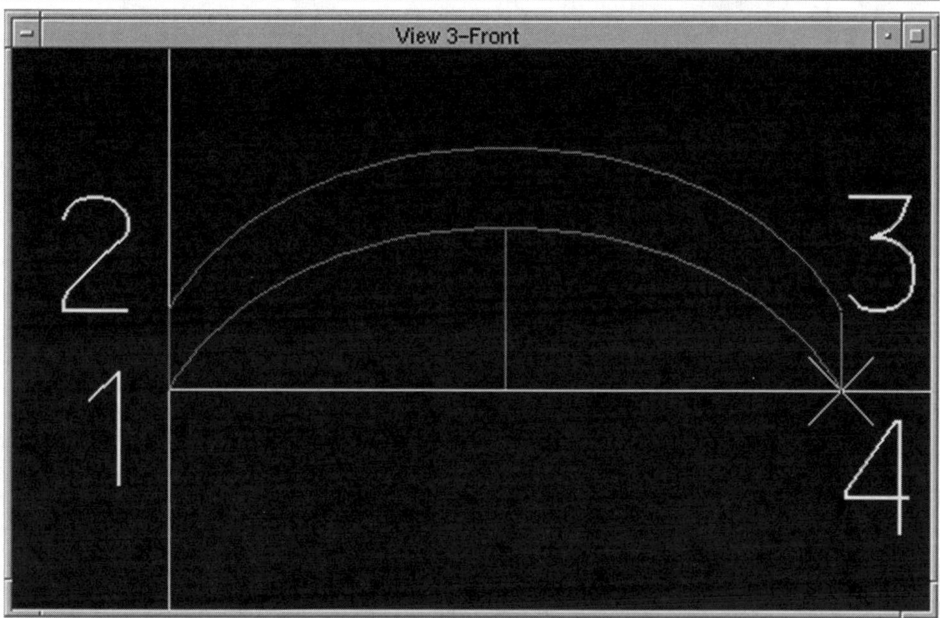

Place Line command between the two arcs with placement points.

9. Use the *Create Complex Shape* command with Manual and Solid settings in the Tool Settings window. Select the first arc in the Front view, then the line connecting the two arcs, then the next arc, and finally the next line connecting the two arcs. Reset to complete the command. You can now delete the construction elements (the first two lines we created, the shape and the line used for projection).

10. Select the *Place Line* command and in the Front view snap to the end point of either arc of your newly created shape. Key in the following:

```
AD=,1:6
```

A line 1.5 feet long is created. You will use this line for creating a solid tile from your arced shape, using the *Construct Surface or Solid of Projection* command.

11. Select the *Construct Surface or Solid of Projection* command and select the arc we just created, then snap to the endpoint of the line you just created (point 1) in the Front view, and data point to accept. Use the *Change Element Attributes* command to change the color to red and the level to 2.

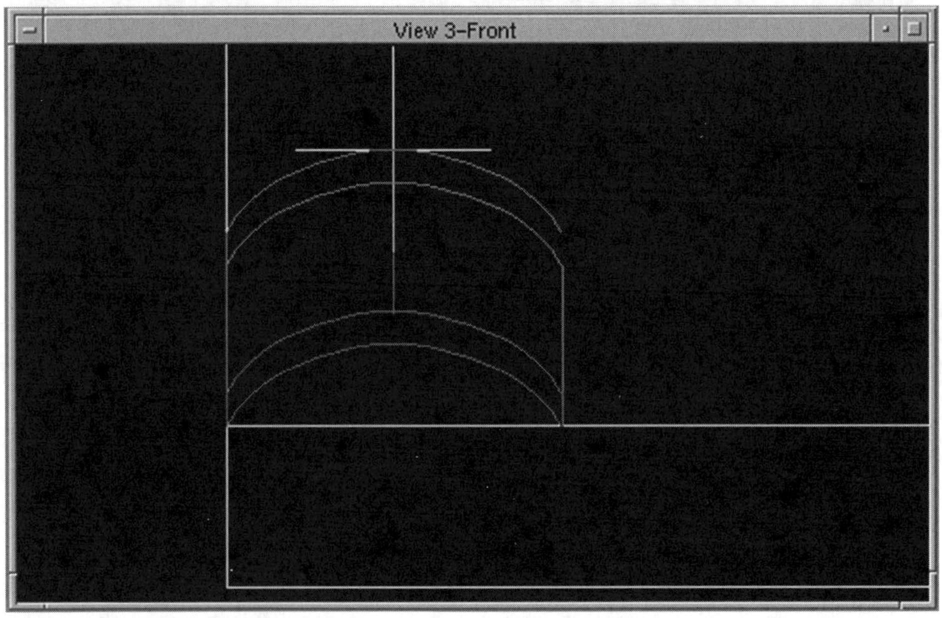

Creating a Solid of Projection from the shape and the line.

Make sure your tile aligns with the roof. Select the *Render View* command from the 3D View Control palette with the Phong setting and render the Front view, and then compare it with the following image.

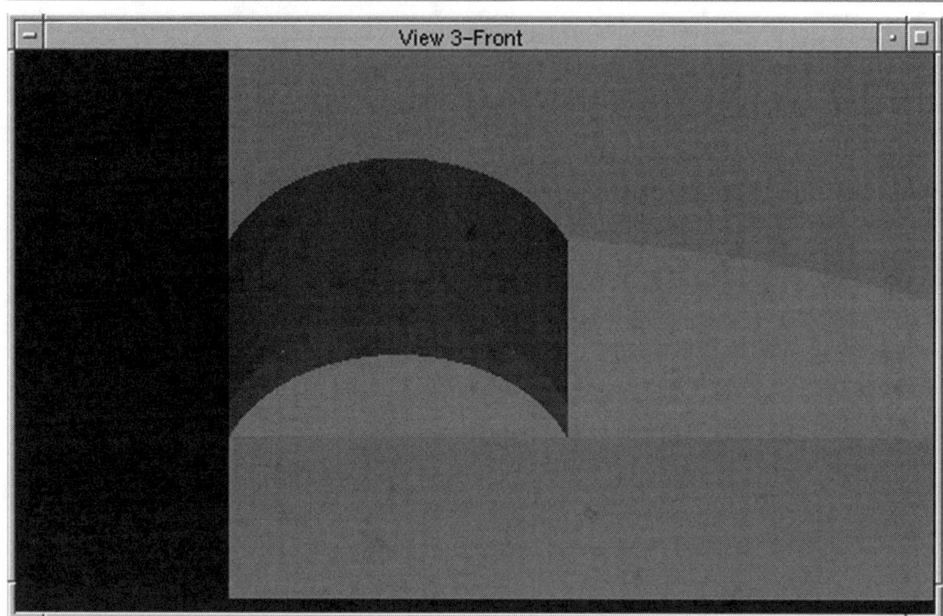

Rendered view of roof and tile.

12. Finally, use the *Rectangular Array* command from the Manipulate Element palette with the following settings:

Type	Rectangular
Rows	13
Columns	11
Row Spacing	1:2
Column Spacing	1:6

Select the tile and then data point to accept the command. The roof should be covered with tiles. Render the Front view; it should look like the following image.

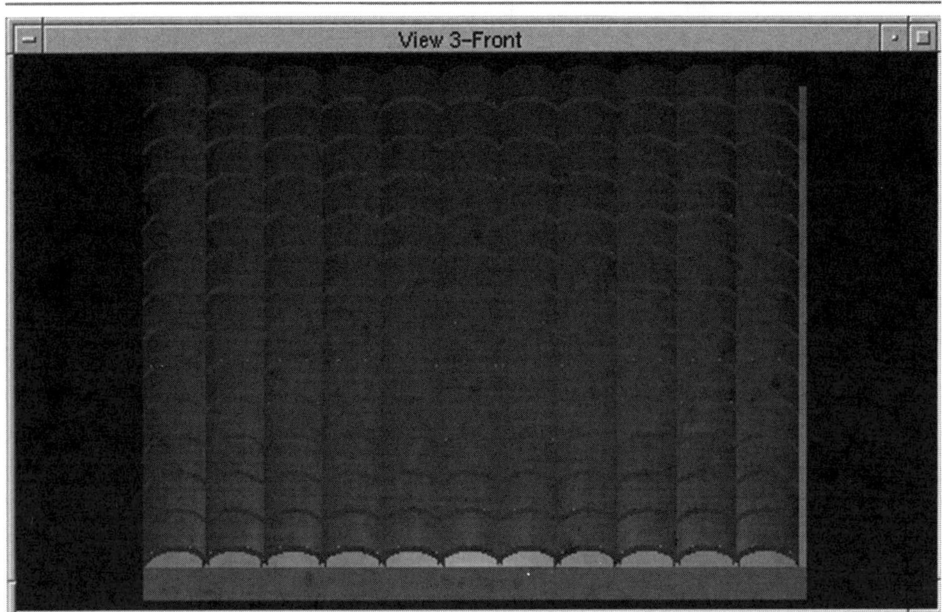

Rendered Front view of the tiled roof.

To further enhance the realism, use specifications from a manufacturer to create a more accurate tile.

Spherical Coordinate Systems

Spherical coordinates and cylindrical coordinates are special cases of the more general curvilinear coordinate system. These coordinate systems also specify points in space, but do it with angles and distances (unlike rectangular coordinates, which only use distances). Remember, you will still use the three axes you used before, the X, Y, and Z axes; however, you will refer to them differently.

When would you use spherical coordinates? The answer lies in the geometry you need to build. If the predominant geometry of your object is spherical or conical, then you should consider using spherical coordinate systems. Some examples of this geometry are geodesic domes (which you will build), cylinder heads, lenses and lighting design, and many other small applications within a larger design. You can also use spherical coordinates to create cones and planes.

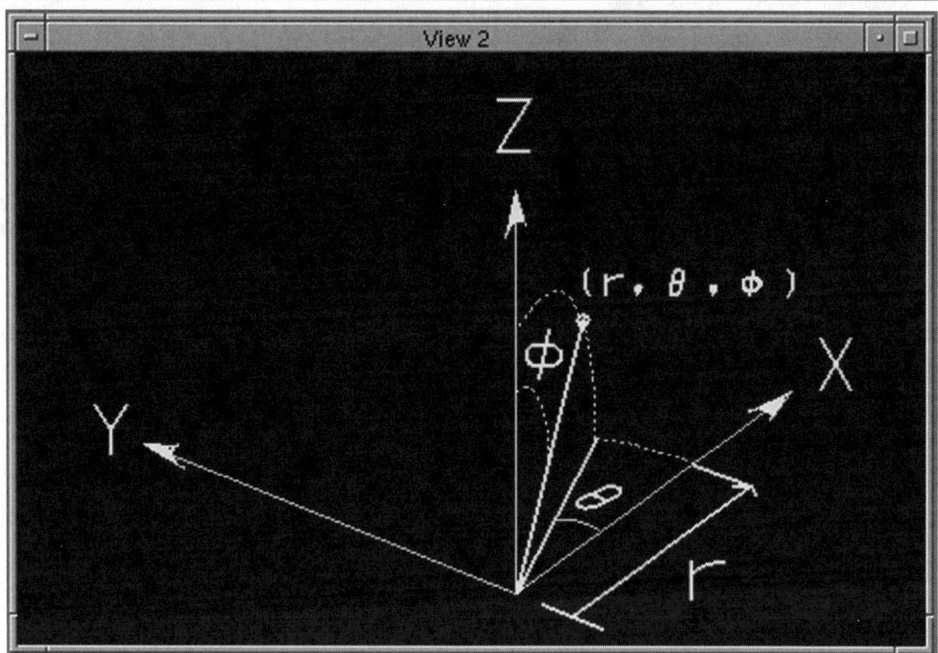

The spherical coordinate system and its three coordinates: radius (r), angle 1 (θ), angle 2 (φ).

You will notice that there are three spherical coordinates needed to specify a point in space. These are the radius, given the symbol *r*; the angle measured counterclockwise from the positive X-axis about the Z-axis called θ (the Greek letter theta); and the angle from the positive Z-axis (the angle between the Z-axis and the radius) called φ (the Greek letter phi).

Sometimes, you will need to convert from spherical coordinates to rectangular coordinates or vice versa. Here are some handy transformation formulas:

```
x = r * sin φ * cos θ

y = r * sin φ * sin θ

z = r * cos φ
```

where X, Y, and Z are rectangular coordinates. Using simple trigonometry these formulas can be calculated and used for conversions. The following are the inverse equations:

$$r = (x^2+y^2+z^2)^{1/2}$$

$$\theta = \tan^{-1}(y/x)$$

$$\phi = \tan^{-1} [(x^2+y^2)^{1/2}/z]$$

Now let's see how you can use these new tools and explore another fork in the road.

Exercise 3: Creating and Using a Spherical Coordinate System

The road leads to a place where spheres reign supreme. So forget X, Y, and Z and say hello to *r*, θ, and ϕ, our new guides. You will now build a geodesic type dome. Once again, this dome is not to specifications for construction but is an exercise to illustrate the use of spherical coordinate systems.

1. Select *SCS Definition* from the Inside 3D pulldown menu *Chapter 8* heading. The file you have just opened has a sphere of radius 1 working unit (architectural working units), with no ACS defined. A 20-foot diameter circle was drawn to keep the display depth beyond the edge of the elements to be drawn. The sphere and circle are on level 62.
2. You begin by defining your ACS. Select the *Define ACS (By Points)* command and set the type to Spherical in the Tool Settings window. Snap to the center of the sphere to define the ACS origin (point 1) in the Iso view, and data point to accept. Then snap to point 2 to define the X-axis, and then to point 3 for the Y-axis. Turn on ACS triad from **View → Attributes** and confirm your axes' orientation.

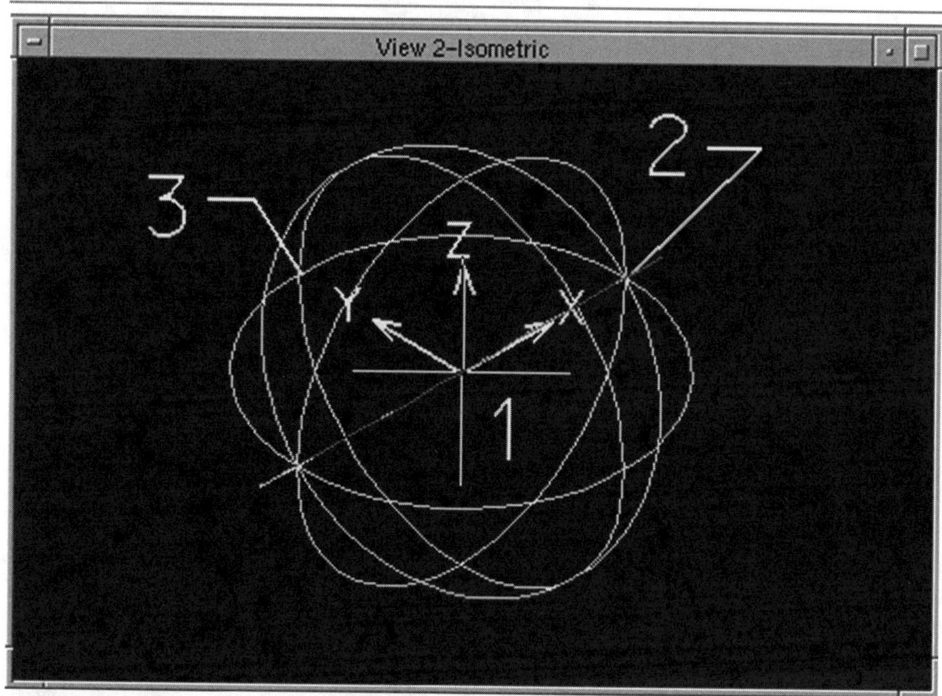

Defining the spherical coordinate system using points.

3. Save your ACS and name it SPHER1 and then Attach it, using the **Settings → Auxiliary Coordinates** dialog box.
4. Select the *Place Line* tool and enter the following:

   ```
   AX=15,0,90
   ```

5. The keyin tells you that you have a 15-foot radius, at an angle of 0 degrees with the X-axis and 90 degrees down from the Z-axis. Reset to complete the command.
6. Use the *Place Shape* command to create the perimeter of the dome and snap to the end of the line you just drew (the beginning of this line will be the center point of the dome). Key in the following once you have chosen the right point:

   ```
   AX=15,10,90
   ```

   then,

   ```
   AX=15,20,90
   ```

7. Continue to place points at 10-degree intervals all the way to 360 degrees.

NOTE: *Using the up arrow key will recall your last keyin.*

The perimeter should look like the following figure.

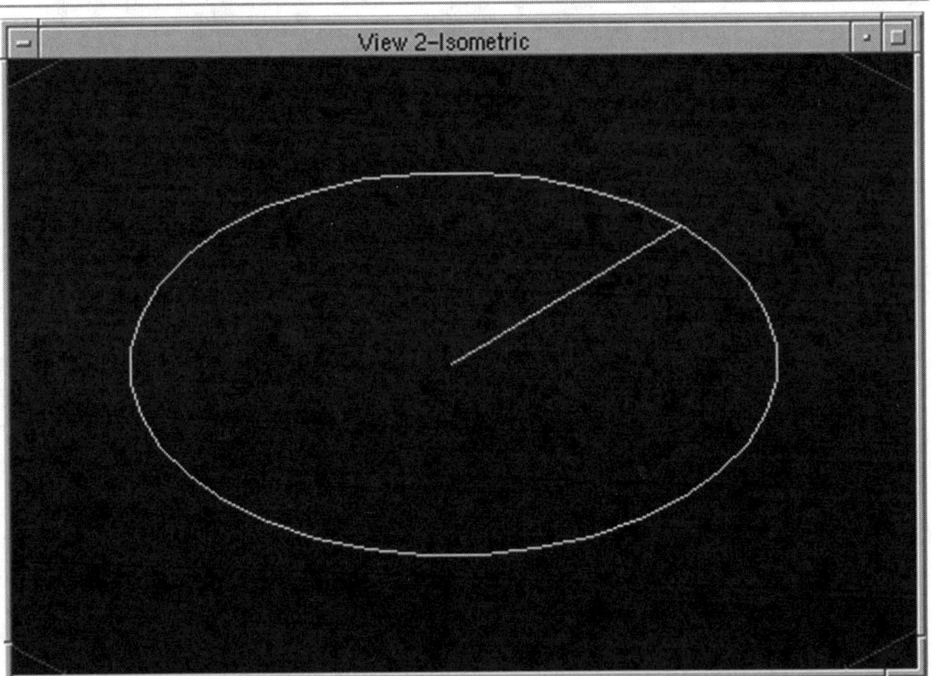

Perimeter of geodesic dome.

8. Select the *Place Shape* tool again and snap to the end point of the first line you drew (the start point of your perimeter), data point to accept, and then key in the following:

   `AX=13.3333,5,80`

   which places a point above, behind, and to the left of the X-axis. Next, enter the following keyin:

   `AX=15,10,90`

9. Then choose *Close Element* from the Tool Settings dialog box.

10. Select the *Construct Rectangular Array* command from the Manipulate Element palette. Choose the Polar type from Tool Settings

and set Items to 36 and Delta Angle to 10 degrees. Also set (turn on) the Rotate Items button.

Tool Settings palette for Polar Array.

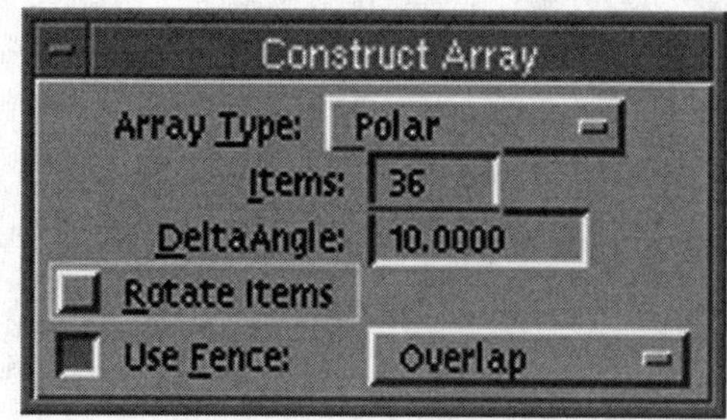

11. In the Top view select the triangular shape that you just created, snap to the beginning of the first line you drew (this will be the center point for your array and dome), and then data point to accept.

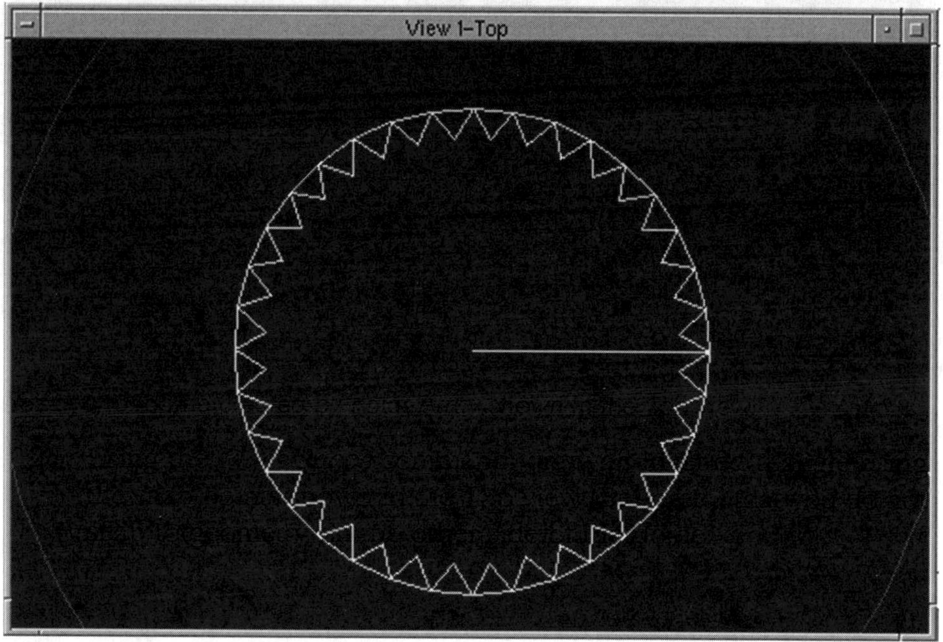

First layer of dome created by Polar Array shown in the Top view.

12. Using the *Place Shape* command, draw an inverted triangle using the snap points shown in the Top view. Be careful that you do not snap to geometry on the other side of the dome.

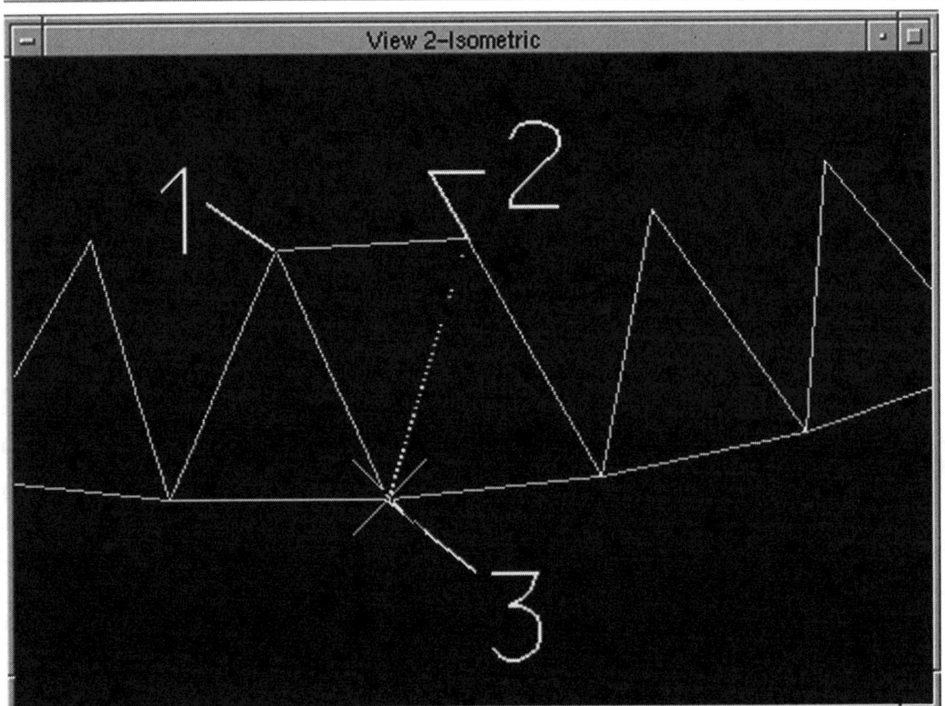

Snap points for Place Shape to create an inverted triangle.

13. Repeat the *Polar Array* command on the inverted triangle with the same settings.
14. Use the *Place Shape* command and input the following keyins:

    ```
    AX=13.3333,5,80
    AX=11.6666,10,70
    ```

    then,

    ```
    AX=13.3333,15,80
    ```

    and finally select the *Close Element* command.
15. Repeat the *Polar Array* command with the same settings as before on the new shape.
16. Create another inverted triangle as before with the *Place Shape* command to fill in the second row.

17. Use the following keyins to create the third row with the *Place Shape* command:

```
AX=11.6666,0,70
AX=10,5,60
AX=11.6666,10,70
```

18. Select *Close Element*, repeat the *Polar Array* command, create the inverted shape, and then array the inverted shape.
19. The keyins for the fourth row are as follows:

```
AX=10,5,60
AX=8.3333,10,50
AX=10,15,60
```

    Repeat Step 18.

20. The fifth row keyins are as follows:

```
AX=8.3333,0,50
AX=7.5,5,40
AX=8.3333,10,50
```

    Once again repeat the process in Step 18.

21. The sixth row keyins are as follows:

```
AX=7.5,5,40
AX=6.6666,10,30
AX=7.5,15,40
```

    Repeat the process to complete the row.

22. For the seventh row enter the following keyins:

```
AX=6.6666,0,30
AX=6.1666,5,20
AX=6.6666,10,30
```

    Repeat the process to complete the row.

23. The eight row keyins are as follows:

```
AX=6.1666,5,20
AX=5.9,10,10
AX=6.1666,15,20
```

Repeat the process, and the dome is complete. The dome is meant to be placed on a foundation of some sort, and the center can be capped by a hemisphere.

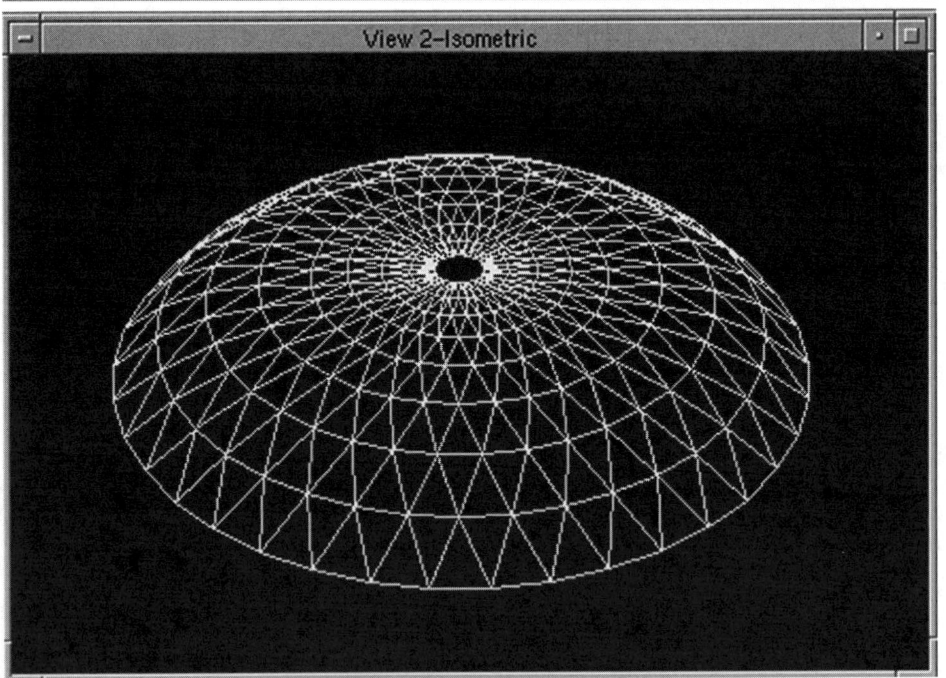

The geodesic dome, with all eight rows.

By keeping one of the three coordinates constant, you can generate different objects. If *r* is constant, then you produce spheres, centered about the origin. If theta is constant (the angle measured counterclockwise from the X-axis about the Z-axis), then you produce planes that rotate around the Z-axis, and if phi (the angle down from the Z-axis) is constant, then you can construct conic sections, whose point is at the origin and angle is determined by theta, with height, r. By mixing these constants you can produce many different wedges and sphere sections.

Cylindrical Coordinate Systems

In the previous section we used the *Polar Array* command to create each level of the geodesic dome. The Polar system is two dimensional and uses radius and angle. If you extrapolate this system in the Z-axis you have the cylindrical coordinate system (CCS).

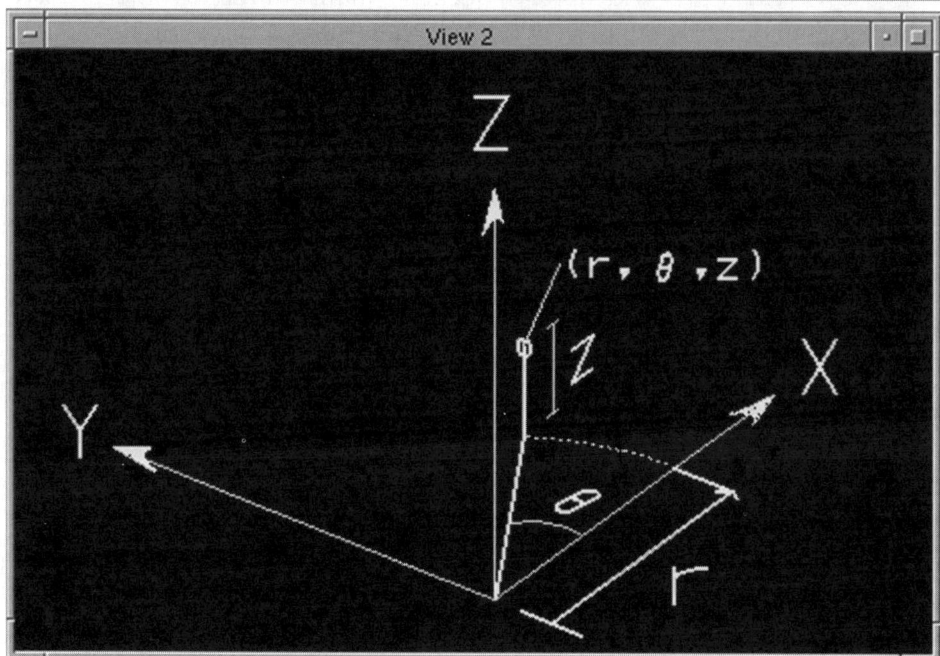

The cylindrical coordinate system and its coordinates: radius (r), angle (θ), and height (z).

So remember, in cylindrical coordinates you can specify any point in 3D space using the radius from the origin, an angle measured counterclockwise from the positive X-axis about the Z-axis, and a height of Z.

The transformations are as follows:

```
x = r * cos θ

y = r * sin θ

z = z
```

Almost invariably the positive Z-axis from the RCS will align with the Z-axis from the CCS; this is the default alignment.

The inverse transformation is the same for *r* as in spherical coordinates, and Z is equal to Z. For theta:

$$\theta = \sin^{-1}(y/(x^2+y^2+z^2))$$

The use of CCSs are widespread because many different objects have cylindrical geometry, such as wheels, axles, tires, conveyor belt parts, engine parts, and airplane fuselages. Generally speaking, the study of hoop stress is a very important subject in engineering. CCSs help you control the placement of geometry so that you can maintain cylindrical symmetry.

To find a point in space you move out along the positive X-axis, a distance *r*, and then rotate about the Z-axis counterclockwise by an angle theta, and finally move up the Z-axis (or parallel to the Z-axis), to your height.

You can use the same ACS precision input tools and lock tools in the CCS.

Exercise 4: Defining and Using Cylindrical Coordinates

In this exercise you will create a spiral staircase.

1. From the Inside 3D menu select the *Define CCS* command from the *Chapter 8* heading. This will open a file of a cylinder with a 1-foot radius, and a height of 1 foot. The center of the bottom face is at XY=0,0,0. Working units are architectural. Begin by attaching a CCS. Notice there are two lines drawn to facilitate proper attachment for the CCS.
2. Select *Define ACS (By Points)*, snap to the center of the bottom face (point 1) or the beginning of both lines (same point) and data point to accept. Snap to point 2 and data point, and then snap to point 3 and data point to accept.

   The idea here is to reproduce an Iso image with the axes (make sure ACS triad is turned on) of the ACS aligned properly.

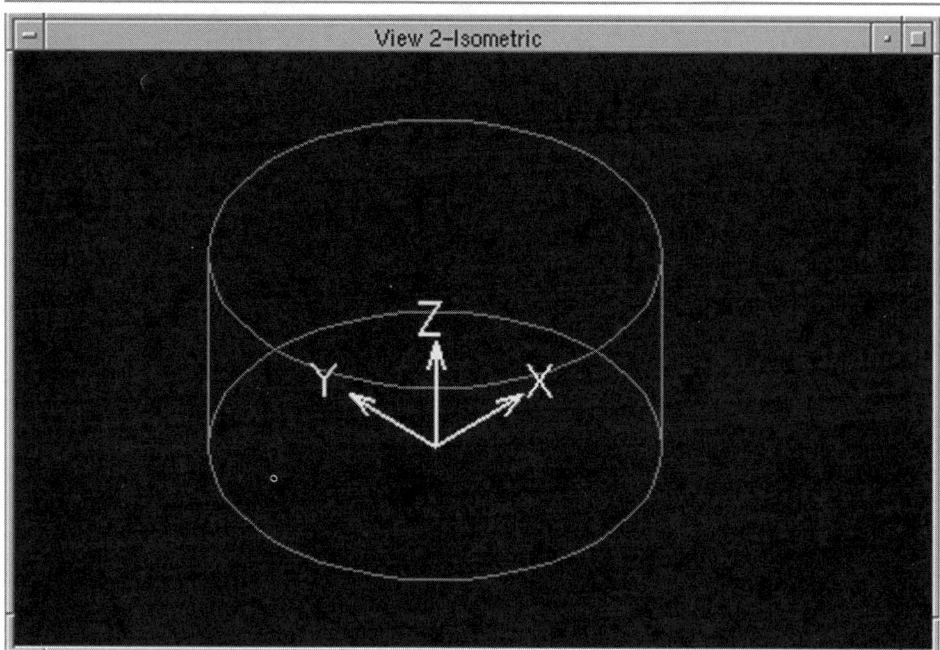

Iso view with ACS Triad turned on after CCS definition.

3. Save the CCS using the **Settings → Auxiliary Coordinates** dialog box. Name the coordinate system CYL1 and give it a description of "Center of staircase." Save it and Attach it.
4. You can now turn off the ACS Triad option. Select the *Scale Element* command and set the Y scale *only* to 11, while leaving X and Z at 1. In the Front view select the cylinder and snap to the beginning of one of the two lines you had from the start (this way you ensure the cylinder grows in the positive Y-axis). You should now have an 11-foot cylinder, with a base center at global 0,0,0.
5. Select the *Place Shape* command and key in the following:

   ```
   AX=1
   ```

   then,

   ```
   AD=3
   AX=4,20,0
   AX=1,20,0
   ```
6. Finally, choose *Close Element* to finish the first stair.

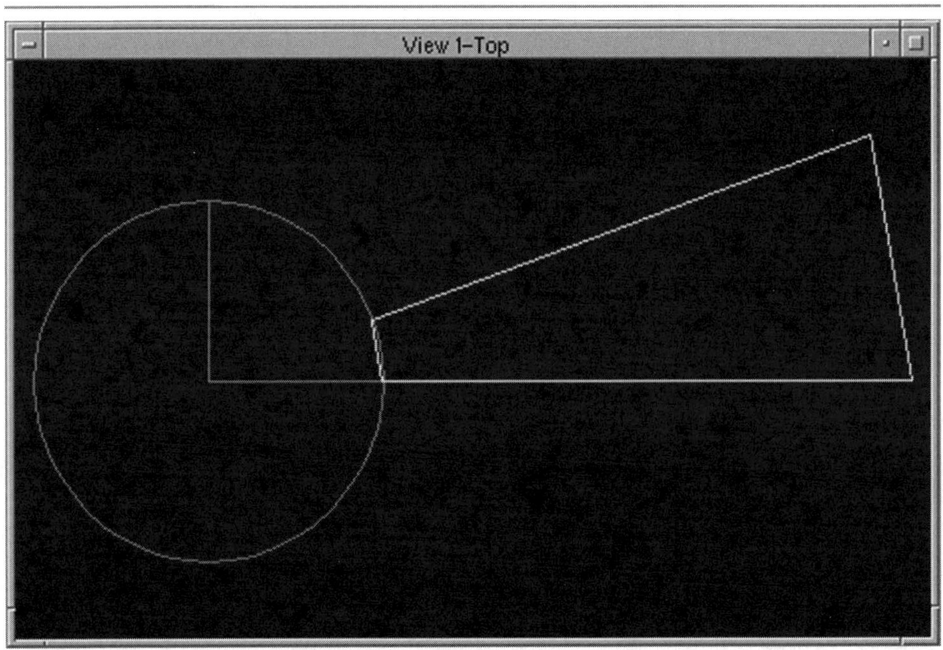

First stair and center support.

7. Select the *Polar Array* command and set the Items to 12, set the Delta Angle to 20 degrees, and turn on Rotate Items in the Settings dialog box. Make sure the type is Polar. Select the stair and snap to the intersection point of the predrawn lines, and data point to accept.
8. Select the *Move Element* command and select the second stair. Create (point 1) in the Top view and key in the following:

```
AX=,,:11
```

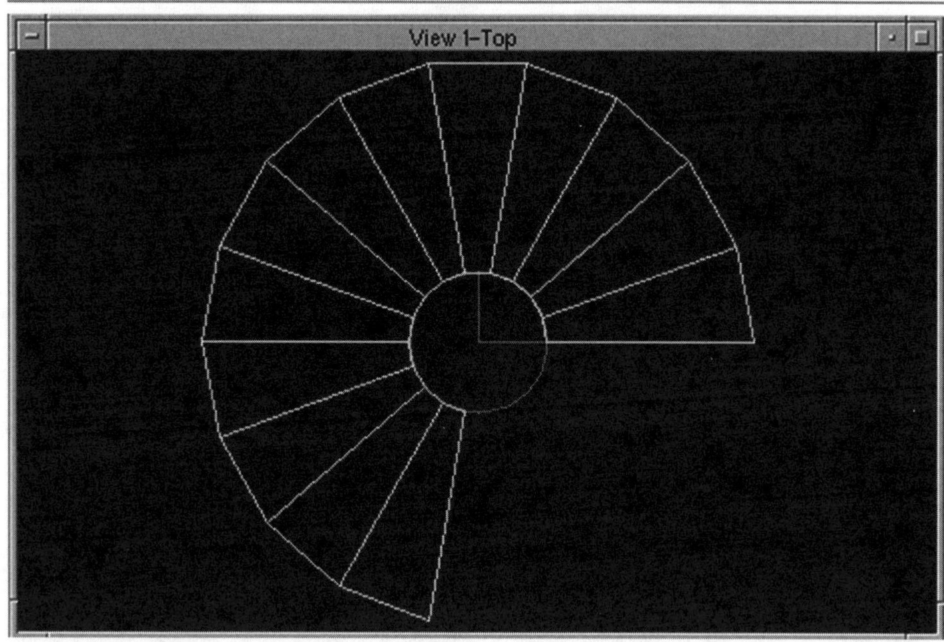

Moving each stair to the right height.

9. Continue to move each stair up by increments of 11 inches. So the next keyin for stair 3 would be `AX=,,:22`, followed by `AX=,,:33`, and so on.

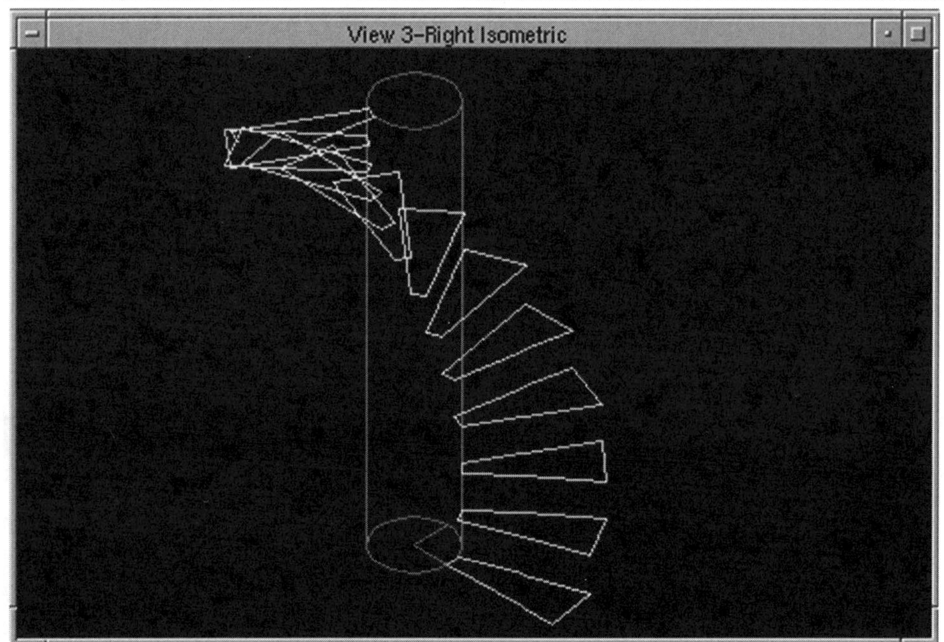

Staircase after the stairs have been moved.

Summary

Using auxiliary coordinates can be a challenge, but a very useful tool. Don't be afraid to wander down these forks in the road, because they lead to interesting places. Transforming from one to another is possible, but not recommended. Instead you should change coordinate systems. Remember, you still have your absolute coordinates (which are rectangular) available even if you attach a different coordinate system, so you have nothing to lose.

Introduction

Most MicroStation adventurers have heard of the B-spline (bicubic spline), usually overheard in conversations between guys with thick glasses and pocket protectors. Not to worry, MicroStation is your "nerd-in-a-box." You will learn how these functions behave and then tame their unruly nature.

B-spline curves are more complex mathematically than the point or stream curves. Hence, they are more accurate in representing real-life objects, such as the curve of a car or the human body. Because they are mathematically complex, most MicroStation adventurers stay away from them. Safari Sam will give you the tools to tame these powerful beasts, without having to understand the mathematics behind B-splines. Then you will learn about composite and predefined curves and how to manipulate the curves. Finally, you will cover B-spline surfaces.

Let's talk about what splines are and what B-splines are. Splines are used to model nonuniform curves. The spline is a polynomial function. The most general form of a polynomial function (in this case the function is in x) is as follows:

$$P(x) = p^0 + p^1x + p^2x^2 + \ldots + p^nx^n$$

where *n* is the degree of the polynomial and the *p*'s are the coefficients.

NOTE: *The order of a polynomial is equal to degree + 1.*

Here are some examples of polynomials:

```
y = ax + b, Linear, Second Order
y = ax^2 + bx + c, Quadratic, Third Order
y = ax^3 + bx^2 + cx + d, Cubic, Fourth Order
```

The polynomial is an estimate of the true curve you want to draw. The method of estimation is called interpolation. The general method of interpolation uses a finite number of "nearest neighbor" points. The problem is that the derivatives (the slope [slope = first derivative] of the polynomial curve at any point) are not always continuous because the slope sometimes goes to infinity (e.g., a vertical line has a slope of infinity). The reason you want to stay away from polynomials like that is that those polynomials have kinks and spikes in them, which do not provide a good estimate of the true curve.

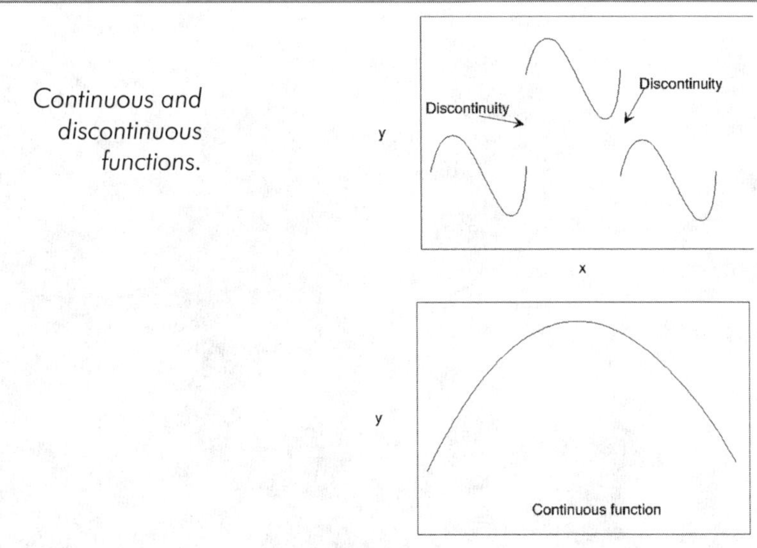

Continuous and discontinuous functions.

In situations where continuity of derivatives is a concern, you must use a "stiffer" interpolation provided by the spline function. A spline is a polynomial between each pair of poles, or control points, but the coefficients are determined to be slightly "non-local," so as to give global smoothness to the curve. Cubic splines are the most popular, since they produce a continuous function through the second derivative (the second derivative is the rate of change of the slope). So which order polynomial is appropriate when? Look at the following figure:

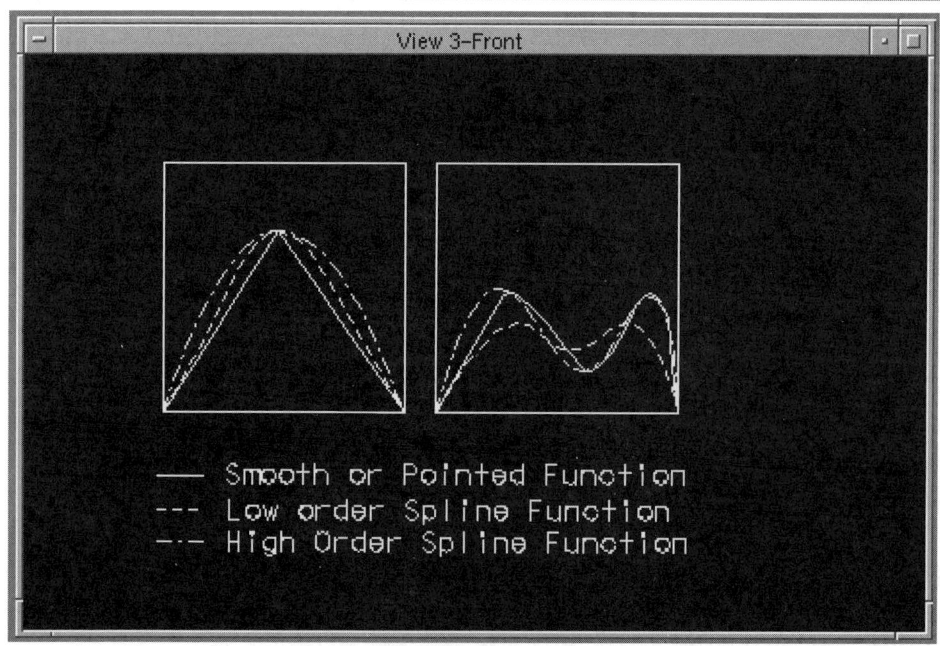

Smooth and pointed functions and high and low order spline functions.

The higher order spline is useful in interpolating more complex curves, whereas the lower order function is better at modeling surfaces with sharper edges.

Because the coefficients of the spline are non-local, adding more poles increases the accuracy up to a point. The recommended number of poles to use is four. Using three, five, or six can be tolerated, but adding more than that can lead to oscillations in the curve globally. Once again the best value for poles is four.

Moving right along, you now can start to talk about bicubic splines or B-splines. Bicubic interpolation requires that there be continuous first and second derivatives for every point on the curve or surface. This forces the curve or surface to be like a baby's bottom (smooth all over). The B-spline uses bicubic interpolation. We will not discuss the formulas for bicubic interpolation because they are beyond the scope of this book.

B-splines were traditionally drawn using a French curve tool. The basic shapes of that French curve are already defined in MicroStation and can be used to place data points that define the curve. These tools allow you to use MicroStation to build complex curves such as non-uniform rational B-splines (NURBS) and composite curves. You will also find the mathematic equations for controlling your curves, as well as ways to control the curves without knowing the math. Starting with simple curves and then moving to surfaces is the best way to proceed. Using a piece of simple geometry helps in understanding how curves are created and changed, usually placing a line string. B-splines can be adjusted after they have been placed.

Understanding Curves and B-Splines

Drafting adventurers have jousted with the curve. Many an hour has been spent trying to pin down the right one. The 3D curve is an unruly beast, but using Control Polygons and other tools, the beast can be tamed. The Curves palette is one you have used before. Curves can be drawn in both two and three dimensions.

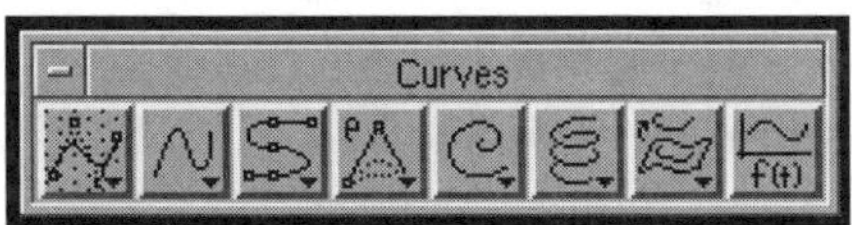

The Curves palette.

The control of B-splines is achieved through use of the Tool Settings palette and the B-splines dialog box, which is located in the Element menu. The dialog box has essentially two sections: Curves and Surfaces. The tolerance setting changes the accuracy of placement in working units. So, a tolerance of 0.01 means that an error in distance between the data points (poles) and the curve, 5 working units apart, is 0.05 working units.

NOTE: *Using the Least Squares Method tool for B-spline creation will display the error in distance if it exceeds the tolerance.*

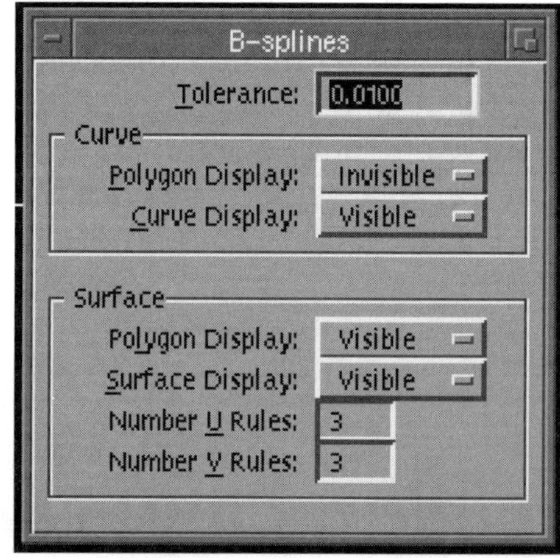

B-splines dialog box from the Element menu.

The curves section of the dialog box has two buttons to control polygon display and curve display. The options are Visible and Invisible. The polygon display shows or hides the Control Polygon of the B-spline. The control polygon is made up of lines connecting data points or poles. You must have the same or a greater number of data points than poles specified in the Place B-spline Curve (Tool Settings) palette. A two-pole curve is a straight line. Remember, each vertex of the control polygon is a pole.

The Curve display option lets you control whether the B-spline curve is displayed or not. Both of these settings are in effect only when you create the B-spline.

The first tool in the Curves palette is *Place B-spline Curve.* This command has many variables and we will spend a lot of time on it here.

Place B-Spline Curve

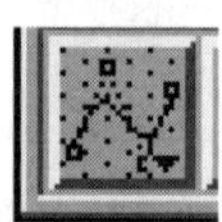

Place B-spline Curve icon.

You will learn about the four different types of NURBs MicroStation can create using the *Place B-spline Curve* command.

Upon selecting this command, the Tools Settings window displays these buttons: Method, Define By, Closure, Order, and Poles.

NOTE: *The Poles button is only available for the Least Squares Method of placement.*

There are four Methods of curves:

- Define Poles
- Through Points
- Catmull-Rom
- Least Squares Method

The availability of Method, Closure, Order, and Poles change as the Type changes.

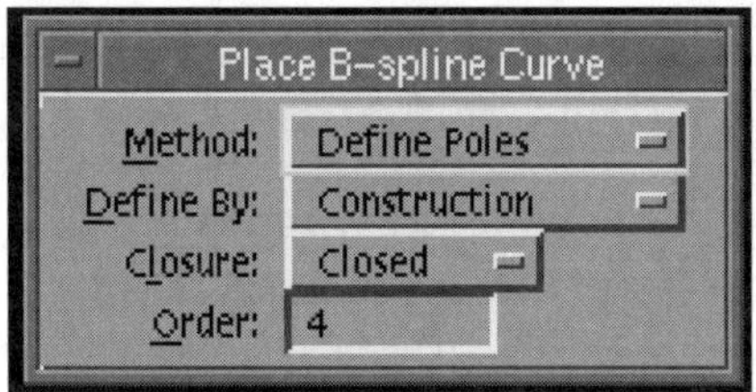

Place B-spline Curve Tool Settings.

Define By defines the way the B-spline is created. Our choices are Placement and Construction. Placement allows you to enter data points to define the poles of the control polygon. Several data points (usually along a preconstructed line string or other geometry) must be entered to define the curve. Construction allows you to select a shape or line string to be your predefined control polygon for the curve. Only one data point is needed (to select the element).

Closure defines whether the first point of the curve is also the last point of the curve. The options are Open and Closed. The Open setting means the ends of the curve do not meet, or the first pole is not the same as the last pole, hence, you have created an open B-spline. The Closed setting creates a closed B-spline, or one where the last pole is the same as the first pole. A closed B-spline encloses an area. A closed B-spline is also either periodic or nonperiodic. A periodic closed B-spline has a smooth curve at the first/last pole, whereas a nonperiodic closed B-spline has a kink in the curve at the first/last pole, but more about that later.

Order defines the order of the polynomial used to draw the curve. In practical terms, Order defines the distance from the control polygons poles. The higher the order, the farther from the poles the curve is. You have seen the Pythagorean equation before:

$$x^2 + y^2 = z^2$$

Here the degree of the equation is two, because the highest exponent (power) is two, and the order is three or degree plus one. So,

```
Order = degree + 1
```

Poles are the vertices of the control polygon.

Let's tame our first jungle beast in the next exercise.

Exercise 1: Place B-Spline Curve

1. From the Command window open the **Inside3D → Chapter 9 → Place B-spline** command. Inside this file are a series of line segments (not a string) at an active depth of Z=0 in the Front view. The line segments will help you to create B-splines. Please notice that there are seven poles (vertices).

Define Poles

2. Select the Place B-spline Curve icon.

   The first Method of B-spline you will create will be a Define Poles type of B-spline curve. Make sure your Method is set to Define Poles, set Define By to Placement, the closure is Open, and set the Order to 3.

3. Snap to the lower left point of the far left line, and data point to accept once you have. Go to the next point in succession moving right, using snapping and a data point to accept. Your curve should look like the following image:

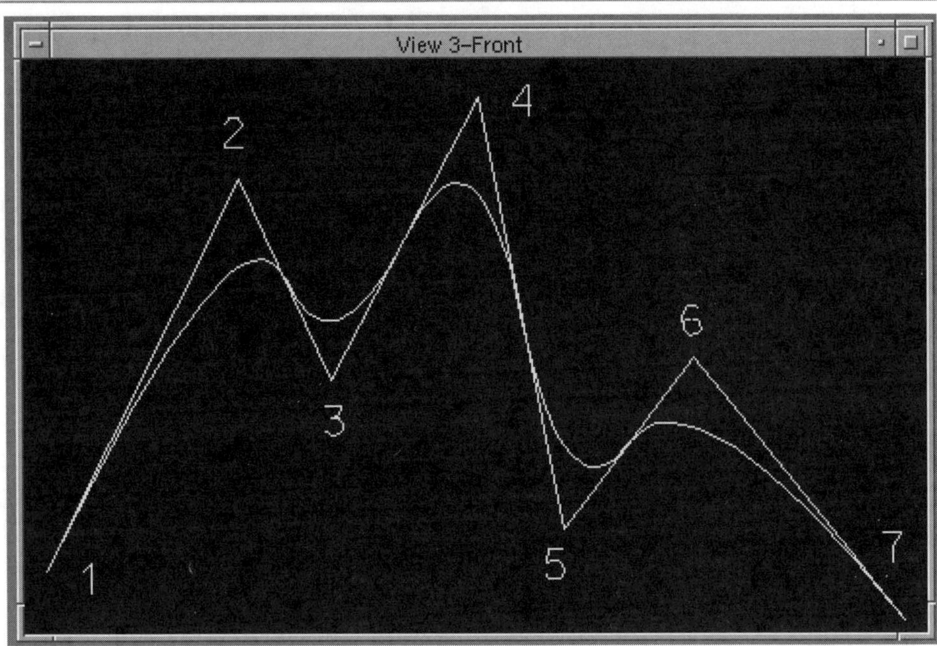

Define Poles, Order = 3, open B-spline curve.

The Define Poles method must have a greater or equal number of poles than the Order. So, for Define Poles:

```
Poles >= Order
```

4. Now go back to your Tool Setting windows and increase the Order to 4 and create a new B-spline using the same control points as before.
5. Repeat the process for Order = 5 to 7, and change the color for each one so you can see the difference. The family of Define Poles curves should look like this:

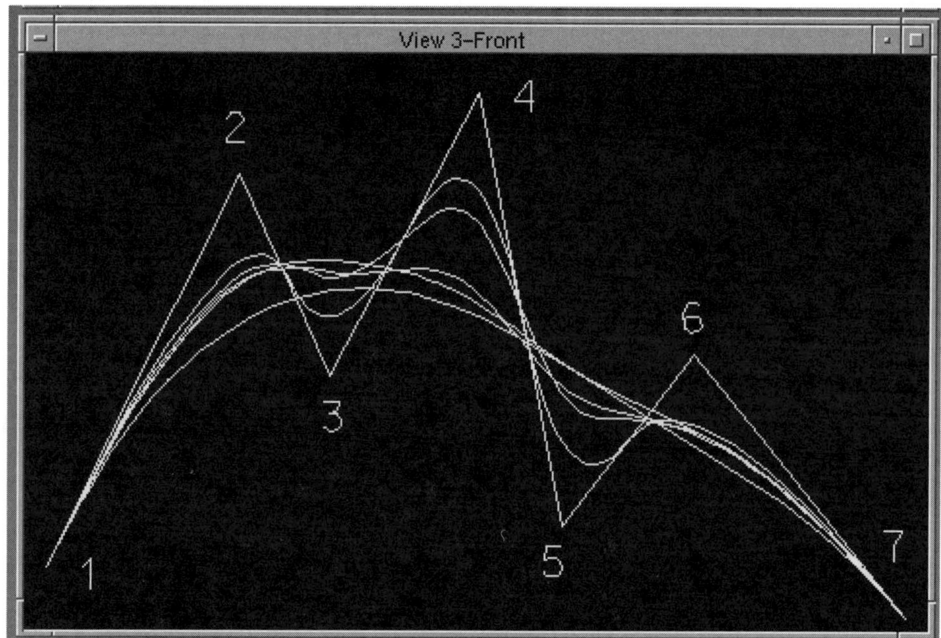

Family of Define Poles, open B-spline curves for Order = 3 to 7.

6. Now, move all the curves to an unused level, and then turn that level off. Leave the control polygon on.

Through Points B-Spline

The next NURB you will learn about is the Through Points B-spline. The Through Points method is a curve that, as the name implies, goes through the data points or line string or shape. The curve is cubic (order = 4) and has continuous second derivatives (the rate of change of tangents is unbroken, which means the curve has no spikes or dropouts), which minimizes the curvature. The relationship between poles and data points is:

```
Number of Poles = Data points + 2
```

Note that using the Through Points type leaves only Define By (Placement or Construction) as an option.

1. Using the same control polygon as before, tentative point to the same points to draw your curve.

The curve should look like the following image:

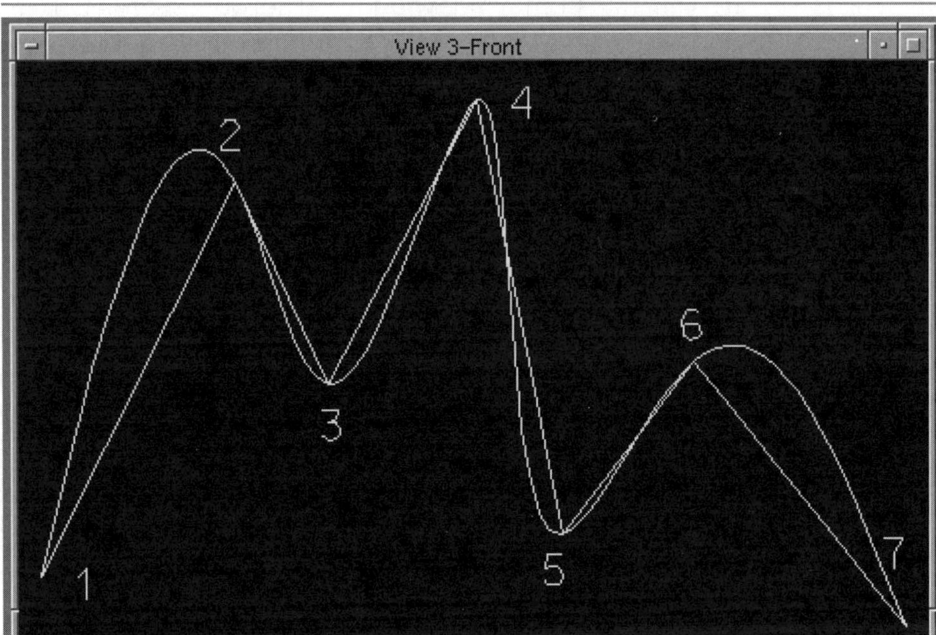

Through Poles Non-Uniform Rational B-spline.

Catmull-Rom Method

The next type of B-spline you will learn about is the Catmull-Rom method. The Catmull-Rom options in the Tool Settings windows are limited to Define By, Placement, and Construction. The Catmull-Rom method uses a cubic polynomial to interpolate the curve. Poles are added automatically to help create a more realistic curve. Catmull-Rom curves are excellent for creating smooth curves and surfaces.

```
Number of poles = (3 * (Data points -1) + 1)
```

1. Using the same control polygon as before, create a Catmull-Rom NURB.

Least Squares Curve

The final type of NURB is a Least Squares curve. This method minimizes the sum of the squares of the distance from the data points (or vertices of the line string or shape) to the curve. If this error exceeds the tolerance, then it is reported in the Command window. The options for a Least Squares NURB are Method, Closure, Order, and Poles. If the order is the same as the number of the poles, then the curve passes through all data points.

1. Using the same control polygon, with Order set to four and Poles set to four, draw a Least Squares B-spline.

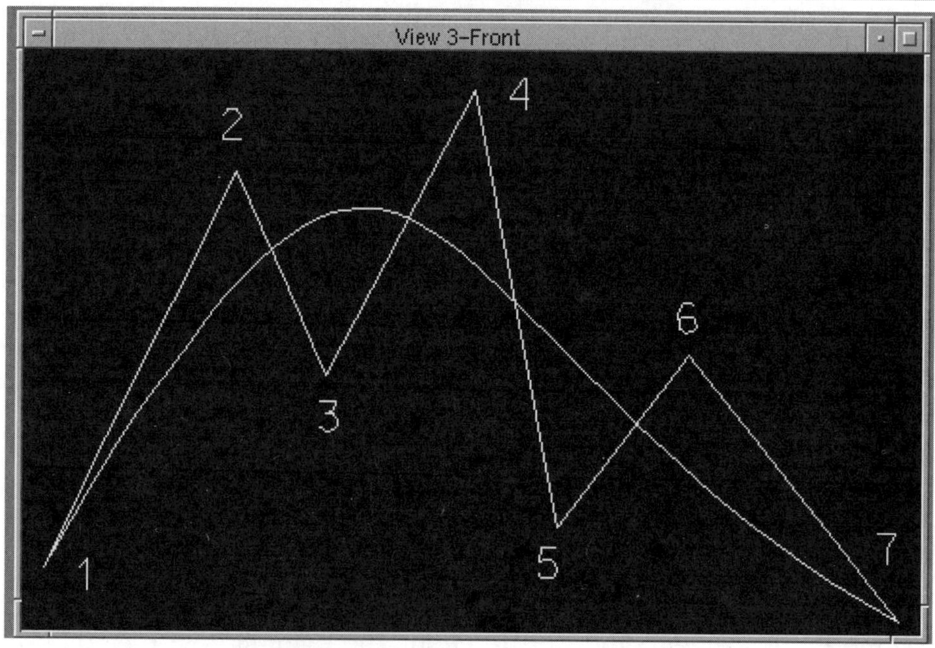

The Least Squares B-spline, with Order set to 4 and Poles set to 4.

2. Repeat the process for Order = 3, Poles = 4 to 7.

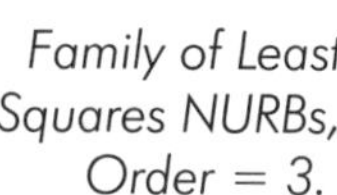

Family of Least Squares NURBs, Order = 3.

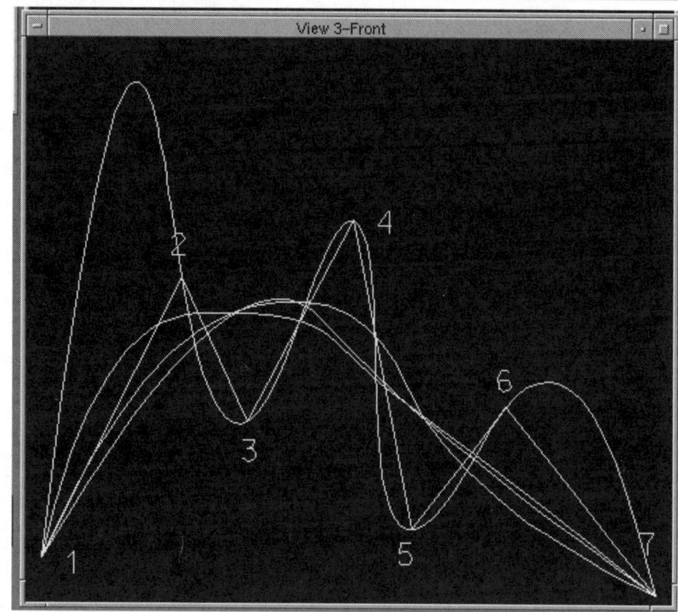

3. Repeat the process for Order = 4, Poles = 3 to 7, and note the differences.

Family of Least Squares NURBs, Order = 4.

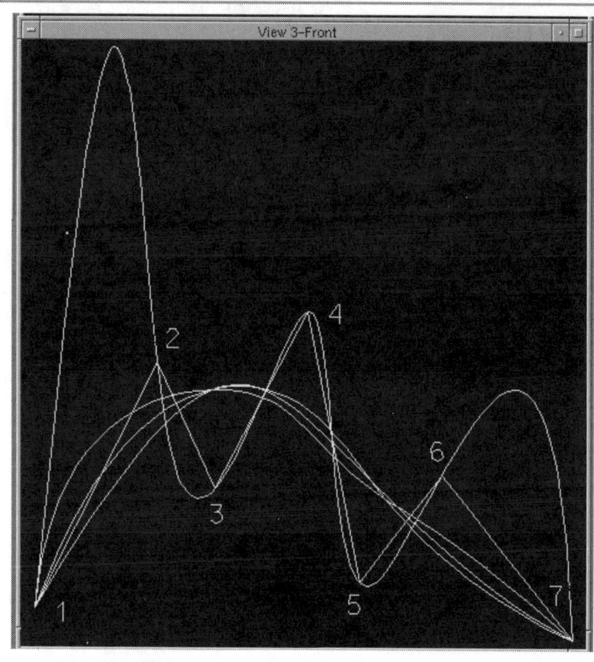

➜ **NOTE:** *Remember, you only have seven poles on your control polygon.*

The next icon is the point or stream curve. These are less mathematically complex, but have their uses.

The point curve is probably the simplest of all MicroStation curves. There are no settings to control the point curve's shape. As the curve is placed by data points, the curve is displayed dynamically. You can also enter data by precision input. On the Tool Settings palette you have the ability to place either a point or a stream curve and decide whether it will be nonplanar or not. If you turn on the nonplanar button, then you can draw the curve in three dimensions.

The stream curve is used sometimes in digitizing. A data point is entered and then the mouse or puck is moved and a stream of data points is entered automatically. Please refer to the MicroStation manual for its settings. Once again you have the ability to place points in a planar or nonplanar fashion.

Place Conic and Place Spiral tools.

The next two icons are special purpose 2D curve tools that can be used to make parabolas, hyperbolas, and other conic sections, as well as clothoid, Archimedes, and logarithmic spirals. These commands behave the same way in 3D as they do in 2D.

Composite Curve

The final type of predefined curve you will learn about is the composite curve. A composite curve is one made up of line segments, arcs, or a special B-spline curve, the Bezier curve. Any B-spline curve with the numbers of poles equal to its Order is a Bezier curve. The advantage of a Bezier curve is that you can control the starting and ending points of the curve. In addition you can also control the tangents at that point. The Bezier curve passes through the control points as a tangent. The line between the first and second data points gives the direction of the first tangent and the line between the last two poles defines the final tangent direction. The length of the distance between the poles controls the size of the tangent at each pole.

Exercise 2: Place Bezier Curve

1. Using the same control polygon as before, create a fourth order (poles = 4, order =4) Bezier curve.

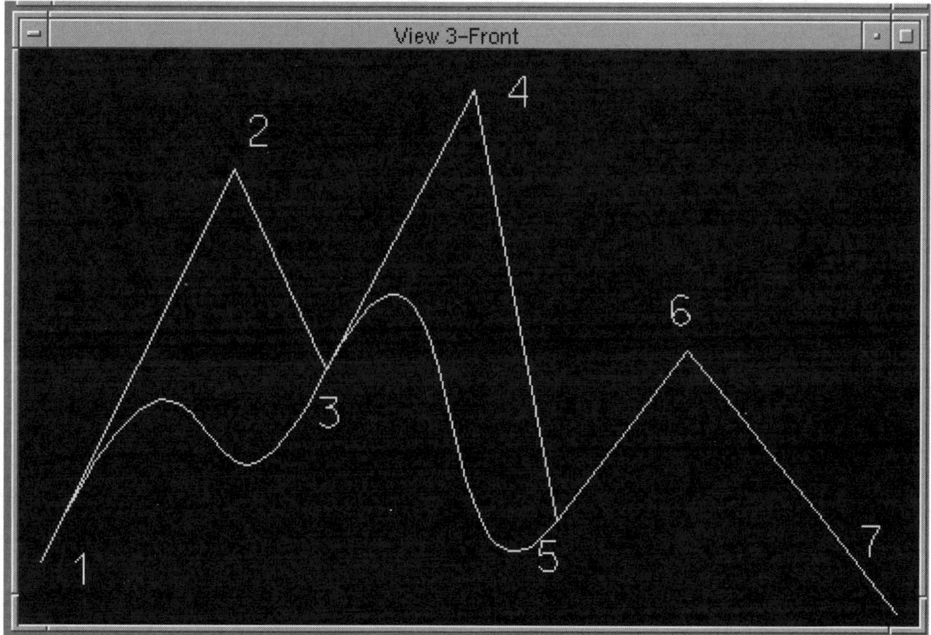

A fourth order Bezier curve.

The Helix Command

The *Helix* command allows the user to place helical or spiral object, such as a spiral for a staircase. MicroStation allows you to create different styles of helices. Even further control of helix generation can be achieved through use of the curve calculator, as you will see in the section ahead.

You will now get a chance to model perhaps the most famous helix of all: the double helix of DNA (deoxyribose nucleic acid). Here are the vital statistics for your DNA model (you will model the B-form of DNA). First, you have two left-handed helices that are intertwined. The helices will start and finish at the same height, but are rotated from each other by 100 degrees. The helix makes one complete turn every 34 angstroms (10^{-10} meters) and has a diameter of 20 angstroms at both the top and bottom of the helix. Now you know all you need to know to be a genetic engineer (not a very good one, but you're learning).

The Helilx command tool.

Exercise 3: Creating a DNA Molecule

1. Select the **Inside3D → Chapter 9 → Place Helix** command, Working Units are metric, scaled up by a million; you will use nanometers instead of angstroms. There are 10 angstroms per nanometer.
2. The helix is defined first to be right- or left-handed; this is set through the Thread button on the Tool Settings palette. You also have to define the axis of the helix, the starting and ending radii, and the pitch of the helix and whether it will be Orthogonal to the axis or not. You will achieve these parameters by using construction elements.
3. Select the *Place Circle By Center* command and enter the following:

   `XY=0,0,0`
4. Place the center of the circle at the global origin, key in a radius of `1.0`, giving you a circle with a diameter of 2.

5. In the Front view, use the *Place Line* tool to snap to the far right side of the circle and after you have accepted, key in:
   `dx=,3.4`

Your elements should look like the following figure:

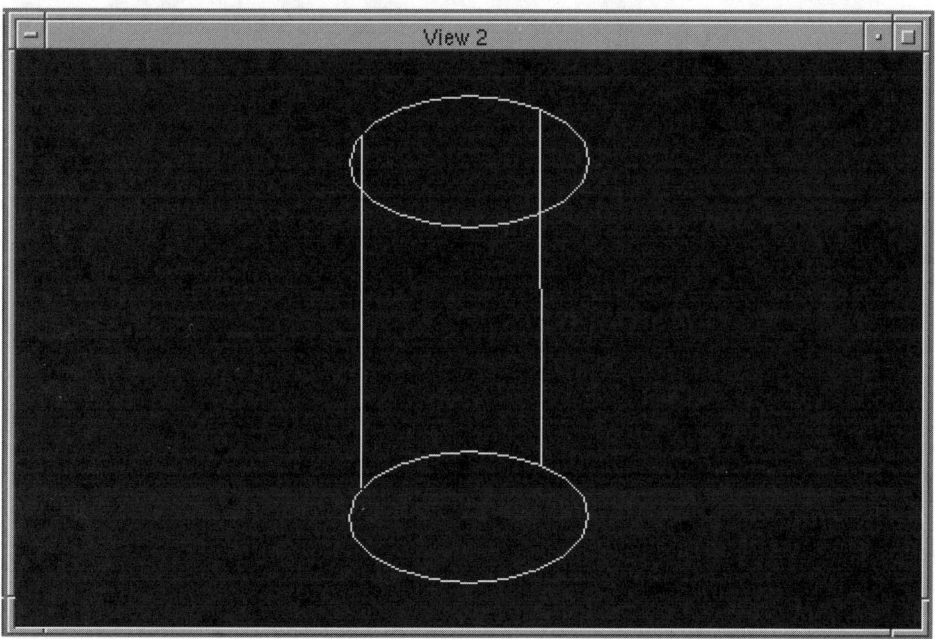

Circle and Line construction elements for a helix in the Iso view.

1. Copy the circle up 3.4 nm in the front view.
2. Next select *Place Line* and set the Length to 1.0 and Angle to 100 degrees in the Tool Settings palette. Snap to the center of each circle and data point to accept. Finally, draw a line that connects the ends of the two new lines on the perimeter of the circles. The final form of the construction elements should look like the following figure.

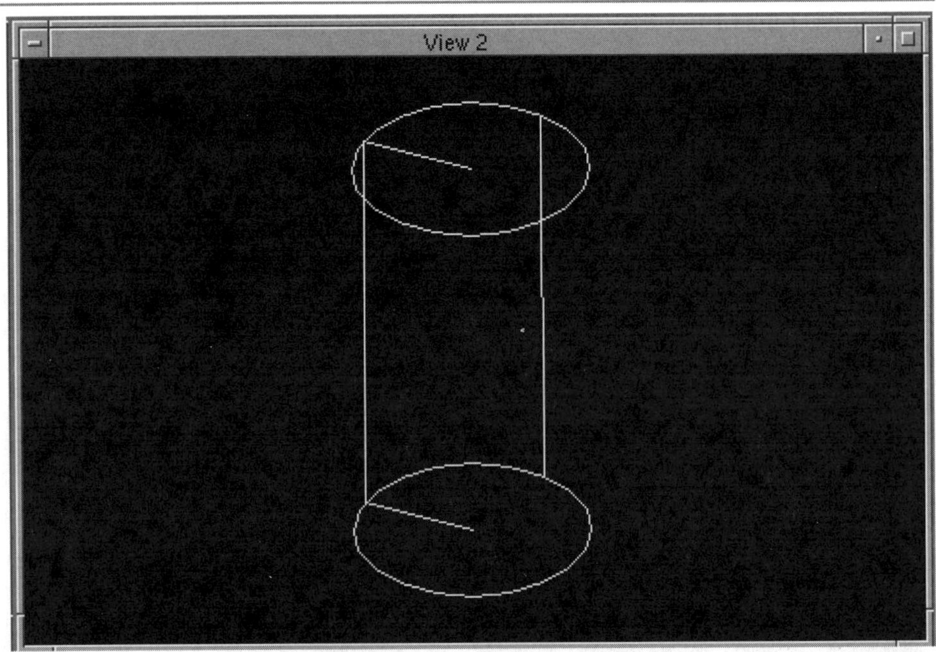

Final form of construction elements for Place Helix.

3. You can now delete the radius construction elements used to create the second line.
4. Set your level to 2, and color to 1. Select the *Place Helix* command. Set Thread to Left, Axis to Points, and turn Orthogonal on. Leave the other settings off. In the Top view select the center of the bottom circle as the center point of the helix. Make sure you have selected the right circle by confirming in the Iso view. Data point to accept the center point.
5. To define the radius of the helix, come down to the Front view and snap to the endpoint of the first vertical line you drew. Data point to accept. Next, snap to the top of this line to define the axis, and data point to accept. Again snap to the end of this line to define the height, and data point to accept. Finally, (you guessed it) snap to the end of the line to define the final radius.
6. Repeat the process for the second helix using the second line. Copy both helices in the Front view up by 3.4 nm (`dx=,3.4`), a couple of times. Turn level 1 off, and there is your double helix.

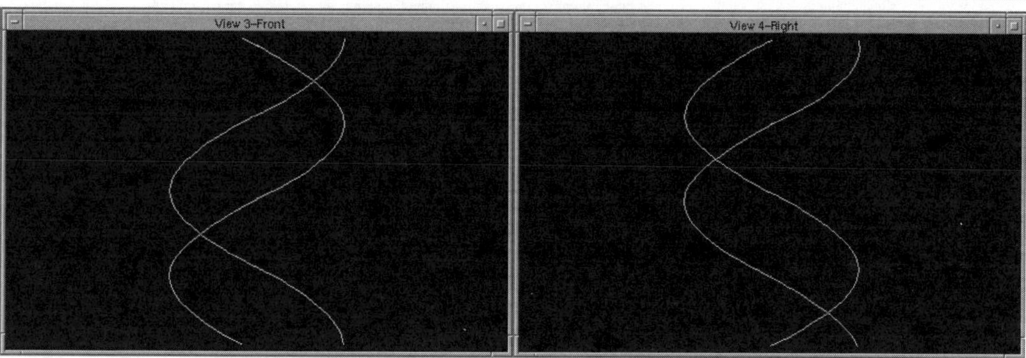

The double helix of DNA.

For those of you who are interested, there are 10 base pairs of deoxyribose sugars for each complete turn. Each base pair is rotated by 10.58 degrees and is separated from the next pair by 0.34 nm vertically.

The Curve Calculator

The Curve Calculator icon.

The Curve Calculator lets you define any conceivable planar curve based on a mathematical formula. Since things do not really change in the 3D world, Safari Sam will just review the basics. Now, you're probably wondering why we spent so much time creating 2D curves. Well, you have already learned how to make a surface from an element; now you can make complex surfaces because you have mastered complex elements. The most complex of all elements is one you can define, and the curve calculator lets you do that.

MicroStation comes with libraries of curves that are predefined. In addition, you can define and save any curve you create.

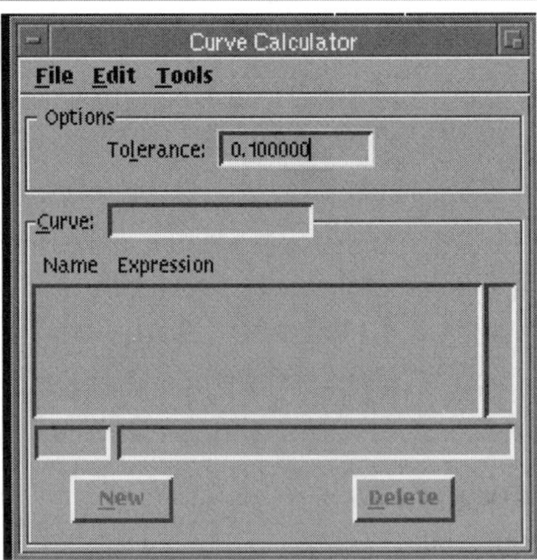

The Curve Calculator dialog box.

There are three pulldown menus here: File, Edit, and Tools. The File menu has the standard file manipulation commands.

Exercise 4: Using the Curve Calculator

1. Choose the *Open File* command. This opens a dialog box for choosing a resource file (extension *.rsc) with the equations in them. Choose the *CURVE3D.RSC* file

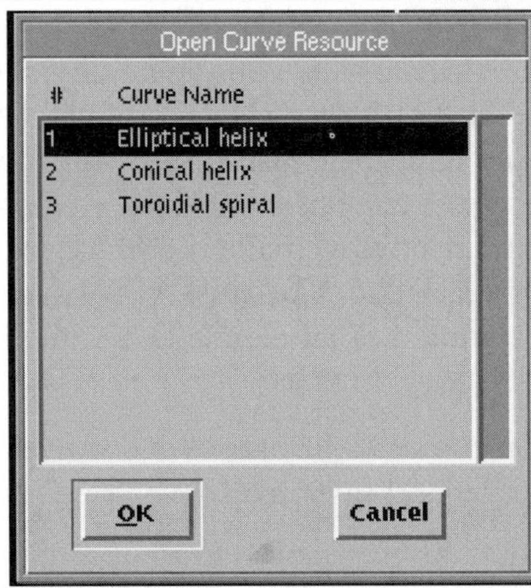

The Open Curve Resource dialog box.

MicroStation also has a 2D Curve library that has 10 predefined curves in it. These curves are defined by equations and are categorized by the method of curve generation. The *curve3d.rsc* file has three curves in it: Elliptical Helix, Conic Helix, and Toroidal Helix.

1. Select Elliptical Helix in the list and click on OK.
2. Looking in the Curve Calculator dialog box, you can set the various parameters of the curve and change the way it is defined. Only those variables that are black are changeable; those that are grayed out are not editable (they are if you read the next section). In the case of the Elliptical Helix, you first define the major and minor axes of the ellipse and then pitch angle and number of revolutions. To place the curve, select **Tools → Place Parametric curve** and then select a center point for your new curve.
3. If you change any parameter, then you can save the new curve and give it a new name by using the *Save* command from the File menu. Use Save As if you want to create a new *.rsc* file.

Take a look at all the curves available to you and place them to see how they appear. Then try changing values and see what happens to the curves.

The next step is define curves of your own creation. This obviously requires some decent mathematical knowledge; stop frowning, not everyone has to do this (insert applause here). First, some ground rules:

1. Changing these formulas is a power user feature. *Do not* attempt if you are not sure about your math; best to do some review first.
2. Variable names are limited to eight characters, and the right side of the equation can only have 40 characters. You have a limit of 25 formulas to define a curve.
3. Each coordinate is defined in terms of a parameter *t*, which can only have a value from zero to 1 or, 0 <= t <= 1.
4. The following constants are valid in formula creation:

   pi - pi

   e - natural logarithm

   Hence, you cannot use *e* as a variable.

The curve calculator understands the C operators (+,-,*,/,** and others).

To prevent you from doing too much damage, MicroStation has a feature to lock any equation used to define a curve. The keyin is:

```
FORMULA LOCK number
```

where *number* is the number of the equation in the list box.

NOTE: *The count for the numbers starts at zero.*

To unlock a formula use

```
FORMULA UNLOCK number
```

The formulas defining the X, Y, and Z coordinates of the curve can be defined using logarithmic, exponential, power, trigonometric, and hyperbolic functions. Here is a list of valid functions:

Functions	Description
sin(var)	sine of var
cos(var)	cosine of var
tan(var)	tangent of var
asin(var)	arc sine of var
acos(var)	arc cosine of var
atan(var)	arc tangent of var
atan2(var1,var2)	arc tangent(y) / x
sinh(var)	hyperbolic sine of var
cosh(var)	hyperbolic cosine of var
tanh(var)	hyperbolic tangent of var
exp(var)	e^{var}
ldexp(var1,var2)	$2var1^{var2}$
log(var)	natural log of var
log10(var)	log base 10 of var
ldexp(var1,var2)	$var1^{var2}$
sqrt(var)	square root of var

To define the formulas you start with the x variable, then y, and then z. If no z variable is input, it is then assumed to be 0, creating a 2D curve.

The formulas for a cosine curve of amplitude 10 and wavelength 12 would then be:

```
x=12*u
y(t)=10*cos(u)
u(t)=2*pi*t
```

The letter *u* is used because *t* can only run from 0 to 1.

Adding a z term of :

```
z(t)=10*cos(u)
```

between y and u yields a 3D cosine curve. By giving 12 and 10 a variable name, you can alter the cosine curve later on.

B-Splines in 3D

In the previous exercise you used a 2D control polygon to create your B-splines. What would happen to the curves if the control polygon was a 3D object? Let's find out.

Exercise 5: 3D B-Spline Curves

Open **Inside 3D → Chapter 9 → 3D Control Polygon 1**.

The control polygon here is a line string that is three dimensional. You will find that the Define Poles method of curve generation yields the most realistic curve of all.

1. Select Place B-spline Curve, using Define Poles, Placement, Open, and Order=4. Use the endpoints of the line segments as your vertices.
2. Repeat the process for the other curve types and note the differences between them.

Notice how the Catmull-Rom and Through points types formed loops to get through the control points.

Modifying B-Spline Curve Attributes

To modify B-splines use the *Change to Active Curve Settings* command found on the Modify Elements palette.

Change to Active Curve Settings icon.

There are four variables in this command. They are Polygon, Curve, Closure, and Order. The first two allow you to change the display of the control polygon or the actual curve.

Closure lets you close a open B-spline or open a closed one.

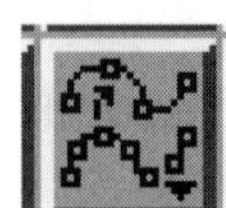

The Reduce Curve Data icon.

Another handy tool is the *Reduce Curve Data* command. Also found on the Modify Elements palette. This command allows you to reduce the number of poles used to define a curve, hence lowering the file size. The tolerance is the maximum distance the new curve can be from the old curve.

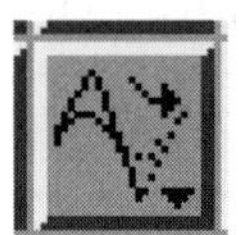

The Extend Curve icon.

Another tool is the *Extend Curve* command, again on the Modify Element palette. The *Extend Curve* command can be used to extend lines, line strings, arcs, ellipses, complex chains, and B-splines by a certain scale. There are two settings for this command: Continuity and Extension Scale. Continuity defines the smoothness of the extension. Your options are Position, Tangent, and Curvature. Position means that the extension is a straight line. Tangent means that the extension is tangent to a continuous line. Finally, Curvature means that the extension is a curve that is continuous in order and poles to the curve being extended.

The Extension Scale is the scale relative to the length between the two poles of the curve at the end of the extension. The scale is greater than zero and less than one.

The *Change Element Direction* command is useful in reversing the element's (same elements as in the *Extend Curve* command) direction or start point.

The Change Element Direction icon.

First, the direction of an element is defined as follows:

- An open element's direction goes from its start point to its end point.
- A closed element's direction is counterclockwise when placed.
- To change an element's direction, select the element and reset.
- To change an element's start point, select the element, select a new start point, and reset.

The Convert Element to B-spline icon.

The final modification tool is the *Convert Element to B-spline* command. This command is used to convert elements (the same as in the *Extend Curve* command, including complex shapes, surfaces of projection or revolution, and cones) to a B-spline curve with the same shape. There are three settings or options in the Tool Settings palette. These are Make Copy, Closed to Surface, and Tolerance. The Make Copy command is self-evident (I hope). The Closed to Surface option allows closed elements to be converted to B-spline surfaces instead of closed B-spline curves. Finally, Tolerance sets how closely the element is converted or copied if Closed to Surface is turned on.

B-Spline Surfaces

The NURB surface is the most flexible of all surfaces. It is flexible because each pole of the control net (versus control polygon before) can be modified, changing only a small part of the whole surface. The command you will be spending a lot of time with is the *Place Free-Form Surface* command. So grab some java, and pay attention!

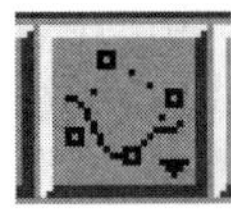

Place Free-Form Surface icon

The options on the Tool Settings palette look the same as the palette for *Place B-Spline Curve*, except there is an extra field for Order and Closure.

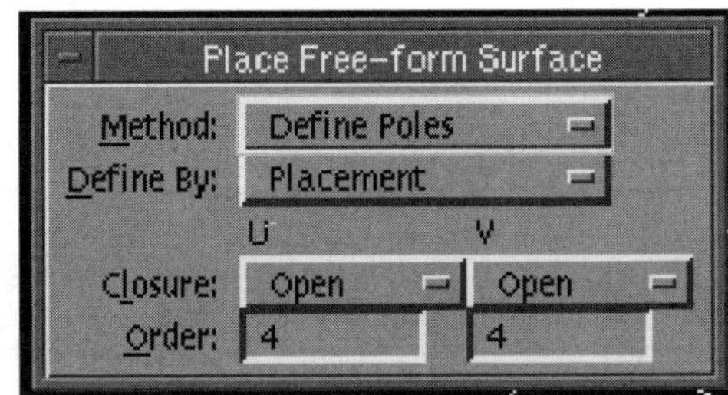

The Place Free Form Surface Tool Settings palette.

You have the same methods available to you: Define Poles, Through Points, Least Squares, and Catmull-Rom. Also Define By is the same as before, with Placement and Construction as your options. Closure sets whether or not the surface is closed or open in the u or v direction. Order sets the order of the equation that creates the surface in the u or v direction. Poles is the number of Poles in the u or v direction if Least Squares is used.

The u and v variables define the control net. The u variable specifies the number of control points in each row of the net. Remember, the number of control points (poles) must be greater than or equal to the order.

The v variable specifies the number of rows of u B-splines. You must specify at least as many rows as the order set for v. Now let's try to create a free-form surface.

Exercise 6: Creating Half a Ship Hull with a Free Form Surface

1. Select the **Inside3D → Chapter 9 → Ship Hull** command. The Working Units are in feet and inches. The only elements in the file are line segments where the poles of hull are located. The keel is 100 feet long.
2. Begin by selecting the *Place Free Form Surface* command. Select Define Poles as the type, Placement as method, set Closure to Open

for u and v, and set u and v order to 4. In the Iso view start by snapping on the far lower left side of the line (point 1). Continue snapping to points as labeled.

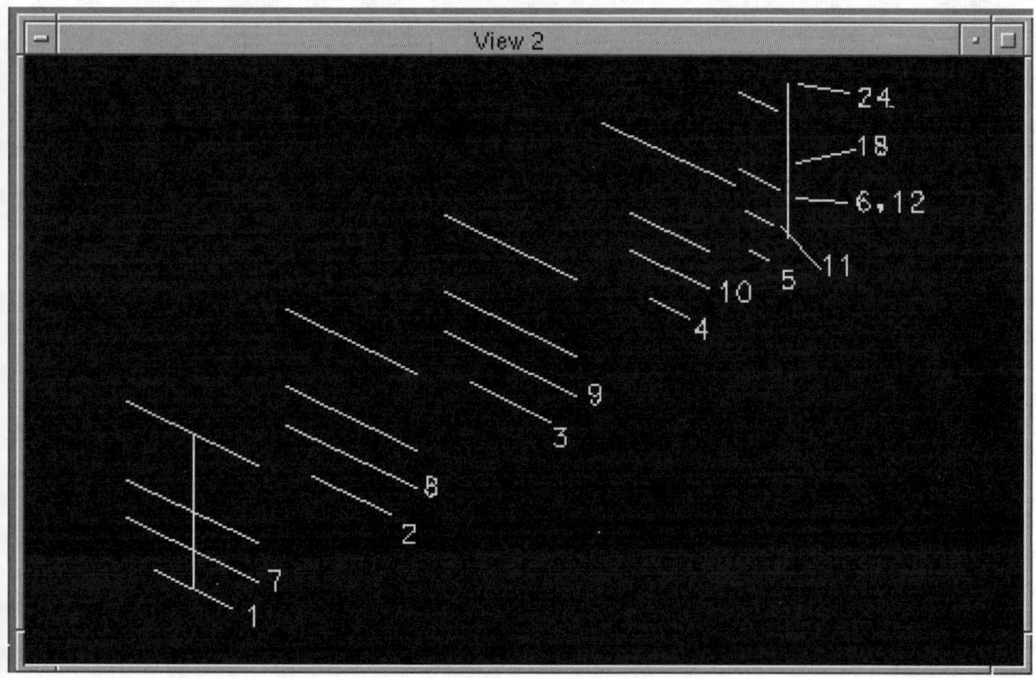

Snap points or poles of the ship's hull.

3. At point 6 you should reset. When you reset at point 6, MicroStation allows you to begin the next row (points 7 to 12), a data point at point 12 lets you begin the next row (13 to 18), and a data point at 18 lets you start the last row (19 through 24). Finally, a data point at point 24 completes the B-spline surface.
4. Redo the B-spline surface, using Catmull-Rom and the other methods. When you try the Least Squares method, use orders and poles of 3 (for both u and v).
5. Finally, redo the hull by adjusting the poles (the line segments that were already there).

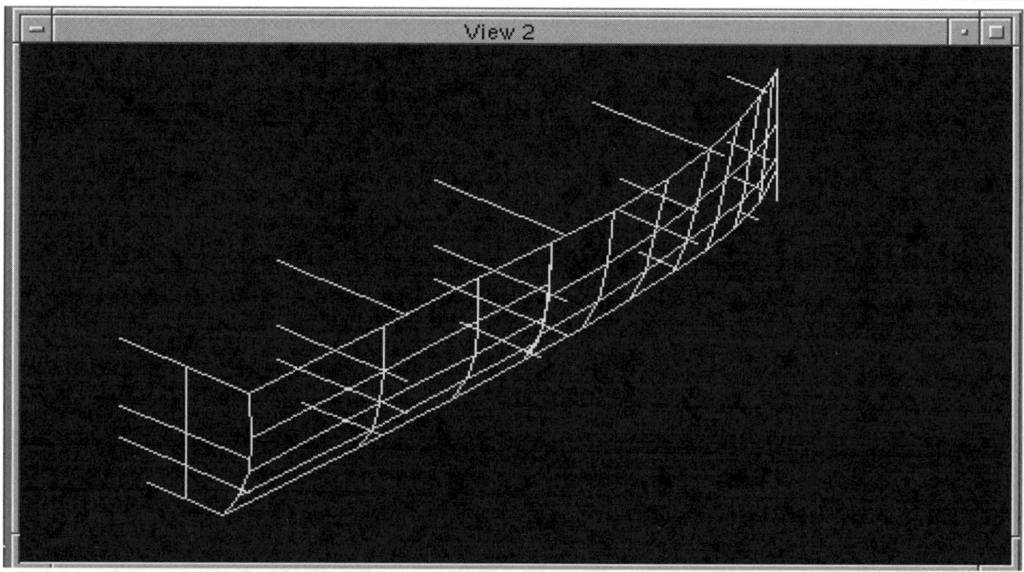

Half the ship's hull, using a Define Poles Free Form Surface.

You may be asking, how did Safari Sam know how to place the hull poles where he did? They poles were placed from dimension data taken from a real ship and then adjusted to create a unique design. The poles were placed approximately where one would imagine structural elements of the ship's chassis to be. Notice how each set of lines looks like a cross-section of that part of the ship. Using this kind of cross-sectional analysis helps in constructing a free-form surface.

Summary

You have now learned how to tame some very nasty beasts. Remember, to be a real master of these elements you must practice further and understand all of the options for each curve or surface. Make sure you tried every option in each exercise, since each one is important. B-splines can be intimidating creatures, but the best way to approach them is to take it one slow step at a time, especially when you first start.

Understanding how to set up the placement geometry is critical to your success, so you must think in a framework fashion, that is, what do I need to have in place before I build this curve or surface? So understanding your placement geometry can be more important than understanding the curves. The only way to truly master B-splines is to practice over and over again, until you can climb into the ring with the beast without breaking a sweat.

Introduction

It was not so long ago that a journey like this would have caused great hardship and frustration. Back in the old days, the 3D tools were primitive and awkward to use. Today, a journey into the 3D wilderness has been reduced to the click of a button (well, almost). Knowing what tools are available and how to use them can make a difficult job easy. Who knows, you might even have some fun doing the work.

In this chapter you will learn how to perform Boolean operations and 3D surface modifications. These tools will empower you to conquer the most formidable of 3D jobs. In this chapter you will learn through discussion and exercises how to use these tools as they apply to various disciplines. Enough said, let's get started.

Modifying 3D Surfaces

Boolean Operations

In the 1850s, a mathematician named George Boole developed the principles of mathematical logic that we now call "Boolean algebra." This mathematical logic is employed in the Boolean tools in MicroStation. Fortunately, the computer has to do all the math. All you have to do is select the elements. Boolean operations can also be described as additive and subtractive modeling. Using additive and subtractive modeling you can create complex geometry by combining or subtracting one solid element from another. Using these techniques, you can quickly create geometry that several years ago would have taken hours of work. The following illustration will help you get a summary of how Boolean tools work.

This is the wedge and a slab geometry before any Boolean operations have been performed. Both the wedge and slab are solid elements.

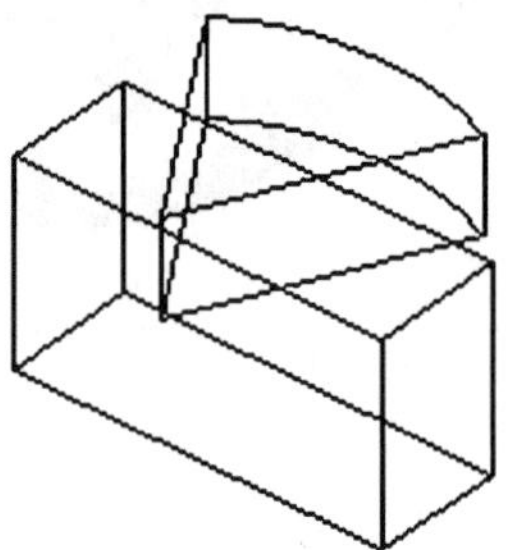

The following illustrations graphically show the three different Boolean operations.

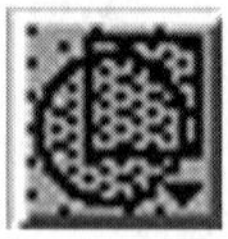

The Construct Union Between Surfaces icon.

The *Construct Union Between Surfaces* tool will trim two elements to their common intersection. This will combine the wedge and slab to create one element. The intersecting region will be removed.

This is what the wedge and slab will look like after the Construct Union Between Surface tool has been used.

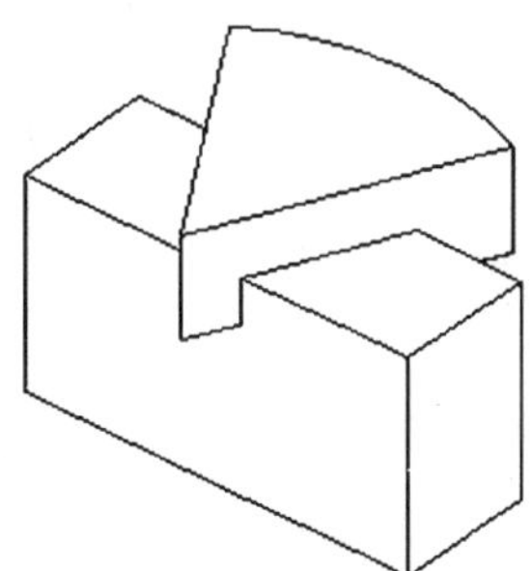

The Construct Intersection Between Surfaces icon.

The *Construct Intersection Between Surfaces* tool will trim two elements to their intersection. This will create one element from the intersecting regions of the wedge and the slab.

This element was created from the intersecting regions of the wedge and the slab.

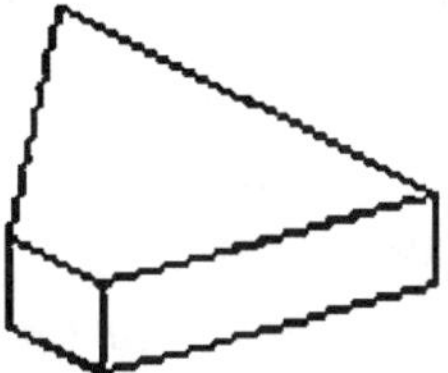

The Construct Difference Between Surfaces icon.

The *Construct Difference Between Surfaces* tool will trim the intersecting region of one element from another element. The first element selected will remain, with the intersecting region of the second element removed.

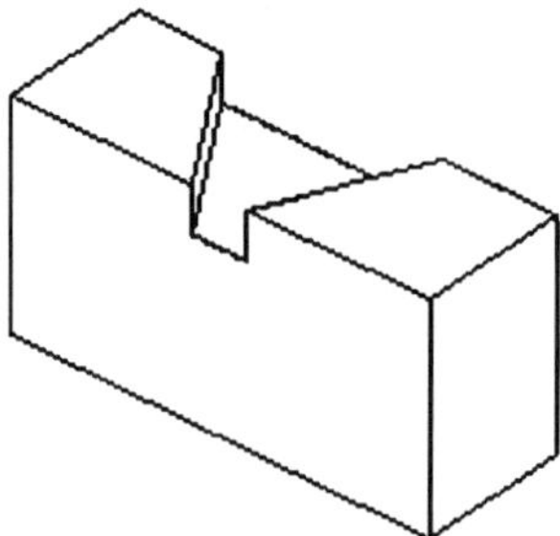

This element was created by removing the wedge from the slab.

Exercise 1: Boolean Operations

In the following exercises, you will perform the previously mentioned Boolean operations. The 3D elements shown in the previous illustrations will be used to practice these operations. At this point, since you now know how to create 3D slabs and wedges from Chapter 5, you can recreate these 3D elements yourself, or just use the ones provided with this exercise. If you choose to create your own 3D elements, remember to use a 3D seed file to start with.

1. From the Command window, select **Inside3D** → **Chapter 10** → **Boolean Exercise**.
2. Select the *Construct Union Between Surfaces* tool from the Modify 3D Surfaces palette.
3. Data point on the blue slab and then data point on the blue wedge. Both of the elements should be highlighted. Select one more data point to accept the second surface you selected. You should now see what the final element will look like. This is the point that you would accept or reject the results. Data point to accept the results. These two elements are now one element.

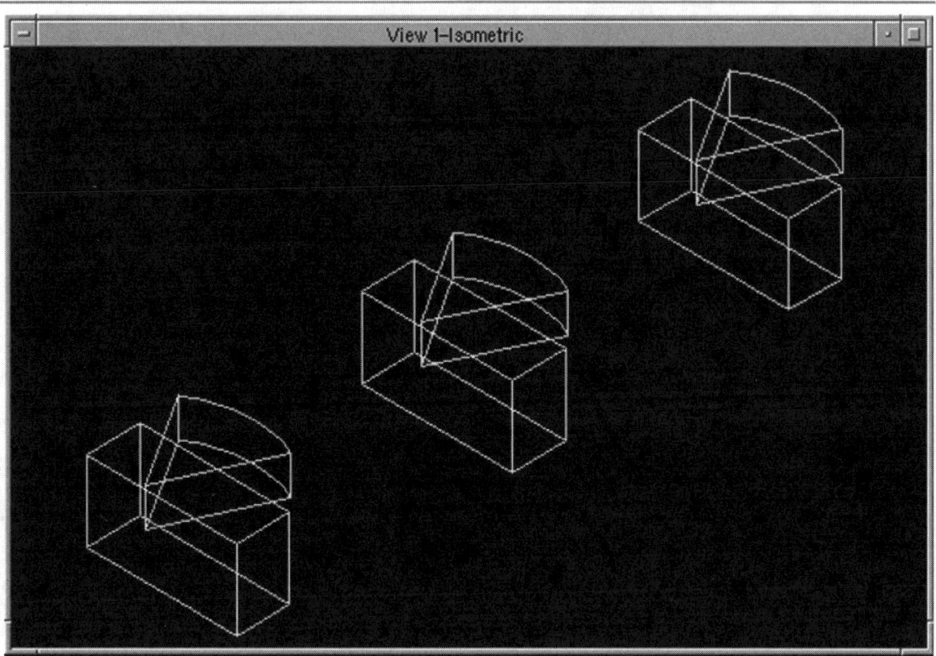

Your View 1 should look like this.

4. Select the *Construct Intersection Between Surfaces* tool.
5. Data point on the yellow slab, and then data point on the yellow wedge. Again, both elements should be highlighted. Data point one more time to accept the second surface you selected. You should now see an element that represents the intersecting region of the two elements. Data point to accept the results.

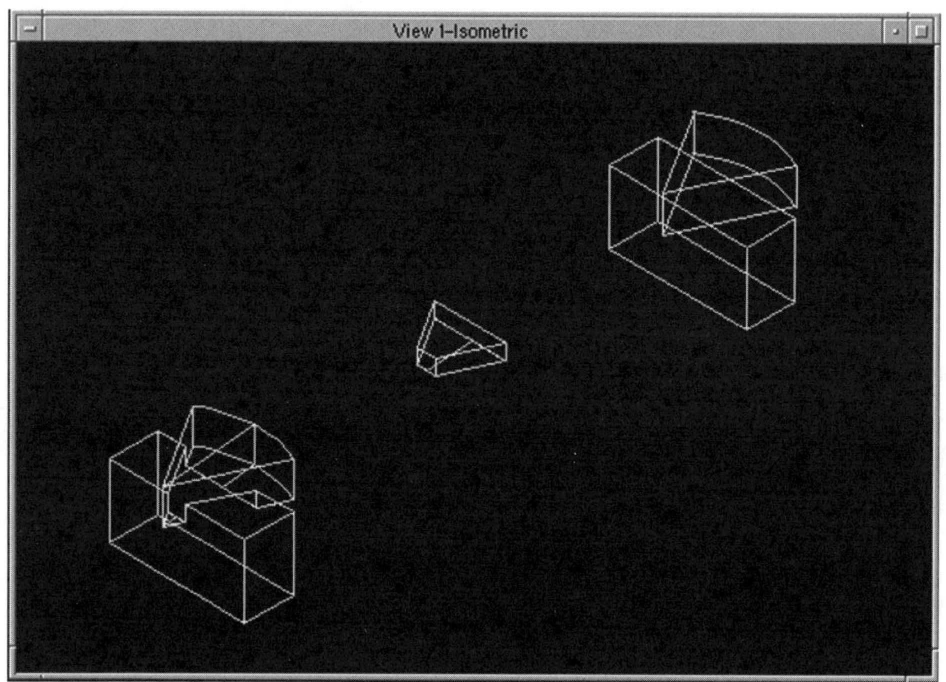

Now your View 1 should look like this.

6. Select the *Construct Difference Between Surface* tool.

NOTE: *When using this tool, the order in which you select the elements determines which element is subtracted and which element remains. Select the element you want to keep first. Then select the element that you want to remove. It's that simple.*

7. We are going to remove the wedge from the slab. Data point on the slab first, and then data point on the wedge. Data point one more time to accept the slab as the second surface selected, and then data point again to accept the results.

8. Now you will want to see your handy work under the bright lights. Use Phong shading, which you learned about in Chapter 7 to render View 1. Being the helpful guide I am, I already placed a light source in this file.

NOTE: *Performing multiple Boolean operations on the same group of elements can produce unexpected results. Even though MicroStation is a powerful 3D software, the level of complexity for an element that has had multiple Boolean operations performed on it can go beyond MicroStation's capabilities. This is not to say that complex 3D objects can't be created. You may have to experiment to get what you want. If you keep it simple, you will be fine.*

Exercise 2: Trim Surfaces

The Trim Surfaces icon.

Modifying 3D surfaces can at best be described as tricky. Looking at the pictures on the icons can induce grand ideas of what the tool does. Many travelers have become disoriented and lost in the highlands of 3D. With Safari Sam as your guide, you can avoid the frustrations of not knowing why things don't work the way you think they should. The *Trim Surfaces* tool is a classic example. Getting the results you want is as simple as knowing which part of the elements to select. The following exercise will demonstrate how this command works. Again, I've taken the liberty of setting things up for you. No thanks required; it's my job as a guide.

1. From the Command window, select **Inside3D → Chapter 10 → Trim Surface Exercise**.

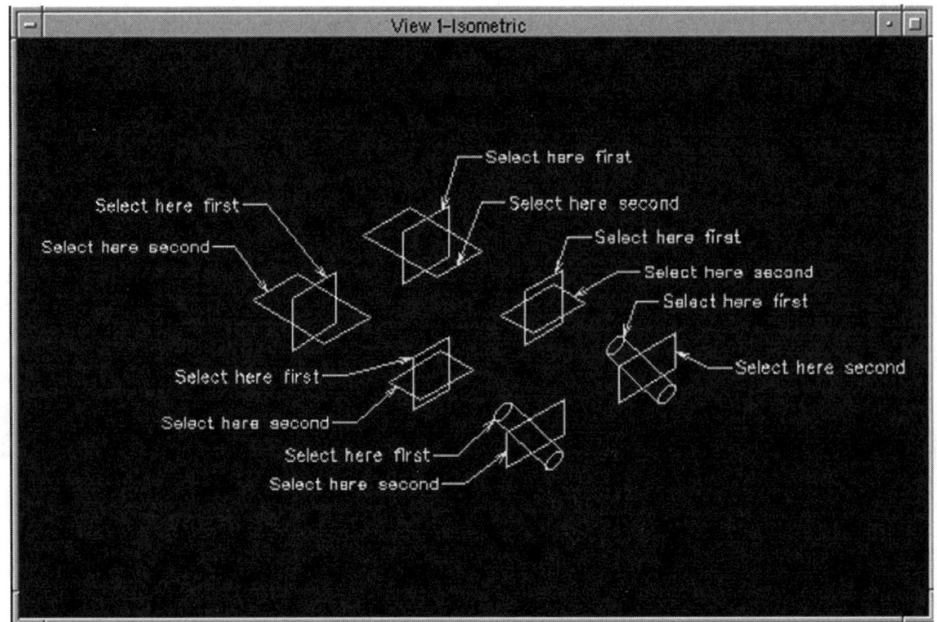

Your screen should look like this. These elements will help to illustrate how to use the Trim Surfaces tool.

2. Select the *Trim Surfaces* tool from the Modify 3D Surfaces palette. Set the truncate method to Single. This truncate method will trim the first element selected to the intersection of the second element selected. I've provided you with three sets of identical geometry that have different selection points. By selecting the elements at different points, you control which part of the element is trimmed.
3. Start with the yellow and brown elements in the left part of the screen. Read the text from the screen to know where to data point first and then second.
4. After you have selected the second element, data point again to accept it. The top part of the first surface selected has been trimmed at the intersection of the second selected surface.
5. Repeat this operation with the other yellow and brown elements. Read the text from the screen for instructions on where to select the elements.
6. Repeat this operation on the other sets of geometry. Again, read the screen for where to select the elements.

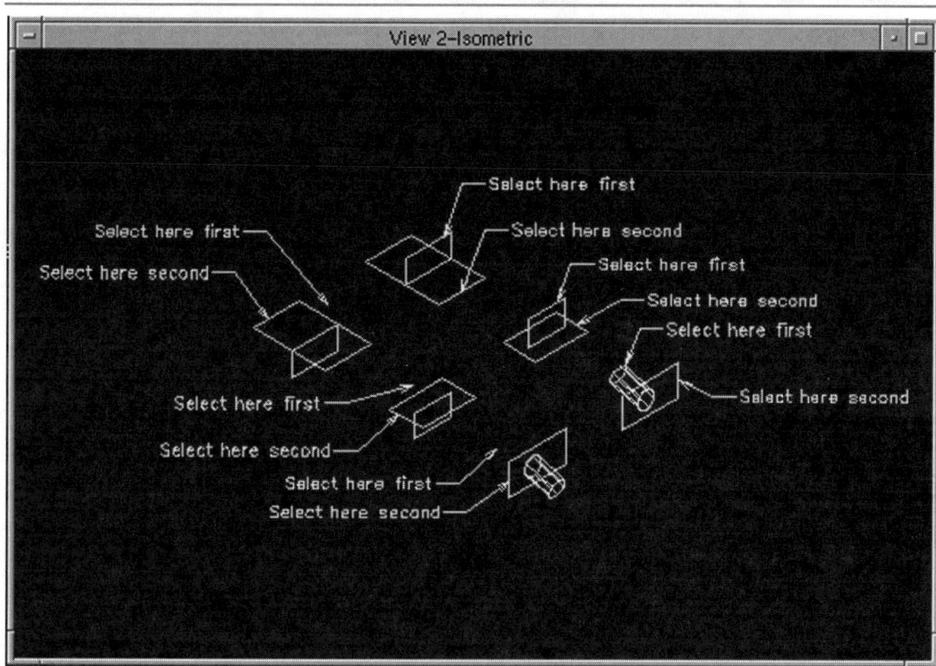

This is how the surfaces should look after being trimmed.

7. For the first part of this exercise you had the truncate method set to Single. Now repeat this exercise from step 1, except at step 2, set the truncate method to Both. This will trim both elements to their intersection if the intersection can be trimmed. Not all intersections can be trimmed, as you will see with the cyan and magenta surfaces. They are trimmed as if the truncate method was set to single.

NOTE: *The truncate method None does not trim either surface. It just creates a curve at their intersection.*

8. This would be a great opportunity to practice your new rendering skills. Phong render View 1. This always helps to visualize your work. If you want to practice this again, you can Undo what you have done, or repeat step 1. Otherwise, let's see what's on the other side of the next ridge.

Exercise 3: Extend Surface

The Extend Surface icon.

The *Extend Surface* tool is not a tool that is used often. But, when you need to extend one edge of a surface, here is your solution. This tool will allow you to extend surfaces such as surface of projection or revolution, B-spline surfaces, and cones along one of their edges. This means that if you created a cylinder, only to find out later that the height is not what it should be, you can use this tool to extend the cylinder to the new height. The *Extend Surface* command has a tool setting called Continuity. There are three continuity methods. They are as follows:

Position: This method is the most commonly used because it works on popular surfaces like slabs, cylinders, and blocks. It will extend one edge of a surface parallel to the edge selected. The edge of the extension will have a sharp edge. If the surface is a curved B-spline, the extension will be planar. You will see what I mean in Part 2 of this exercise.

Tangent: This method will extend one edge of a surface tangent to the selected surface. The extension will not be planar, nor will it project normal to the selected surface edge. You will have some flexibility as to where the extended edge will be.

Curvature: The functionality of this setting is similar to the Tangent method. The results are sometimes hard to notice also. It depends on the surface you are extending. For example, this method takes into account curves on the surface of a B-spline and will mirror them onto the extended surface.

Extension Scale: This setting only applies to Tangent and Curvature. It controls the distance between the poles (refer to Chapter 9 for a description of poles) at the edge of the selected surfaces, and the new poles of the extension. The value must be between 0 and 1. The default value of .5 is what you are going to use in the exercises.

Part 1: Extend Surfaces Exercise

In Part 1 of the Extend Surfaces exercise, you will extend the surfaces of three elements. The Continuity setting will be set to Position for all three elements. As you can see, in this exercise you have a blue planar surface, a cyan slab, a green cylinder, and a yellow line. First, you are going to extend the blue planar surface toward the slab. Then you will extend the slab to the yellow line. Finally, you will extend the cylinder in the Z-axis 7 units. Take a deep breath, and let's go for it.

1. From the Command window, select **Inside3D → Chapter 10 → Extend Surface Exercise 1**.
2. Select the *Extend Surface* tool from the Modify 3D Surfaces palette. The Continuity setting is set to Position. That's where you want it.
3. In View 1, data point on the right side of the blue surface with a data point.
4. Now key in the following:

   ```
   DX=5
   ```
5. Press Return to enter the precision input.
6. Notice how the surface was extended to the right a distance of 5 units.
7. In View 2 (the isometric view), data point on the top of the cyan slab. Notice how the slab is dynamically extending from the side that you selected.
8. Move your cursor into View 3 and tentative and data point on the yellow line. The slab has been extended to the Z-depth that the line exists on.
9. In View 3, data point on the top of the cylinder.
10. Now key in the following:

    ```
    DX=,10
    ```
11. Press Return to enter the precision input.
12. Just for the fun of it, Phong render View 2. I always like seeing my work rendered. Good job, traveler, now it's time to move on.

Part 2: Extend Surfaces Exercise

In Part 2 of this exercise, you will work with B-spline surfaces. These surfaces are flat, as you can see. The Tangent and Curvature settings will be used in this exercise. First, you use the Position setting for the green B-spline surface, the Tangent setting for the magenta B-spline surface, and the Curvature setting for the cyan B-spline surface. I have provided two lines to help define the distance to extend the B-spline surfaces.

1. From the Command window, select **Inside3D → Chapter 10 → Extend Surface Exercise 2**.
2. Select the *Extend Surface* tool from the Modify 3D Surfaces palette. The Continuity setting is set to Position. That's where you want it for the green B-spline surface.
3. In View 1, data point on the bottom of the green B-spline surface.
4. In View 1, move your cursor down to the red line and data point. Notice how the surface projected straight without any lateral play. In View 2, notice how the extended surface is planar.
5. Change the Continuity setting to Tangent. Leave the Extension Scale setting of .5 unchanged.
6. In View 1, data point on the bottom of the magenta B-spline surface.
7. In View 1, notice how the surface now moves laterally. It does not matter where you select the surface. It has to do with how the surface was created, and where its poles are located. Move your cursor down to the red line so that the surface is extended straight down. Data point close to the red line so it looks even with the extended green surface.
8. In View 2, notice that there is slight bump on the extended surface. The Tangent setting will mirror surface features onto the extended surface. Zoom in if you want to get a better look.
9. Change the Continuity setting to Curvature. Again, leave the Extension Scale setting of .5 unchanged.
10. In View 1, data point on the bottom of the cyan B-spline surface.
11. In View 1, again, notice how you need to move your cursor down and over to the left to extend the surface straight down. Data point

close to the red line so that it is even with the other extended surfaces.

12. In View 2, notice how the Curvature setting mirrored more of the surface features. This would be a good time for another Phong rendering.

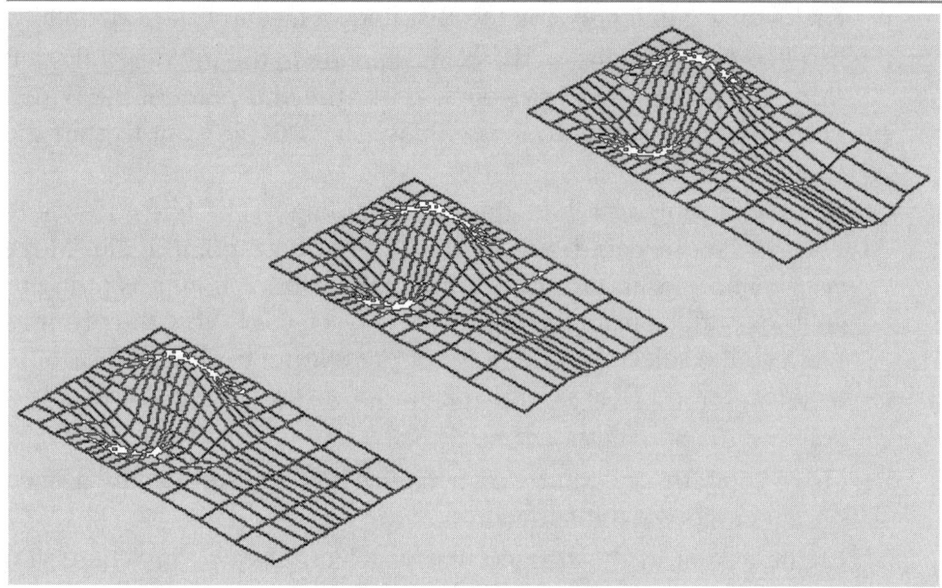

The results: your B-spline surfaces should look like this.

Exercise 4: Split Surface

The Split Surface icon

The *Split Surface* command works in 3D, much like the *Delete Part of Element* command works in 2D. It allows you to break one element into two elements or break a continuous element such as a slab or cylinder. In this exercise, you will split the surface of three types of elements. You will split a B-spline surface, a slab, and a planar block. Reading the prompts for this command will explain part of what's going on, but not all of it.

That's where I earn my pay—I will tell you what the prompts don't. Are you ready? Here we go.

1. From the Command window, select **Inside3D → Chapter 10 → Split Surface Exercise**.
2. Select the **Split Surface** tool from the Modify 3D Surfaces palette. First, you are going to split the B-spline surface in half horizontally.
3. In View 1, data point on the cyan B-spline in the middle of the left side. The prompt will ask you to select the end point of the partial delete. What it wants to know is, where do you want to start the split?
4. Move your cursor a little above the middle of the left side where you selected a data point in step 3, and data point again. Move your cursor around to the left and right. Notice how it is trying to split the surface from the edge that you selected. Now the prompts want you to select the end point or press Reset to change direction.
5. Press Reset to change the direction of the split. Now you are splitting the surface horizontally.
6. Move your cursor up and down. Notice how the split can be moved on either side of the surface.
7. Create a split in the surface that you like. Data point where you want to end the split. Wasn't that easy?
8. Now let's move on to the next element. This is where the prompts don't tell the whole story, so pay close attention. In View 1, tentative and data point on the lower right corner of the blue slab. This identifies the surface that you want split and the point at which you want to start the split.
9. Tentative and data point on the lower right corner of the blue slab again to select what the prompt is asking for as the "end point of the partial delete." This is going to be the start of the split.
10. Now the prompt is asking for the "direction of the partial delete." Tentative and data point on the middle of the left side of the blue slab.
11. Update all views. If you want, you can repeat this part of the exercise to get it straight.

12. Next, let's split the magenta slab into two surfaces. Select the *Split Surface* tool if necessary. In View 3, data point on the middle left side of the magenta slab. This selects the point at which the slab will be split.
13. In View 3, data point on the same spot again on the magenta slab. This selects the direction of the split. By moving the cursor around you can see that it wants to split the slab like the other slab you split.
14. Press Reset to change the direction of the split.
15. Notice in View 2 (the isometric view) how the slab is being split into two slabs.
16. In View 3, you can adjust the split by moving your cursor up and down the slab. Create a split you like and data point to accept it. Again, update all views to see your handiwork.
17. Finally, you will split the yellow block. Select the *Split Surface* tool if necessary. In View 1, data point on the top middle part of the yellow block to select the start of the split.
18. In View 1, data point again to select the direction of the split. If you move the cursor up and down, you will see that it wants to split the surface from the edge.
19. Press Reset to change the direction of the split.
20. Now the split is going in the left, right direction, splitting the block into two elements.
21. Again, split the element to your liking, and then Phong render View 2 to see how they look. This was a long exercise, but you did wonderfully. Grab something to eat from your pack and take a break. You deserve it.

Exercise 5: Stitch Surface

The Stitch Surface icon.

Now, here is a helpful tool. The *Stitch Surface* tool allows you to join two surfaces, creating one new surface. This command works with B-spline surfaces and surfaces of projection or revolution. What really makes this command great is that the surfaces you want to stitch together don't have to be close to each other. In fact, they can be on other sides of the design cube. As you will see in the following exercise, this command can save you much time.

If you are going to create free-form surfaces, there is something you should know. Both of the surfaces have to be *open*. There are two settings when creating free-form surfaces, *open* and *closed*. If you get the error message, "unable to perform operation," use the *Change to Active Surface Settings* command to make them both *open*. In this exercise, you will stitch B-spline surfaces to curved surfaces, curved surfaces to planar surfaces, and planar surfaces to planar surfaces. With that out of the way, let's make our way down the path to the next exercise.

1. From the Command window, select **Inside3D → Chapter 10 → Stitch Surface Exercise**.
2. Select the *Stitch Surface* tool from the Modify 3D Surfaces palette. Take a moment and become familiar with the elements in the file and where they are in 3D space.
3. In View 1, there are two magenta blocks in the top left part of the view. Data point on the right edge of the block on the left. Then data point on the left edge of the block on the right side.
4. Data point to accept the second surface selected. This will create the new surface.
5. Data point to accept the results. Notice in the other views how the new surface looks. Also, notice that the color changed. This is because this is a new element and is created with the active color of cyan.
6. In View 2 (the isometric view) data point on the top part of the green curved surface.
7. Again, in View 2, data point on the edge of the blue B-spline surface that is closest to the green curve.
8. Data point to accept the second surface selected. The two surfaces are now stitched together. Data point to accept the results.

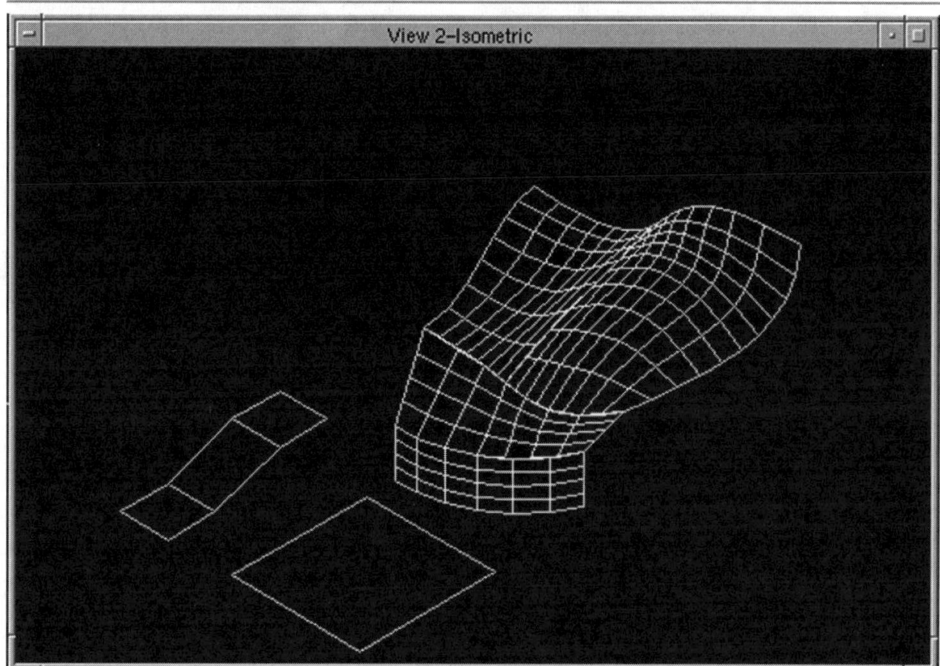

Your isometric view should look like this.

9. In View 1, data point on the right side of the yellow block.
10. In View 2, data point on the bottom of the curved surface that was green, but now is cyan.
11. Data point to accept the results. It's time to Phong render View 2. I don't think that I need to mention again how helpful a tool like this can be in the 3D wilderness.

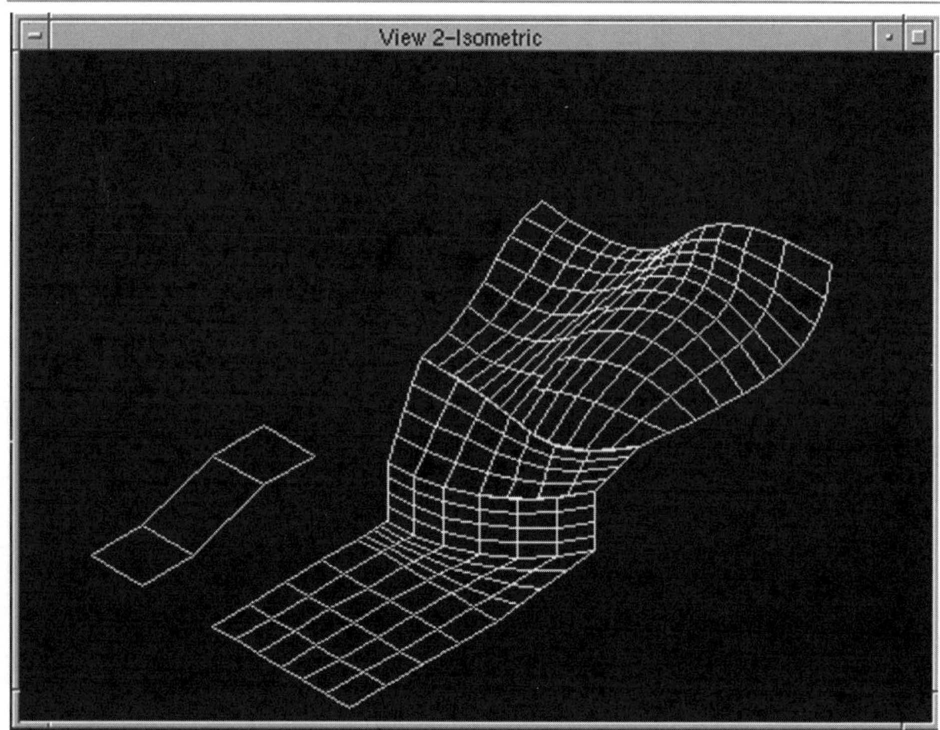

This is what the results look like.

Exercise 6: Change to Active Surface Settings

The Change to Active Surface Settings icon.

The *Change to Active Surface Settings* tool allows you to change certain settings that are unique to B-spline surfaces. Some of these settings relate to how the surfaces are created, whereas others control how the surface is displayed. Since you just went through B-splines in Chapter 9, this exercise will be a breeze. In this exercise, you change the settings of an existing B-spline surface that I dug up from behind some rocks. There will be two surfaces for you. One surface will be kind of a control surface, so you have a before and after picture. I think this will solidify the whole B-spline thing for you. Ready or not, here we go.

1. From the Command window, select **Inside3D → Chapter 10 → Change Active Surf Settings Exercise**.
2. Select the *Change to Active Surface Settings* tool from the Modify 3D Surfaces palette. You can see the two B-spline surfaces that I found. I even marked the U and V for you so you know which way they go.
3. The first thing that you will change will be the Polygon setting. Change the setting to Visible.
4. In View 1, data point on the green surface and data point again to change the settings of the surface. Notice that the control poles are now visible.
5. Now set the Surface setting to Invisible. Then data point on the green surface and data point again to change the settings of the surface. Now all you see is the control poles and no surface outline.
6. Now change the Surface setting to Visible, and the Polygon setting to Invisible. Also change the Rules setting for both the U and V to 10.

Your settings should look like this.

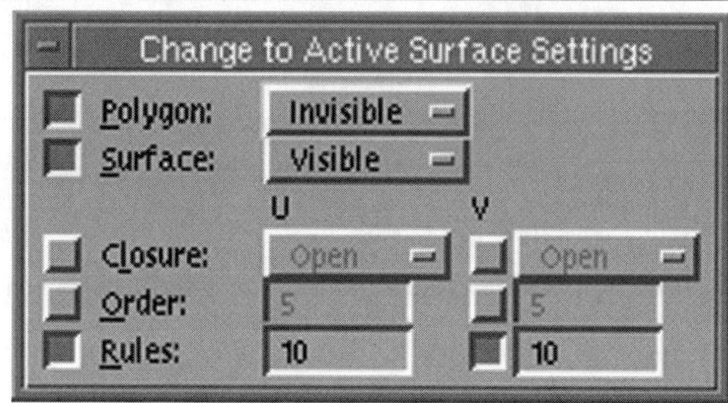

7. In View 1, data point on the green surface and data point again to change the settings of the surface. As you can now see, the Rules for the surface are now 10 for both the U and V. The Poles are also Invisible.
8. Change the Rules setting for the U to be 5. Then data point on the green surface and data point again to change the settings of the surface. As you learned in Chapter 9, this can be helpful when doing complex B-spline surfaces that vary in their complexity.

9. One last change; change the Polygon to Visible, the Rules for U and V to 10, and the Order to 10. Toggle on the Preserve shape setting.
10. Data point on the green surface and data point again to change the settings of the surface. You have just increased the number of control poles for this surface. This will allow greater control when modifying this surface, which you are now going to do.

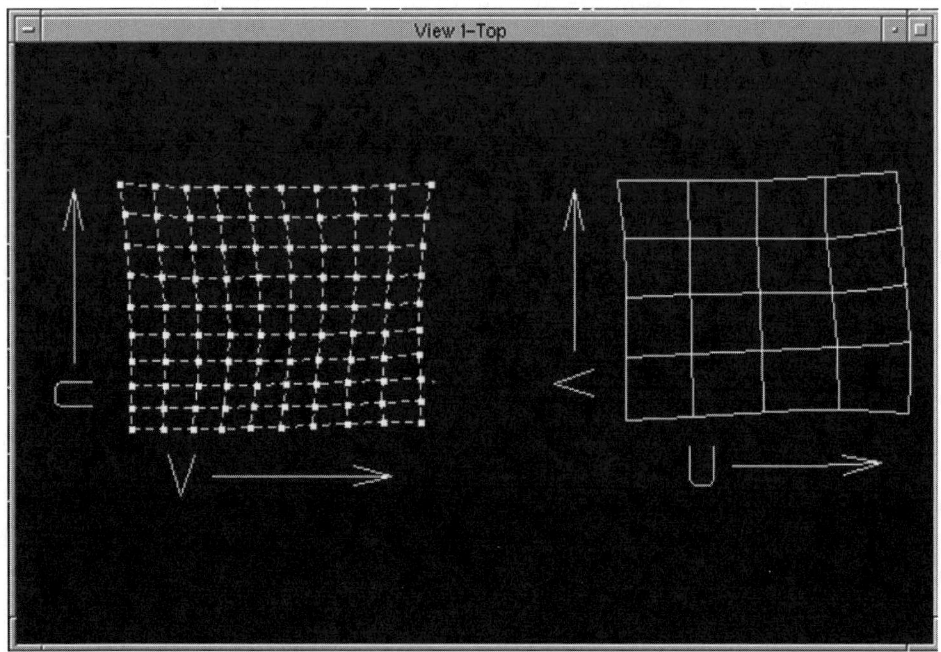

View 1 should look like this.

11. Select the *Modify Element* tool from the Modify Element subpalette. You are going to modify a control pole of this B-spline surface. In View 1, data point in the middle of the green surface.
12. Key in the following:

    ```
    DX=,,600
    ```

13. Press Return to enter the value. You have just modified a control pole for the B-spline surface. You moved the pole 600 units on the Z-axis. The more poles you have, the more precisely you can modify a surface.

14. B-spline surfaces look really cool when they are modified this way. Phong render View 2 and see the effect. With every step you take, you're that much closer to becoming an Indiana Jones of 3D.

Exercise 7: Change Surface Normals

The Change Surface Normal icon.

Being a 3D pioneer in the pristine wilderness often has a spiritual effect on people. If the spiritual world is an unseen world, then *surface normals* can qualify as part of that world. They are invisible to you and me. Yet they can have visible effects on your work in the 3D wilderness. In simple terms, surface normals are the direction of a surface. The normals of a surface will affect the way a surface is rendered, and sometimes modified. As you become an experienced 3D pioneer, you will be able to recognize and know how to change surface normals. You will conquer them and make them bend to your will. The *Change Surface Normal* tool is where you will go to change surface normals. Surface normals are defined by how a surface is created.

In this exercise, you will see how surface normals are defined, how they affect your work, and how to change them. In the first part of this exercise, you will learn that surface normals are defined by how an element is created. You will create two blocks by keyins. One will be defined from the lower left to the upper right. The next block will be from the lower right to the upper left. You will see how the surface normals will be opposite. In the second part of the exercise, you will fillet two surfaces. You will then reverse the surface normals and watch the effects.

Part 1

1. From the Command window, select **Inside3D → Chapter 10 → Change Surface Normals Exercise 1**. Don't panic, this file has no geometry in it.
2. Select the *Place Block* tool from the Polygons subpalette.

3. Key in the following:

   ```
   XY=0
   DX=40,40
   ```

4. Select the *Change Surface Normal* tool from the Modify 3D Surfaces palette.
5. In View 1, data point *just once* on the blue block you just created. Notice in View 2 how the surface normals are pointing in the positive Z-direction. Now press Reset to cancel the command.
6. Now you are going to create another block. This time, you will create it from the lower right to the upper left. Select the *Place Block* tool and key in the following:

   ```
   XY=100
   DX=-40,40
   ```

7. In View 1, data point *just once* on the second blue block you created. Notice in View 2, the surface normals, depicted by arrows at the corners of the block, are pointing in the negative Z-direction. Now press Reset to cancel the command.
8. To change the direction of surface normals, you would data point a second time to accept the element selected. Let's try it now. In View 1, select the first blue block you created and data point twice.
9. Now data point on that block again. Notice that the surface normals are now opposite of what they were. Now let's see how they affect the modification of elements.

Part 2

In this part of the exercise, you are going to fillet two surfaces. This exercise is not to teach you about the *Construct Fillet Between Surfaces* command, so don't worry about that tool for now. You will learn how surface normals affect the *Construct Fillets* command. In this exercise, you will see two sets of surfaces. The surface normals are identical for both sets. You will change the surface normals, and see how the fillets are affected.

1. From the Command window, select **Inside3D → Chapter 10 → Change Surface Normals Exercise 2**.
2. Select the *Construct Fillet Between Surfaces* tool from the Fillet Surfaces subpalette.
3. In View 2, you can see text that indicates where to data point to create the fillets. On the green set of surfaces, data point where it says to select first. Notice the direction of the surface normals is displayed when you select the surface. This is done because the normals affect this command.
4. In View 2, you are being prompted to accept. Data point where the text indicates to select. Again, notice the normals are displayed for that surface.
5. Data point to accept the surface. A fillet is created. Data point again to accept the results. Look at all the other views and see where the filleted surface was created.
6. Select the *Change Surface Normals* tool.
7. In View 3, you are going to reverse the normals for one of the magenta surfaces. Data point on the horizontal magenta surface to select it, and data point again to accept it. Now the surface normals have been reversed.
8. Select the *Construct Fillet Between Surfaces* tool from the Fillet Surfaces subpalette. In View 2, data point on the magenta surface where the text indicates to select first.
9. In View 2, data point on the magenta surface where the text indicates to select second. Data point to accept the results. Notice how the fillet was created on the opposite side of the surface, compared with the green set of surfaces. Just by changing the surface normals, you alter how elements are modified. Your fillets should look like the following illustration.

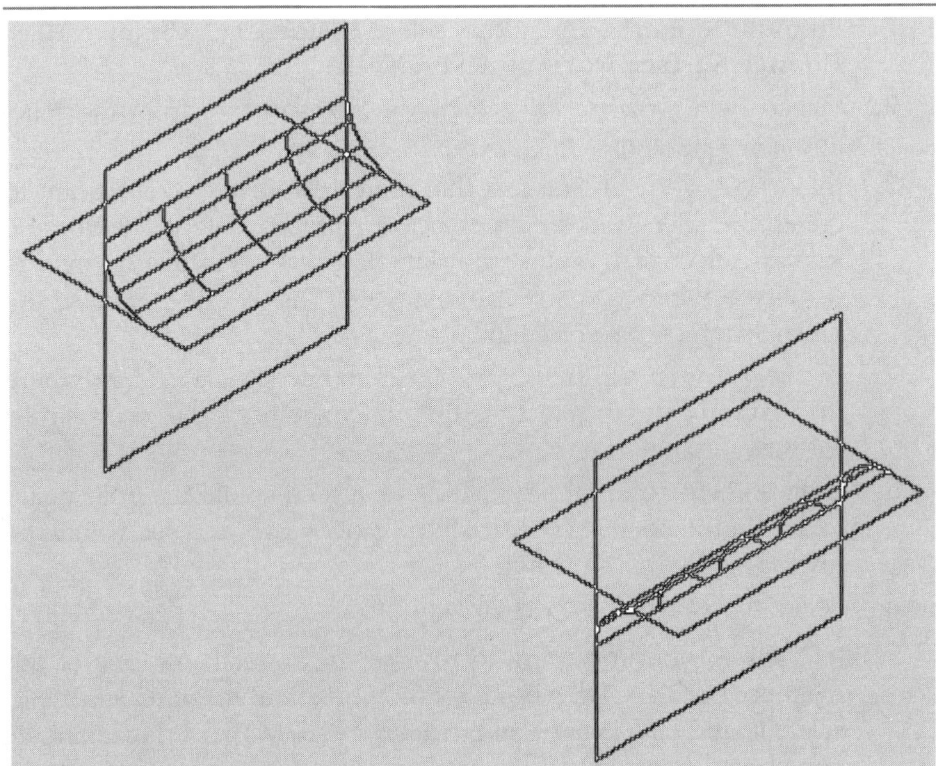

Your fillets should look like these.

Exercise 8: Change Surface Boundary

The Change Surface Boundary icon.

The *Change Surface Boundary* tool is straightforward. If you have a B-spline surface that has a hole in it, this command will allow you to reverse the boundaries of the hole and B-spline surface. It will create a B-spline surface in the area that was a hole and remove the area that was surface. If this is a little confusing, don't worry. The exercise will clear it all up. This will be a short exercise. I just happen to have an extra B-spline surface with a hole in it that I can let you use. What are the odds of that?

1. From the Command window, select **Inside3D → Chapter 10 → Change Surface Boundary Exercise**. This is the B-spline surface with the hole in it that I told you about. Isn't it a beauty?
2. Select the *Change Surface Boundary* tool from the Modify 3D surface subpalette.
3. Data point on the B-spline surface and data point to accept. There you go. You are all done with this exercise. I told you it was a short one.

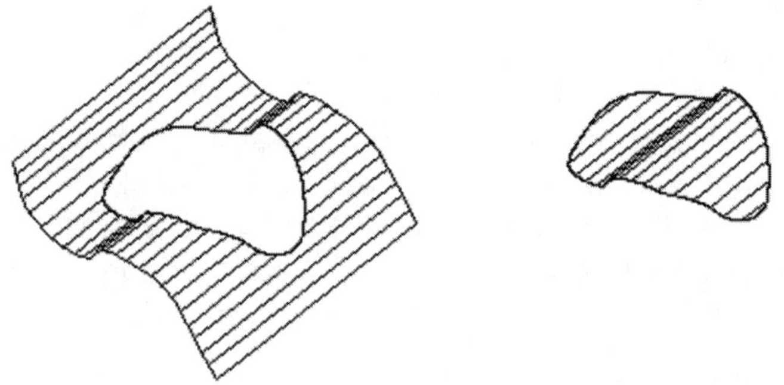

Here is a before and after look at the surface.

Exercise 9: Delete Surface Boundary

The Delete Surface Boundary icon

The *Delete Surface Boundary* tool is just as simple as the tool in the last exercise. This command allows you to remove holes in B-spline surfaces. You can remove them one at a time, or all at once. This exercise will be just as short as the last one. In fact, you can use the same old B-spline that I let you use in the last exercise. It will work perfectly for this exercise.

1. From the Command window, select **Inside3D → Chapter 10 → Delete Surface Boundary Exercise**. Recognize that B-spline surface with the hole in it? It gets around.
2. Select the *Delete Surface Boundary* tool from the Modify 3D Surface subpalette.

3. Data point on the B-spline surface and data point to accept. The hole is all gone. This sure beats having to recreate the B-spline surface minus a hole.

Exercise 10: Change to Active Solid or Surface Status

The Change to Active Solid or Surface Status icon.

When you create a slab or cylinder, there is a setting for Solid or Surface. This setting controls whether or not the element has surfaces on all sides. If the setting is set to surface, only four sides of a slab will be created. It would be like an open box. A cylinder would be like a pipe. It would have open ends. The *Change to Active Solid or Surface Status* tool allows you to change the setting on existing elements. Creating elements with the wrong setting happens to even the most experienced 3D users.

In this exercise, you will be provided with two elements. You will see a slab and a cylinder. The view is set to display Hidden Line. Your job will be to change the existing setting from Surface to Solid. This will update the display immediately. This is also a short exercise.

1. From the Command window, select **Inside3D → Chapter 10 → Change Active Status Exercise**.
2. Select the *Change Active Solid or Surface Status* tool from the Modify 3D Surface subpalette. Notice the setting type of Solid or Surface. You are going to change the elements from Surface to Solid.
3. Data point on the cylinder. Data point again to accept. The cylinder now has a cap on it.
4. Data point on the slab. Data point again to accept. The slab now has a top and bottom. That was real tough, wasn't it? Time to move on.

Exercise 11: Punch Surface Region

The Punch Surface Region icon.

The *Punch Surface Region* tool is very powerful. You can punch a hole through a B-spline surface using the profile of a curve. You can create complex elements using this command. It also will project the curve outline onto the B-spline surface. One problem with tools like this is that they are very CPU intensive. This exercise could take a few minutes to calculate, depending on your system. For this exercise, you will punch a hole through a B-spline surface using a complex shape. The complex shape in this exercise file is on a different level than the B-spline surface. The level the complex shape is on is not displayed in View 2 of this file. This will aid you in viewing the results.

1. From the Command window, select **Inside3D → Chapter 10 → Punch Surface Region Exercise**.
2. Select the *Punch Surface Region* tool from the Modify 3D Surface subpalette. For this exercise, you will use the Project By setting of Perpendicular.
3. The prompt asks you to identify a surface. In View 3, data point on the cyan B-spline surface.
4. Now the prompt asks you to accept the surface or identify the curve. In View 3, data point on the green complex shape.
5. The prompt asks you to accept the curve or identify the direction to impose. What it wants to know is, in which direction do you want to project the curve. In View 3, data point below the green complex shape. It is important that you data point in the correct direction. If you indicate in the incorrect direction, you will get an error message stating "Error: No Bounds." This could take a few minutes, depending on your system. Let it work a little while before giving up.

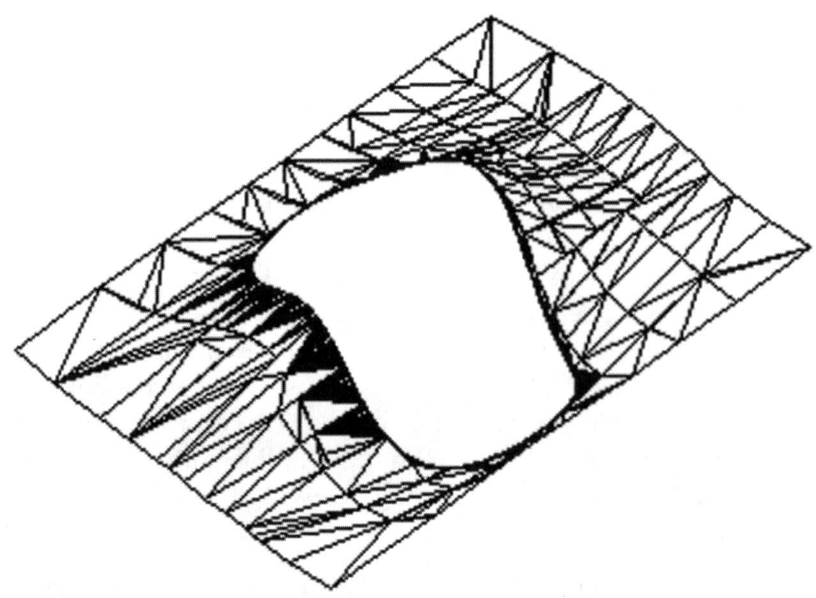

The B-spline surface will look like this when done.

Fillet Surfaces

The Fillet Surfaces subpalette has tools that can make a pioneer's work go fast. One might go as far as saying it's fun. These tools can fillet, chamfer, and blend surfaces at the click of a button. There was a time when filleting two surfaces would have taken hours. Now you can benefit from the tools of technology. In this part of the chapter, you will work with some existing geometry to test out these new tools.

Exercise 12: Construct Fillet Between Surfaces

The Construct Fillet Between Surfaces icon

This *Construct Fillet Between Surfaces* tool is used to construct 3D fillets between two surfaces. It can construct fillets between cones, surfaces of projection or revolution, and B-spline surfaces. This tool is a big time saver for those pioneers who work in the mechanical world. In this exercise, you will fillet two cylinders to a slab. You will vary the radius for the different fillets also.

1. From the Command window, select **Inside3D → Chapter 10 → Fillet Exercise**.
2. Select the *Construct Fillet Between Surfaces* tool from the Fillet Surfaces subpalette. Notice that the Define By setting is set to Constant Radius. This is the setting you are going to use. Also, notice that the Radius is set to a value of .5. You will change this later.
3. In View 2 (the isometric view) data point on the slab. The normals are displayed for the slab.
4. In View 2, data point on the tall cylinder. Notice that the normals are displayed and are pointing inward on the cylinder.
5. In View 2, data point on the slab. Data point again to accept.
6. Notice that you have an error message that states “Error: Unable to perform operation.” This has to do with two things. First, the surface normals for the cylinder and the radius were too large for it to create. With the surface normals pointing inward on the cylinder, the radius would have been created inward on the cylinder, not outward, blending with the slab.
7. From the Modify 3D Surfaces palette, select the *Change Surface Normals* tool.
8. In View 2, data point on the tall cylinder. Data point to reverse the normals for the cylinder. Now you are going to repeat the filleting of the cylinder.

9. In View 2, data point on the tall cylinder again. Notice that the normals are pointing outward this time.
10. In View 2, data point on the slab. Data point again to accept the slab.
11. Data point again to accept the results. That's how the fillet should look.

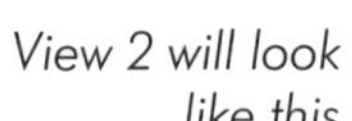
View 2 will look like this.

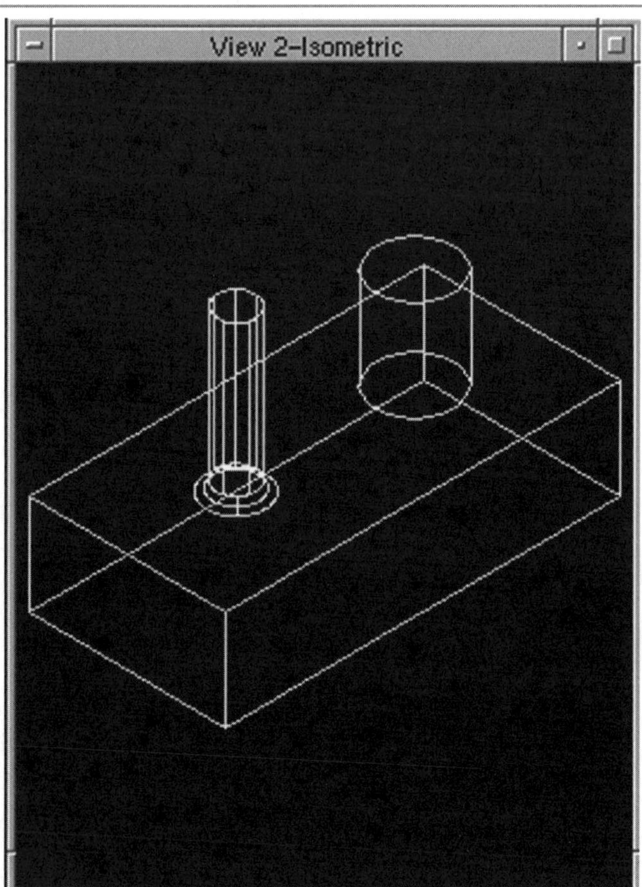

12. Change the Radius value to 1.
13. In View 2, data point on the short cylinder. Notice the normals are pointing outward. This is the way you want them to point for this exercise.

14. In View 2, data point on the slab. Data point twice to accept the slab and the results.
15. Now Phong render View 2. If you want to get an idea of the complexity of the surfaces after you filleted them, do a Wiremesh rendering in View 2. These aren't simple 3D elements any more.

Exercise 13: Construct Chamfer Between Surfaces

The Construct Chamfer Between Surfaces icon

Here is another handy tool that can save 3D pioneers many hours. The *Construct Chamfer Between Surfaces* tool will create a B-spline surface with a user-defined length, at the intersection of two elements. The elements you can chamfer are B-splines, surfaces of projection or revolution, and cones. The settings are similar to the 2D chamfer command. Your options for the truncate method are Both, Single, and None. The user-defined length refers to the hypotenuse of the chamfer. In this exercise, you will create a chamfer between a B-spline surface and a planar surface, a cylinder and planar surface, and two planar surfaces. I think that you will be surprised how the cylinder turns out. Just a few more steps and we will be at the exercise.

1. From the Command window, select **Inside3D → Chapter 10 → Chamfer Exercise**.
2. Select the *Construct Chamfer Between Surfaces* tool from the Fillet Surfaces subpalette. The truncate option is set to Both, which is where you want it for this exercise.
3. In View 2, data point on the green curved surface. Notice the direction of the normals. This is the direction that the chamfer will be created on the curve.
4. In View 2, data point anywhere on the blue block. Notice the direction of the normals for this surface. Data point again to accept the block.
5. The chamfer is now displayed in the highlight color. Data point to accept the results.

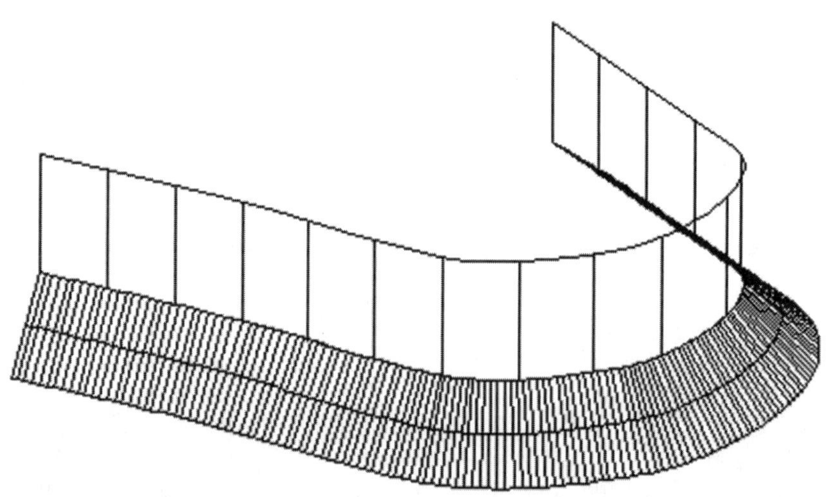

Your curved surface should have a chamfer like this.

6. If you were paying close attention, you would have noticed what direction the normals for the blue block were pointing. Now you are going to reverse them. From the Modify 3D surface palette, select the *Change Surface Normals* tool. Data point on the blue block to reverse its normals.
7. Select the *Construct Chamfer Between Surface* tool. Set the Chamfer Length to a value of 2.
8. In View 2, data point on the cylinder. Again, notice that the normals are pointing outward for the cylinder. This will produce the chamfer on the outside of the cylinder.
9. In View 2, data point on the blue block. The normals will be pointing down from the surface. The chamfer will be produced on the side of the block that the normals are pointing.
10. Data point to accept the block. The chamfer is now displayed in the highlighted color.

11. Data point to accept the results. You just created what mechanical folks call a hole with a "countersink."
12. Select the *Change Surface Normals* tool and reverse the normals for the blue block so that they point up.
13. Select the *Construct Chamfer Between Surfaces* tool.
14. In View 2, data point on the magenta block. Then data point on the blue block.
15. Data point twice to accept the block and the results.
16. Don't you think it's time for a Phong rendering? Render View 2 and see if your work looks like the following illustration.

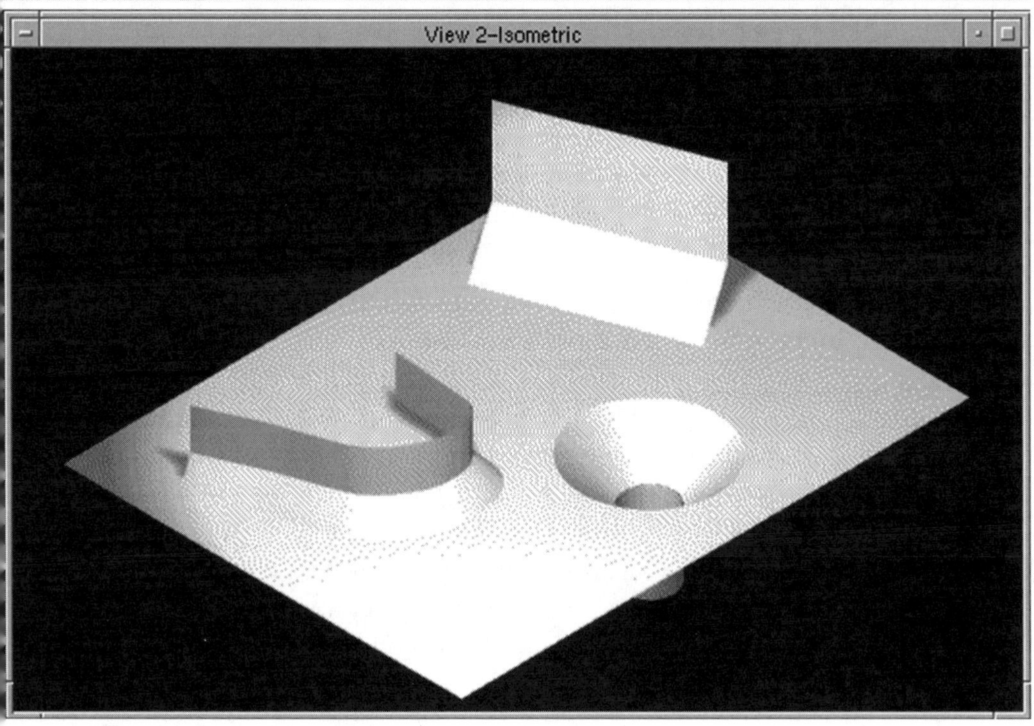

Your Phong render should look like this.

Exercise 14: Blend Surface Between Rail Curves

The Blend Surface Between Rail Curves icon.

The *Blend Surface Between Rail Curves* tool is an important command to take on any 3D adventure. This tool will assist you in putting the final touches on your projects. The smooth blending of surfaces creates a realistic appearance. This realistic appearance can translate into real gold for you. This command is fairly easy to use once you understand how it works. Everybody will think you spent hours or days working to get these results. This can be our little secret.

In this exercise, you will create both types of blended surfaces. The two types of blended surfaces are the round and the chamfer. The round type is like the fillet command you used in a previous exercise. It will produce a rounded B-spline surface. The chamfer type will produce a flat B-spline surface between the two selected rails. I think it's time to start the exercise. The sun is getting low on this chapter.

1. From the Command window, select **Inside3D → Chapter 10 → Blend Surface Between Rails Exercise**.
2. Select the *Blend Surface Between Rails Curves* tool from the Fillet Surfaces subpalette. Notice that the Blend Type is set to Round. This will be the first type you create.
3. The two sets of geometry you are looking at are identical. You will use them to create the two types of blended surfaces. You are going to blend the cyan curved surface with the blue block surface. The yellow point curves will be the rails for this command.
4. In View 2, data point on the cyan curved surface in the set of geometry in the upper left part of the view.
5. You are now being prompted to identify the first rail. Data point on the yellow curve that lies on the cyan curved surface (it's the one in the middle of the cyan curved surface).

6. The prompt now wants you to select the second surface. Data point on the blue block.
7. The prompt now wants you to select the second rail. Data point on the yellow curve that lies on the blue block surface.
8. The message reads "Accept to create blend, or Reset." Data point to accept the blend. The blend is displayed in the highlight color.
9. Data point to accept the results. The rounded blend is now created in the active color of green. Now that's a smooth blend.

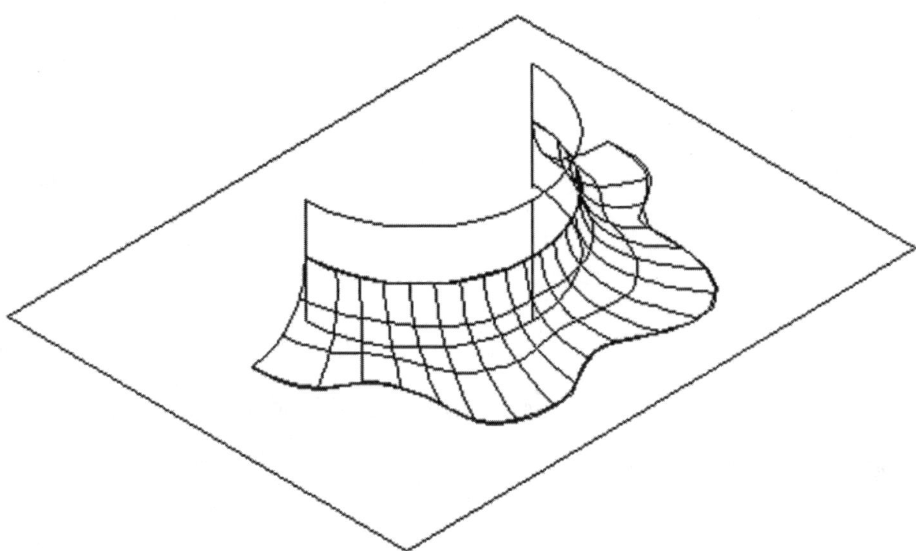

The results should look like this.

10. Change the Blend Type to Chamfer and repeat the operation on the other set of geometry. If the altitude is making you light-headed, and you forgot the steps, look back at step 4. Good job, explorer. You are now checked out on the advanced tools. It's time for us to have a heart to heart on how to use these tools.

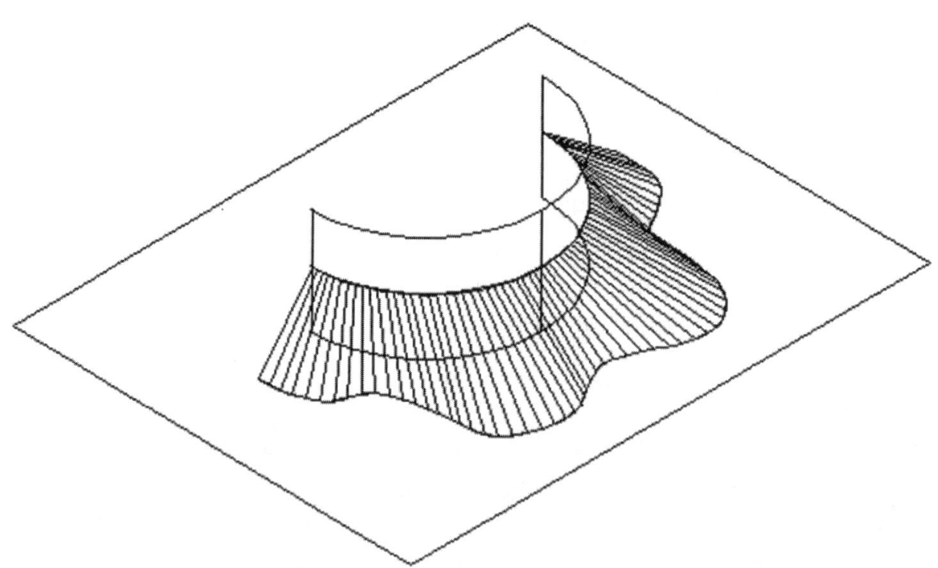

The second blend will look like this.

Did I Mention MicroStation Modeler?

MicroStation Modeler is a parametric, solid modeling MDL extension that runs on top of MicroStation. MicroStation provides some powerful 3D tools to work with, but if you are serious about working in this new 3D wilderness, you should look into buying MicroStation Modeler.

Why use Modeler's parametric solids over MicroStation solids? A parametric solid is stored differently than a nonparametric solid. The parameters used to create a solid are stored with the solid. These parameters can then be modified at a later time, instead of recreating the object. For example, what if you created a slab with a height of 3, a length of 4, and a width of 5? A parametric slab would have these values stored as parameters. What if you had to change the width from 5 to 10? Changing the parametric slab would be as easy as changing the width value from 5 to 10, and the slab would then reflect the change. You would not have to stretch or recreate the slab.

Modeler is a feature-based application. What does that mean for you? Features in Modeler are things such as a hole in a slab or a chamfer on a slab. Let's say, for example, you have a hole through a slab. If you wanted to change the diameter of that hole through the slab, because the hole is a feature, you can change the parameters of that feature, like the diameter. You can also move the hole to a different location on the slab and have the geometry automatically adjust for the change. If the hole was defined as a through hole, it will continue to go through the slab, even if the slab's thickness is changed. This kind of technology will improve productivity on the order of magnitude. We will talk more about Modeler later in the book.

Summary

Look, you have reached the summit of the MicroStation 3D surface mountain. I'm impressed; it's no small accomplishment. From this summit you can look out over the world and see all of the objects made of 3D free-form surfaces. At this point, you can practically roam the 3D wilderness on your own.

Introduction

So, you think you are ready to slay the dragon. Let me give you a bit of advice—slaying the dragon is the easy part; telling the story so others understand it is the tough part. The dragon is your model, and once you have created it, you must pretest it to see if others can understand it. In this chapter I will show you the tools used to tell the story of your dragon slaying.

We will open MicroStation's bag of tools and tricks. Each tool is useful in its own way, for example, placing some text to identify the unique green eyes of your dragon, giving its dimensions, using a pattern to simulate its scales, or using cells and reference files to add detail or show different views.

One of the most useful tools is *drawing composition*. It is basically a method of story telling, which you will learn about later. I will also show you another method of story telling, flattening or exporting your dragon to 2D.

Text in a 3D World

You have decided that leaving bread crumbs for those who follow is not good enough. You want to leave notes for them, or for yourself. Usually text and annotation should be placed in a sheet file and not in your model. The model is referenced into another drawing where the finishing touches are applied. Drawing composition, discussed later in this chapter, is ideal for this purpose. There is no reason that text can't be placed in a model, except that it can get a bit tricky. Unless you have a special situation or are actually modeling text, avoid placing it in a model.

The main reason for this is that text is treated like any other 2D element in a 3D environment. If text is placed in the Top view, it will appear flat in the Front/Back and Left/Right views. In the Bottom view it will appear backward and upside down. The text below was placed by origin in the Top view. Notice how it appears in the other views.

In this figure, text is placed in the Top view.

This is an example of text placed in the Front view. Now the Top and Left views appear flat.

Place Text palette.

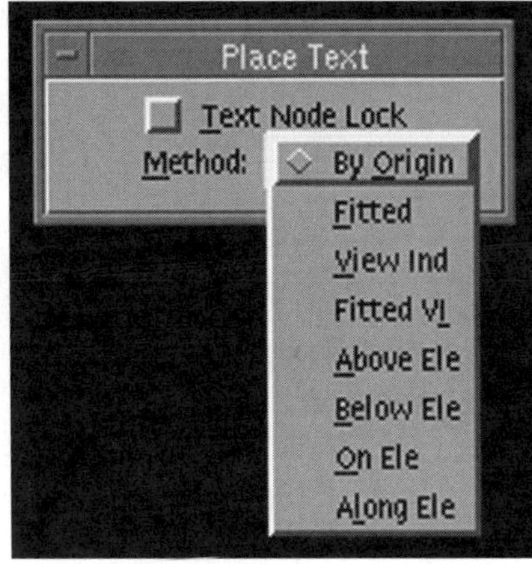

One interesting twist on this is *view-independent* text. You have experience placing text using many 2D methods. View-independent text will appear rotated parallel to every view.

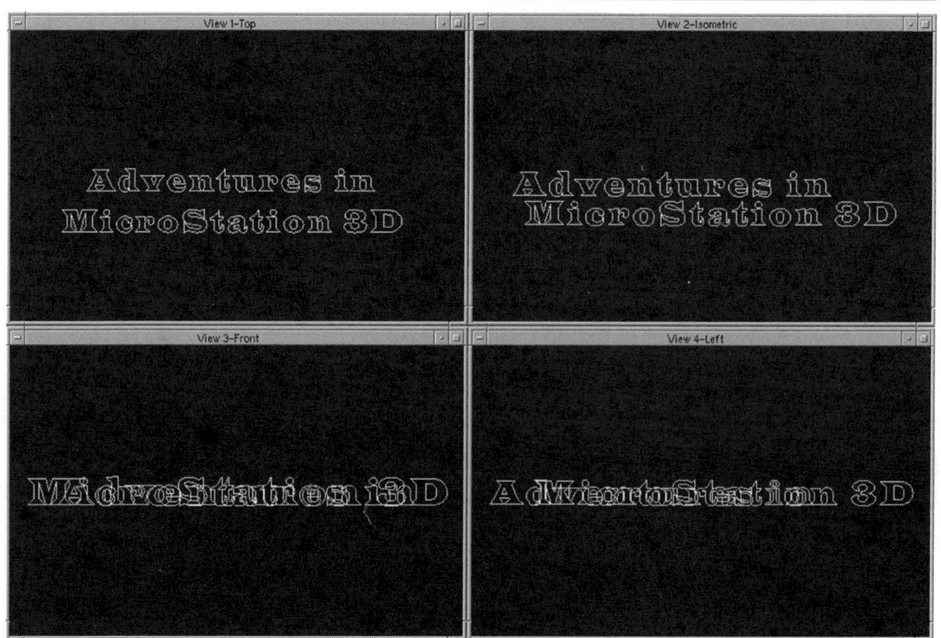

View-independent text.

The one limitation is that text nodes and multiple text strings appear overlapped in all views except the Top view and can result in some odd displays. This feature is especially useful for placing survey or similar types of information.

Dimensioning the Dragon

For obvious reasons this is an important part of the story of your dragon. Dimensions in 3D function exactly like they do in 2D. All of the settings for dimensions that you have spent time with in the 2D world apply here. The standards you have established in 2D should continue to be used.

The difference in 3D is the ability to view the dimensions from many different views. Dimensions are 2D elements in a 3D world. They can be aligned with any view desired. The view in which the dimension is started is the view that the dimension will be aligned with. If it is started in the Front view it will be aligned with the Front view. It functions just like text.

If you encounter a situation where you must begin a dimension in a view other than the one you wish it to be aligned with, MicroStation can accommodate you. The solution is *Change Dimension View.* It is not located on a palette; it must be keyed in in the Command window. The syntax is as follows:

```
CHANGE DIMENSION  <VIEW #>
```

The portion `<VIEW #>` should be replaced with the number of the view the dimension should align with.

Patterning

Patterning is the same concept in 3D as in 2D. Both line and area patterning can be used. The same controls are available, such as pattern angle, pattern spacing, etc. All of the patterning options are also still available, such as flood, fence, union, etc.

As in 2D any closed, planar, solid element can be area patterned. Closed and solid should be familiar concepts, however, planar needs a bit of explanation. To be planar, all vertices of an element must share the same depth. Depending on the view in which the element is created, this could be the X, Y, or Z axis. If all of the vertices are not planar, the element can not be patterned.

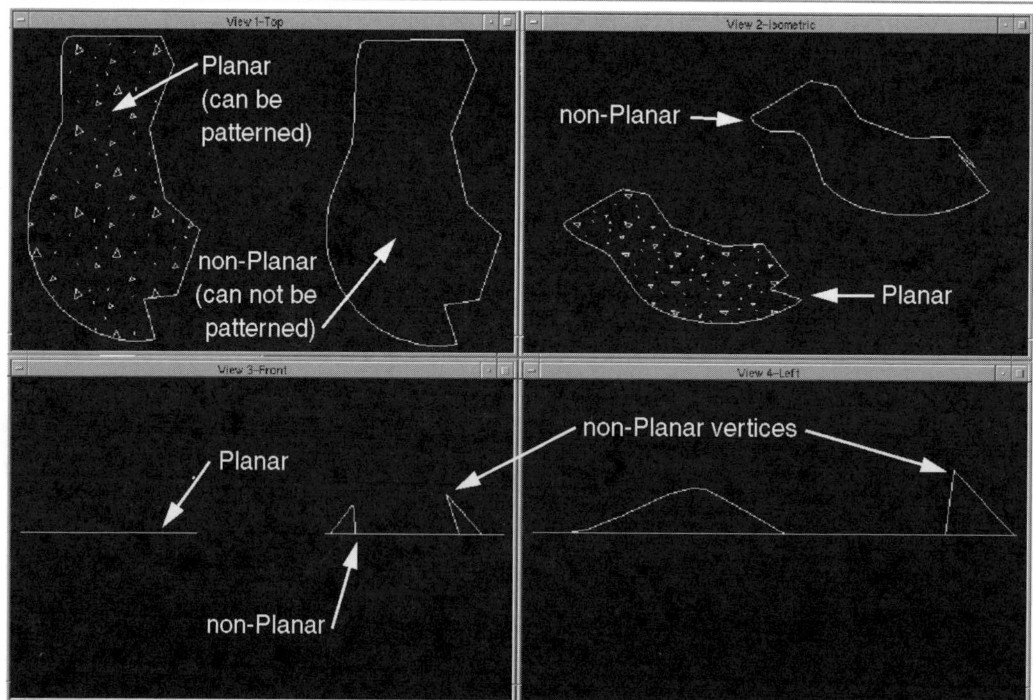

View of planar and nonplanar elements.

Hole elements can also be used the same way in 3D as 2D. They are not patterned and can be used to form voids in area patterns. However, the hole element and the patterned element must share the same plane (be planar).

Line patterns can be used with any type of line or line string. They do not have to be planar. This is because MicroStation does not have to calculate any type of an area, it just uses the vertices and information of the line or line string to place the pattern.

NOTE: *Some interesting effects can be achieved by applying line patterns to nonplanar curved line strings.*

Exercise 1: Text, Dimensions, and Patterns

We will use the following set of shapes to work with text, dimensions, and patterns.

1. Open **Inside3D → Chapter 11→ Ch11-ex1.dgn**.
2. Set the active level to 10 and the active color to 10.
3. Open the text palette **(Element → Text).**
4. Set the font to 3 (Engineering), the text size to 3, line spacing to 1.5, and the justification point for both text and text nodes to upper left.
5. Label the purple slab and the green cylinder similar to the following image.
6. Change the active depth in the Top view to the top of the white slab and place a label on the slab.

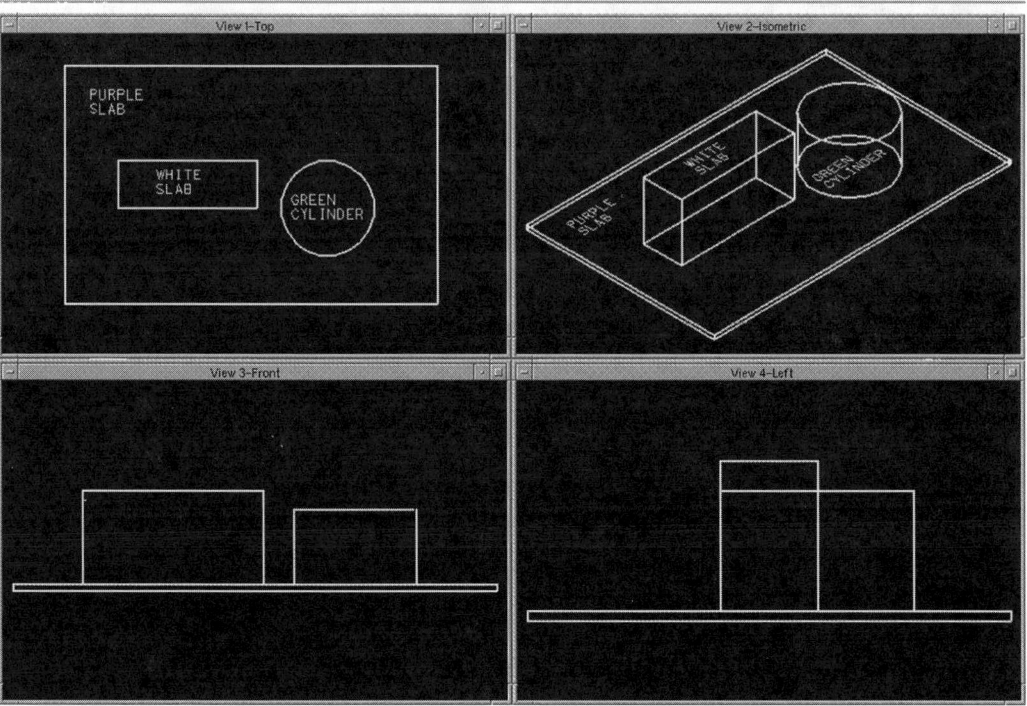

Set the text as shown.

Notice the way in which the text is displayed in the Left, Front, and Iso views.

Now let's do some dimensioning.

7. Change the level to 11 and the color to 11. Open the Linear Dimension palette **(Palettes → Dimensioning → Linear**). Now add the dimension strings shown below.

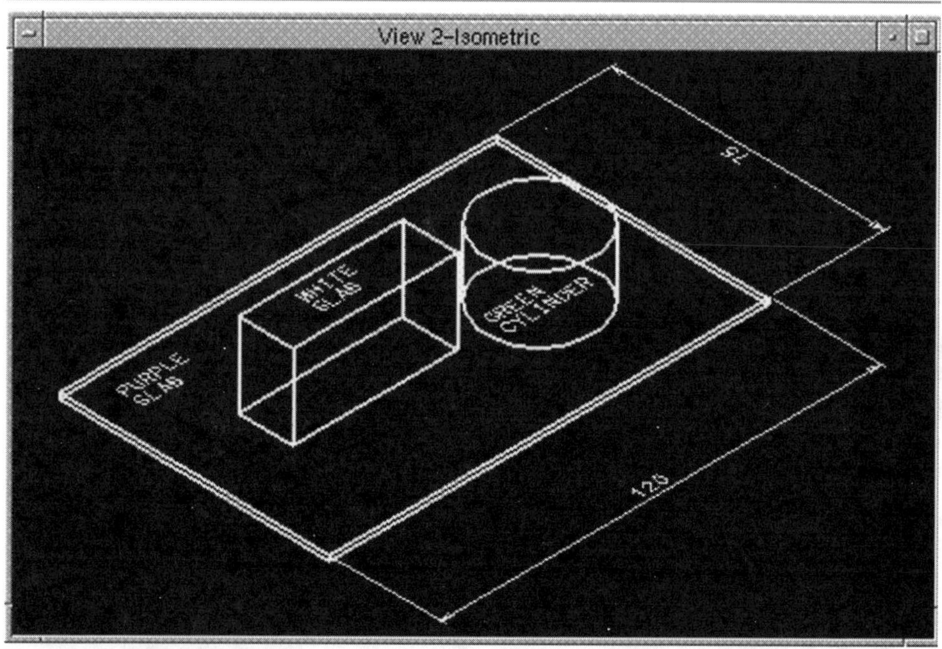

Set the dimensioning as shown.

Notice that the dimensions are legible in the Top and Iso views, but not in the Left and Front views. We will now change this using the *Change Dimension View* tool. Its effects can be clearly seen in the Iso view below.

8. Input `Change dimension view 3` in the Command window. This tells MicroStation that you wish to align a dimension with view 3. Now place a data point on the 120 dimension. It is now aligned with view 3. Input `Change dimension view 4` in the Command window and place a data point on the 75 dimension.

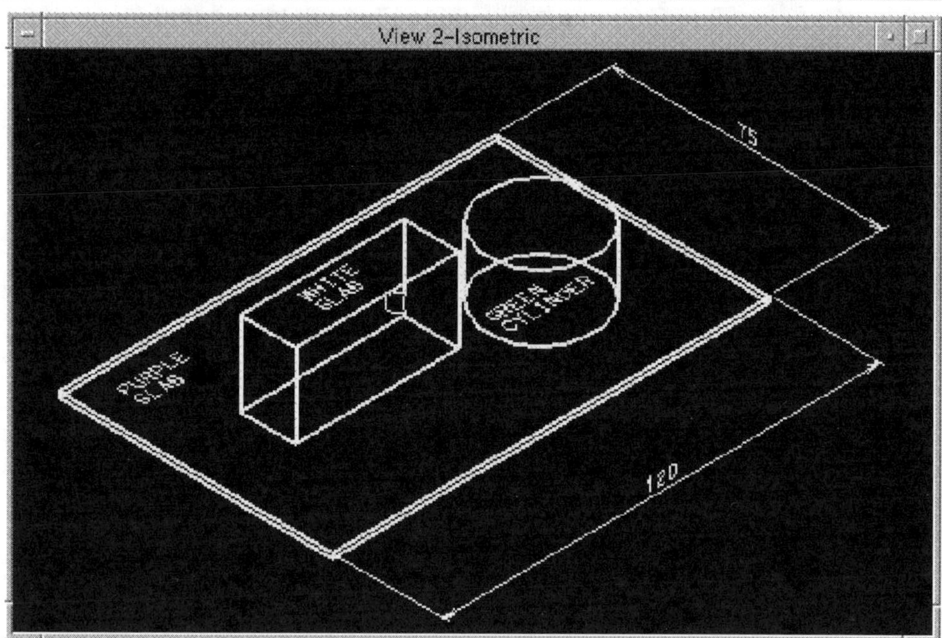

Dimension with changed views.

Let's place some patterns on the shapes. All of the elements in the model are solids and cannot be patterned. So you will have to place a shape to pattern.

9. Change the level to 12 and the color to 12. Activate the *Place Block* tool. Tentative and data point to the upper left corner of the purple slab, and then tentative and data point on the lower right corner of the same slab. You should now have a block sitting on top of the purple slab.
10. Attach the *Areapat.cel* cell library and open the Patterning palette **(Palettes → Patterning)**. Activate the *Pattern Area* tool. Make the active pattern *concrt* and pattern the block that you just created.

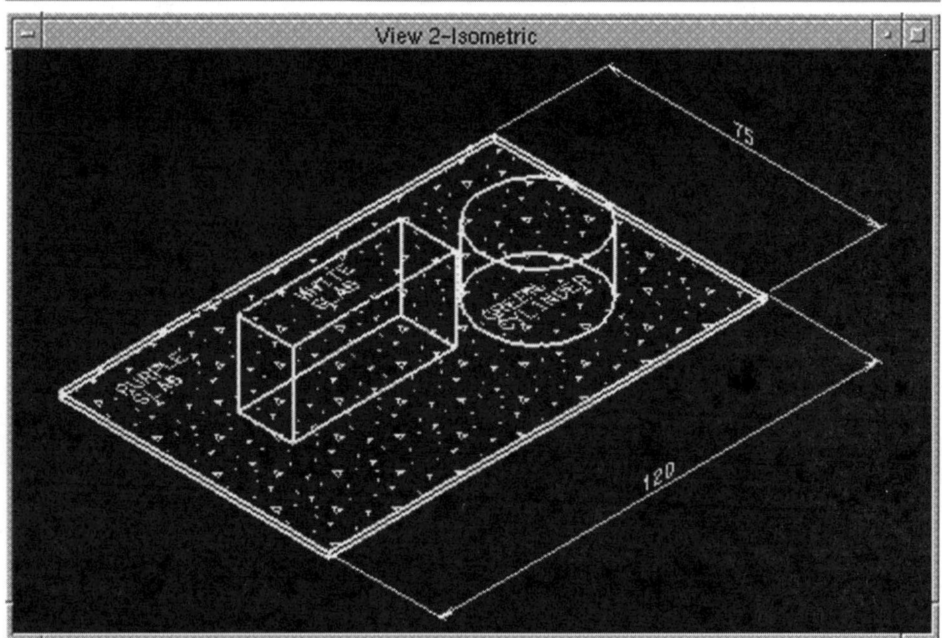

Set the patterns shown.

Notice that the pattern continues under the solids and the text. It would be nice to have the pattern go around them. To do this a hole shape must be placed on the same plane as the element to be patterned. I have already placed some shapes for you to explain this principle.

11. Delete the patterning you did previously.
12. Change the active level to level 60.

    Notice that a number of shapes show up. These are the hole shapes that share the same plane as the top of the purple slab.

13. Place a new block on top of the purple slab and pattern it.

    The pattern now goes around the solids and the text.

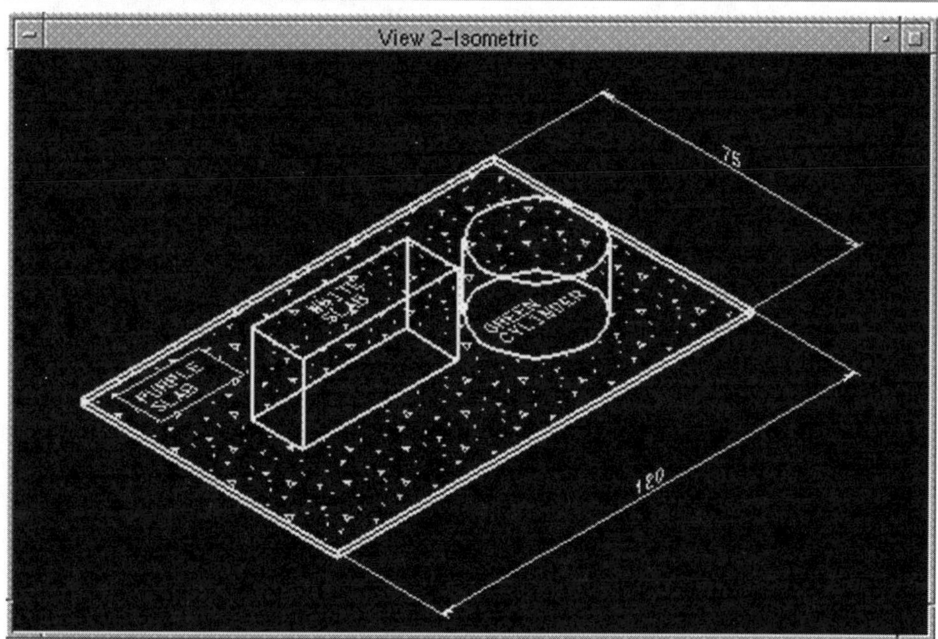

Pattern with hole elements.

We now leave the annotation tools behind and trek into some tools that can make life in the 3D world interesting. Cells and reference file can be used in many different ways. We will cover the basic concepts, but they will really begin to prove their value when you adventure on your own.

Cells and Shared Cells

Cells are one of the major building blocks of the 3D world, just as they are in the 2D world. Cells can contain most any MicroStation element. They can be used in the creation of the model or the annotation of it. There are few tools as versatile and useful. The time required to understand and use cells is well invested.

The major difference in a 3D cell is the fact that it must deal with information in the X, Y, and Z axes, not just the X and Y axes. Because of this MicroStation has a separate type of cell library for each. A 2D cell library can be attached to either a 2D or 3D design file. A 3D cell library can only be attached to a 3D design file. If there is no specific reason to use a 3D cell library, a 2D library should be used. However, cells cannot be added to a 2D library from a 3D design file. MicroStation cannot invent the 3D information.

Three-dimensional cell creation is done in the same process as 2D cells. A fence defines the elements to be in the cell, a cell origin is placed to define the cell's insertion point, and then the cell can be created on the cell dialog palette (*Settings→ Cells*). The major difference is that the view's display depth determines the extent of the cell in the third dimension.

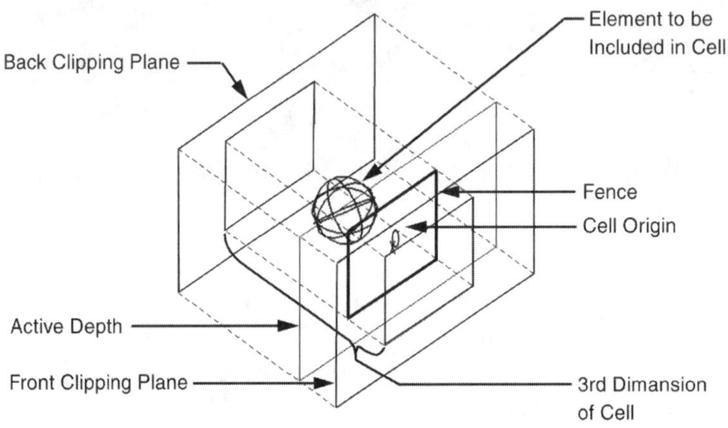

Cells in the 3D world.

> **NOTE:** *Grouping elements (Edit→ Group) is a simple method of including elements into a set for easy movement, duplication, modification, etc. The group can be dropped using the Ungroup command (Edit→ Ungroup).*

Cells can be a quick and effective way of creating a model. They also provide a powerful method for updating or changing your model. With the *Replace Cell* command (on the Cells subpalette) individual cells can be altered or updated with new information or geometry.

Shared cells are just a variation of cells. You should already have a working knowledge of shared cells, but we will do a brief review. When a cell is placed for the first time in a design file with shared cells turned on, a shared cell definition is placed in the file. This definition stores all of the information about the cell. When the same cell is placed in the file, an instance of the geometry is placed, not the actual geometry. An instance is basically a link back to the definition. So you can have many copies of a cell (let's call the cell "Inside3D") in your file. Each copy of Inside3D is an instance; they are linked to the definition that was placed in the file with the first Inside3D cell placed.

There are three main benefits to shared cells. The first is that they reduce the size of your design file and add to efficiency. Second is the fact that once the definition is in the file, the cell library is no longer required for further placement of the cell. The third, and most important here, is the ability to update shared cells. If the *Replace Cell* command is used on one instance of a cell, all instances are updated. If you redefine the Inside3D cell in the cell library and replace one instance of the cell in your file, all instances of Inside3D are replaced. This would be useful in updating a plumbing model, changing the size of a butterfly valve, or changing all columns from round to square in an architectural model.

Reference Files

Reference files are the icing on the MicroStation cake. They can be used in many ways for many purposes in 3D. We will spend some time talking about their uses. I'm assuming that you are familiar with the basics of reference files; if not, spend the time to get friendly with them. There are a number of tools that every adventurer needs before trekking in the 3D world and this is one of them.

Reference files in 3D have unlimited uses. The following examples will give you some insight into specific applications in the forests and deserts that you will encounter.

Many of you will be bringing plans from the 2D world into the 3D world. You might wish to model your piping for interference checking, or you want to create the parts of your widget to check tolerances. I'm sure someone will want to take a floor plan and generate the volumes and shapes of a building before it is built. In each of these examples referencing a 2D plan into your 3D design file is the ideal place to start. The original information is safe from changes; it will be updated with changes that occur over time, not to mention that you can copy and snap to the 2D plan.

Another use for reference files is creating your model in multiple files. Consider a model of a home: the interior walls are in one design file, the exterior in another, and the building site in yet another. This keeps the portions of the model small and easy to work with, while allowing you to use the entire model for reference. It also allows you to display only the portion required. This helps in both modeling and rendering.

Drawing Composition

Another major use for reference files is drawing composition. Drawing composition is a tool that will help you set up a design file that shows different views of your model.

Drawing composition design files are designated with the file extension *.s* (usually *.s01). They are good locations to add annotation (text, dimensions, etc.) to a model. The Drawing Composition palette is located on the Command window *File → Drawing Composition.*

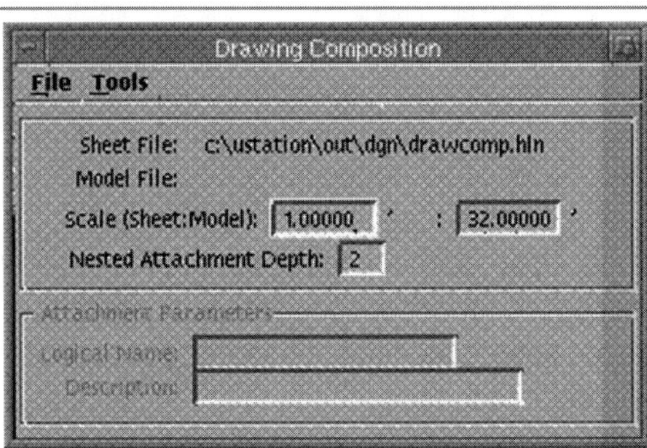

Drawing Composition palette.

Once a sheet file is created, a model (design file) is attached. At this point the first view of the model is attached. Orthographic and saved views can be used. Most often the Top view is used. Once the first view of the model is attached, the others are "folded" off from the first.

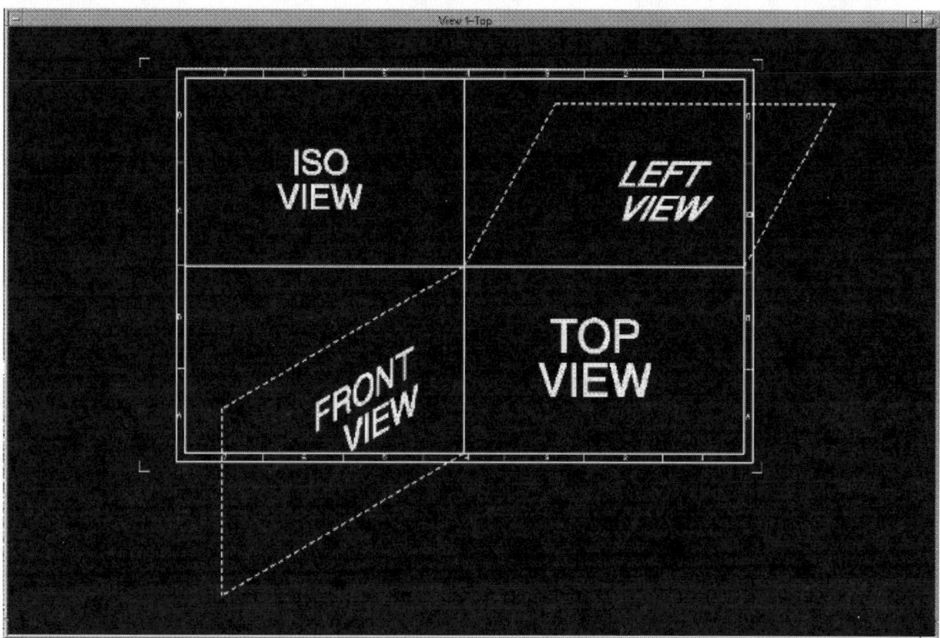

Typical drawing composition layout.

Each of the attached views is actually a reference file. If the model is adjusted, all of the views are updated. On complicated models, the use of the Front and back clipping planes may be inadequate to achieve the desired views. However, on uncomplicated models, drawing composition is a powerful tool.

Exporting Your 3D Model

There are a number of reasons you might wish to export your 3D model. Some are obvious, such as hidden line files, but others are not so obvious. Safari Sam will guide through some of the reasons for exporting and the process.

Let's start with the basic 3D to 2D export. It is often useful or necessary to convert a 3D design file into a 2D design file. A example of this would be topographical information. In a 3D format it can be cumbersome and taxing on a system. By converting it to 2D, it can be simplified and improve efficiency. The process is simple. *File→ Export→ 2D* is the location of the Export 2D palette (you must be in a 3D file for it to be available).

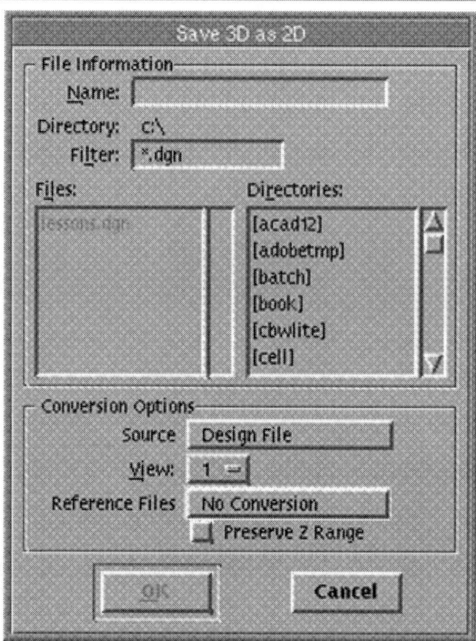

Export 2D palette.

There are three main options on this palette. The first option, Source, allows you to select the design file or the attached cell library to be exported. View is the second option; it allows you to designate the view the 2D design file will look like, that is, Top, Left, Iso, etc. The third option is Reference Files. It controls whether or not attached reference files are merged into the 2D file.

When a model is more complex, exporting a flat 2D file can be unsatisfying. MicroStation has the ability to export the *visible edges* of a 3D file to a new 2D or 3D design file. If you cannot achieve the desired results with drawing composition, visible edges is your solution *(File→ Export→ Visible edges).*

Visible Edges palette.

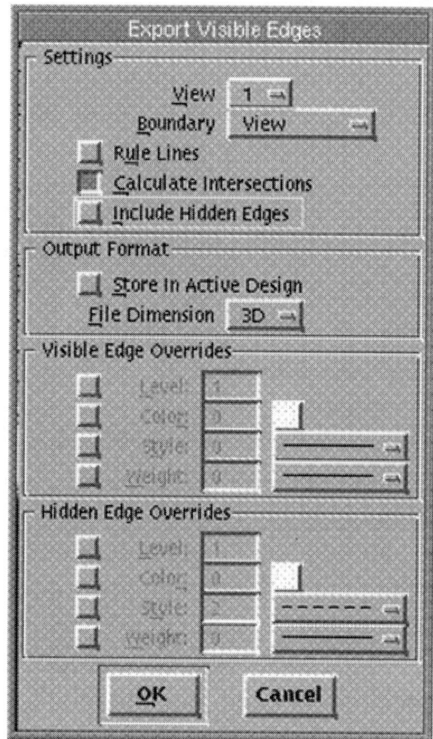

Let's deal with each of the four areas.

- **Settings** controls the overall configuration of the conversion.
- **View** controls which of the open views will be used to export the visible edges.
- **Boundary** controls whether the entire design file, the view (set above), or a fence is used in the conversion.
- **Rule lines**, if set to On, adds extra lines to define surfaces and shapes.

Visible edges without rule lines.

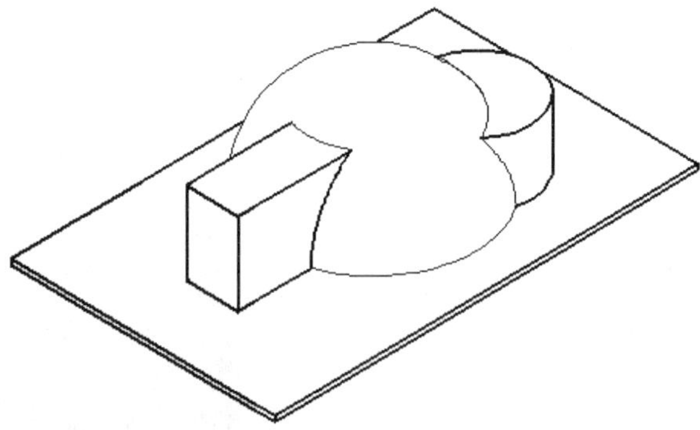

Visible edges with rule lines.

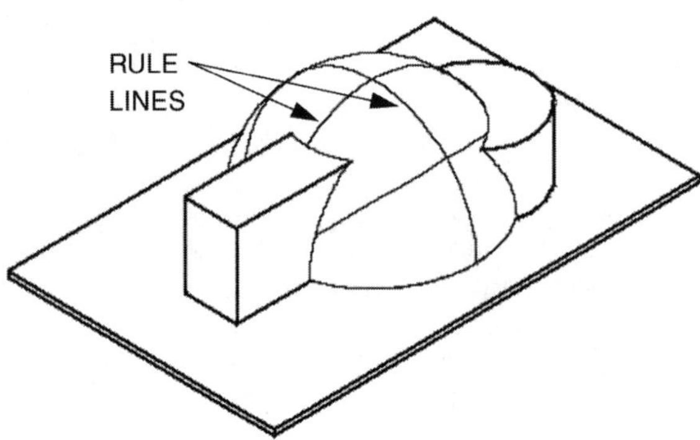

- **Calculate intersection**, if set to On, calculates where elements intersect and defines them with lines or arcs.
- **Include Hidden Edges**, if set to On, includes the edges or elements that would not be visible.

The next section, *Output Format*, has only three options. These options control where the visible edges calculation will be saved, in the active design file, a new 2D file, or a new 3D file.

- **Visible Edge Overrides** gives you control over the level, color, style, and weight of the visible edge lines.
- **Hidden Edge Overrides** gives you control over the level, color, style, and weight of the nonvisible (hidden) edge lines.

Exercise 2: Cells in 3D

1. Open **Inside3D → Chapter 11 → Office.dgn**.

   This office will be the location of our exercise. You have worked hard; you deserve the corner office with the breathtaking view. Now you have it, but let's customize it to suit your style.

   The first thing is to get some furniture in here. We are going to place the furniture in a separate file and then reference it to this file. This is so your interior designer can make some last minute modifications to the furniture plan while you are working on the office.

2. Open **Inside3D → Chapter 11 → Furn.dgn**.

   You will notice that in this file the office is already attached as a coincident reference file, so let's begin to place some furniture.

3. Attach the Chapter 11 cell library **(Inside3D → Chapter 11 → Cell Lib)**.
4. Make "desk" the active cell, active angle 0, and active scale 1. In the Top view place the cell roughly in the location shown below.

Furniture placement.

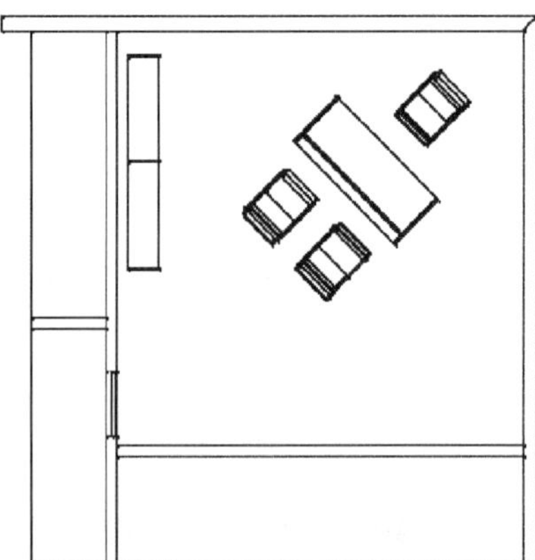

5. Make "case" the active cell and place it in the file as shown above.

   Now let's return to the office and attach *furn.dgn* as a reference file.

6. Open **Inside3D→Chapter 11→Office.dgn**.
7. Attach *furn.dgn* as a reference file, giving it a logical name of "Furniture." The two files are coincidental in location, so there is no need to scale, move, or rotate furn.dgn.

   What good is a corner office if it doesn't have any windows? Let's put in a few. I think that three in each of the exterior walls will be good.

8. To simplify the model turn off the display of the reference file *furn.dgn.*
9. Attach *Chap11.cel* **(Inside3D → Chapter 11 → Chap11.cel)**.
10. Activate the *Place Cell* command. Make "window" the active cell, the active angle 0, and the active scale 1.
11. In the Front view, tentative point to the inner corner formed by the two exterior (yellow) walls (see following figure).
12. Activate the *Copy* command. In the Top view data point on the window and place a copy 6.5 feet to the left (`DL=-6.5,0,0`). Repeat the copy so that there are three windows.

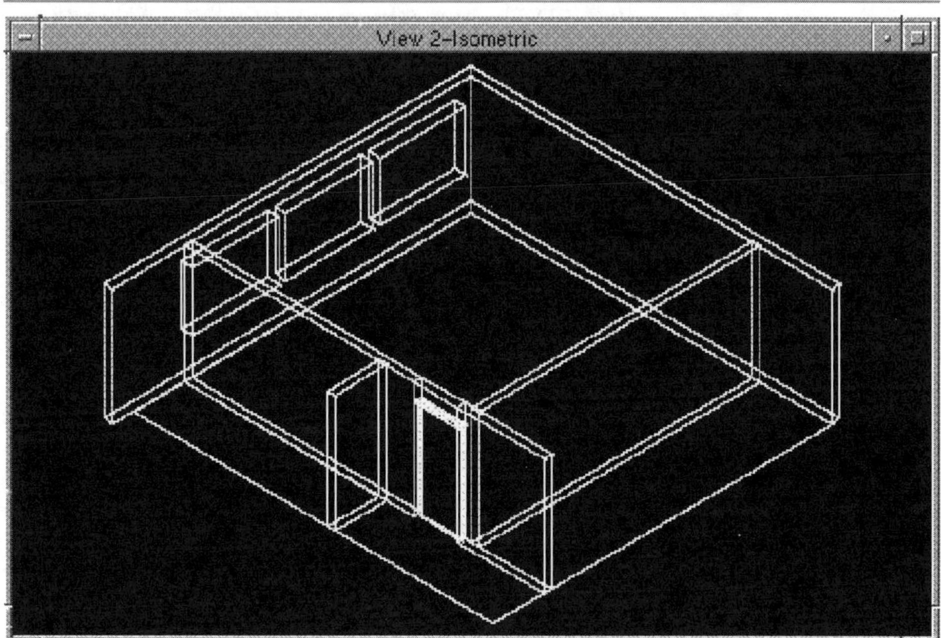

Office with three windows.

Let's put three windows in the other wall.

13. Use the *Selection* tool to select all three windows. Place a data point just to the outside of the office and drag the cursor until all three windows are included in the rectangle and release the data button. The windows should now have the white "handles" displayed at each corner. Activate the *Mirror* command. In the Setting palette, set Mirror About to line and activate the Copy option.

    Now you need to define the line to mirror the windows about.

14. In the Top view, tentative point to the inner corner of the office (make sure you tentative point to the wall, not the window), then place a data point. Now tentative point to the outside corner of the office and place a data point.

15. Deselect the windows (activate the Selection tool and place a data point in the void space of the Top view).

    You now have the shapes for all of the windows in place. Let's use them to actually create some windows in the walls.

16. Use the *Drop Complex Status* command on all six of the windows.

17. Activate the *Construct Difference Between Surfaces* command (see Chapter 12). In the Front view, data point on the exterior wall (yellow) and then on one of the window shapes. Make certain that you select the exterior wall, not the interior wall. Adjusting the front and back clipping plane would be effective here, but not necessary.
18. Continue to subtract all three shapes in the Front view.
19. In the right view, subtract one of the window shapes from the exterior wall.

    You will notice that the wall goes away and the window stays. This is because the surface normals of the window shapes were altered in the mirror process.
20. Use *Change Surface Normals* on the three mirrored windows and subtract the remaining windows from the wall.

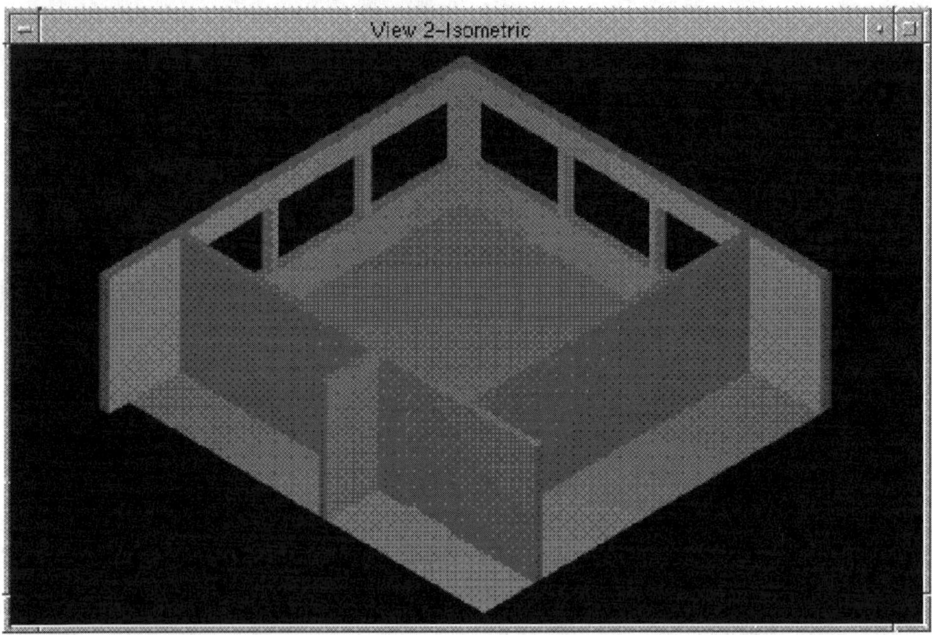

Office with windows.

Now let's place some mullions in the window.

21. Activate the *Place Cell* command. Make "mull" the active cell, the active angle 0, and the active scale 1. Make sure that the shared cells option is also activated on the Cell Dialog palette **(Settings → Cells)**.
22. In the Front and Right views, place the mullions by placing a tentative point in the center of the lower edge of each window.

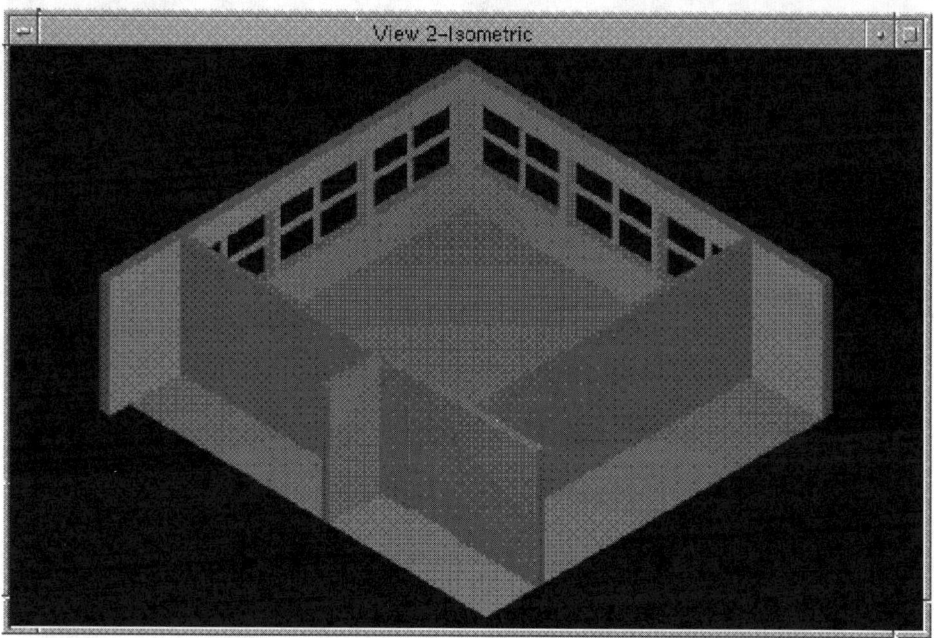

Windows with mullions.

Unfortunately, your architect tells you that some buried code issue requires you to redo your window mullions. It sure is a good thing that you used shared cells (sorry, it's late, the coffee is gone, so the jokes are bad). You will rework the mullion configuration and create a new mull cell. When you replace one existing mull cell, all of the mullions in the office will be updated.

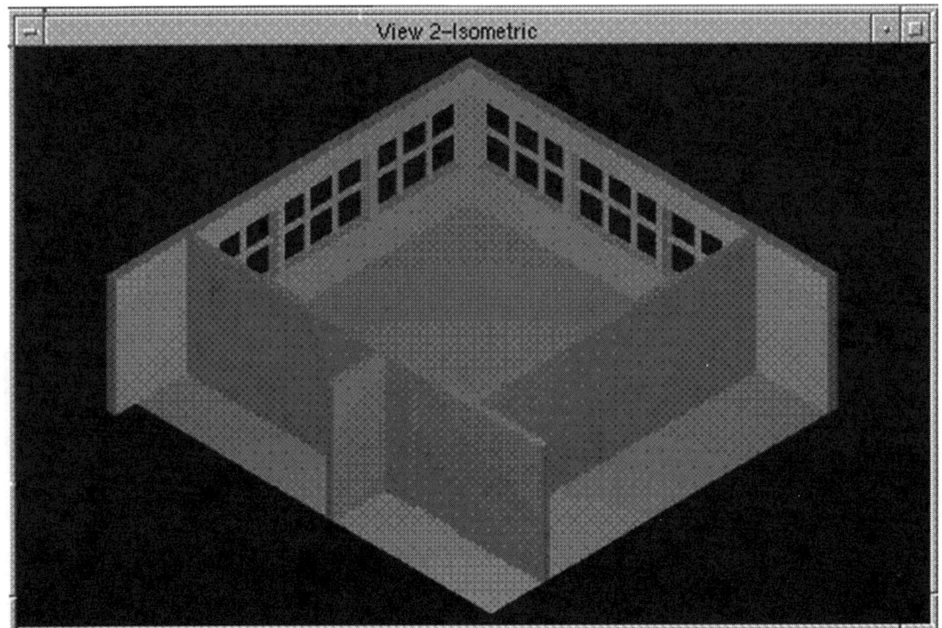

Windows with replacement mullions.

23. Activate the Cell Dialog palette **(Settings → Cells)**.
24. Make "mull2" the active cell.
25. In the Front view, place the cell 10 feet above the right window (tentative point to the center of the window and use precision input `DX=,10`).
26. Drop the cell.
27. On the Cell Dialog palette, delete the cell mull, and turn shared cells off to allow the cell to be deleted.
28. In the Front view, place a fence around the new mullions.
29. Activate the *Place Cell Origin* command. Tentative point to the bottom center of the window below the mullions and use precision input `DX=,10`.
30. Now create a new mull cell.
31. Once the cell is created, delete the element in the fence used to create the new cell.

You now have replaced the mullions in the mull cell and can proceed with the replacement of all the mullions in the model.

32. Activate the *Replace Cell* command and data point on one of the mullion cell in the office.

This really shows the usefulness of shared cells. Now that the design of the office is complete, let's set up a drawing composition to explain the office to the boss. The following image displays what the final sheet will look like.

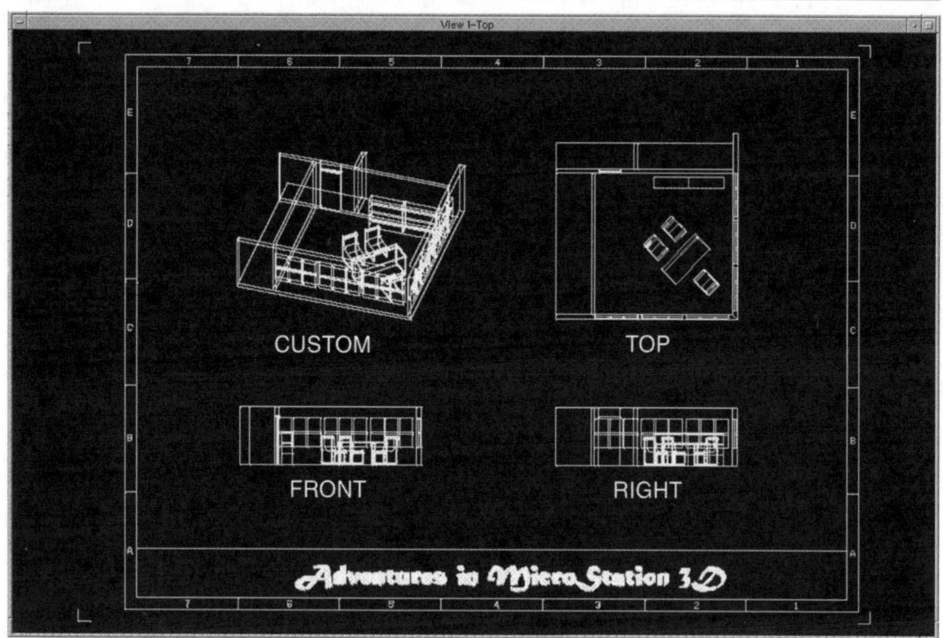

Drawing composition sheet.

We will go through the process one step at a time.

1. Turn display on for the furniture reference file.
2. Open the Drawing Composition palette **(File → Drawing Composition)**.

   The first thing is to create a drawing composition sheet.
3. From the drawing composition palette, open the Create Sheet palette **(File → New → Sheet)**.

4. Make sure that *In3Dseed.sht* is the seed file and name the new file *Drawcomp.s01.*

   You are now in the drawing composition sheet. Let's attach the office model.

5. From the Drawing Composition Palette, activate the open model command **(File → Open → Model)**, and select *Office.dgn.*
6. Change the scale to 1:32 (for a 3/8 scale drawing) and Nested Attach to 2.

   This will attach all your views at the appropriate scale and bring *office.dgn* and *furn.dgn* into the sheet as reference files. You will attach the Right view of the model and then "fold" the Front and Top views from it.

7. From the Drawing Composition Palette, activate the *Attach Right View* command **(Tools → Attach Standard → Right)**. Move the dashed rectangle to the lower right portion of the sheet and place a data point.
8. Activate the *Attach Folded* command **(Tools → Attach Folded → Orthogonal)**.

   MicroStation now needs to know which reference file is the principal attachment.

9. Place a data point on the left edge of the view just attached (the primary attachment).
10. Then place a data point to the left of the model to tell MicroStation that you wish to fold a view to the left.
11. A new dashed rectangle appears (representing the Front view). Slide it to the left portion of the sheet and place a data point.
12. Place a data point above the principal attachment. A new dashed rectangle appears (representing the Top view). Slide it upward and place a data point.

    The last view you will attach is the custom rotated view from the *office.dgn* file.

13. Activate the *Attach Saved View* command **(Tools → Attach Saved Views)**.
14. Select Custom and move the dashed rectangle to the upper left corner and place a data point.
15. Perform a fit all in all four views. This will completely display all of the office.

This layout is a great place to add notes and dimensions. All of the views of the model are reference files. As the model is altered, the changes are updated in the sheet. However, some models are too difficult to understand with all of the line and elements displayed. MicroStation provides two options for this.

- ❒ The first option is to modify the front and back clipping planes. This option would work for the Top, Front, and Right views of your sheet. However, it will not help the custom view.
- ❒ The second option is to export the sheet file as a visible edges file. You will try this option.

1. Activate the *Visible Edges* file tool **(File → Export → Visible Edges)**.
2. The export view should be view 1.
3. The Calculate intersections option should be activated.
4. The output format file dimension should be 3D.
5. All other settings should be deactivated.
6. Keep the default name, but change the directory to the Inside3D chapter11 directory.

   Once the file is created, a preview of the file is display on the screen. Let's actually open the file.
7. Activate the open file palette and change the type to hidden line and select *Drawcomp.hln*.

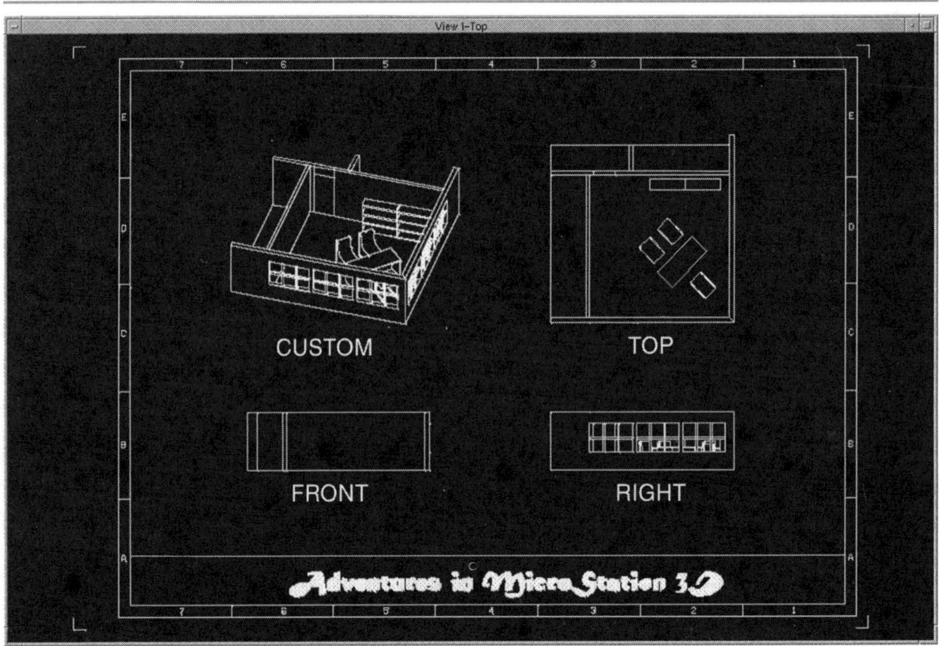

Hidden line file.

This is the ideal place to add your notes for final printout. Best of luck on getting your boss to approve the corner office.

Summary

In this chapter, you have learned the differences and similarities between 2D and 3D text, dimensioning, patterns, and cells. These are tools that can make life so much easier. If you learn how to use these tools in 3D, you will reap the benefits in the future.

Exporting your 3D file back to 2D is easy with the tools on the Export palette. Sometimes dragons are best handled by converting them into 2D.

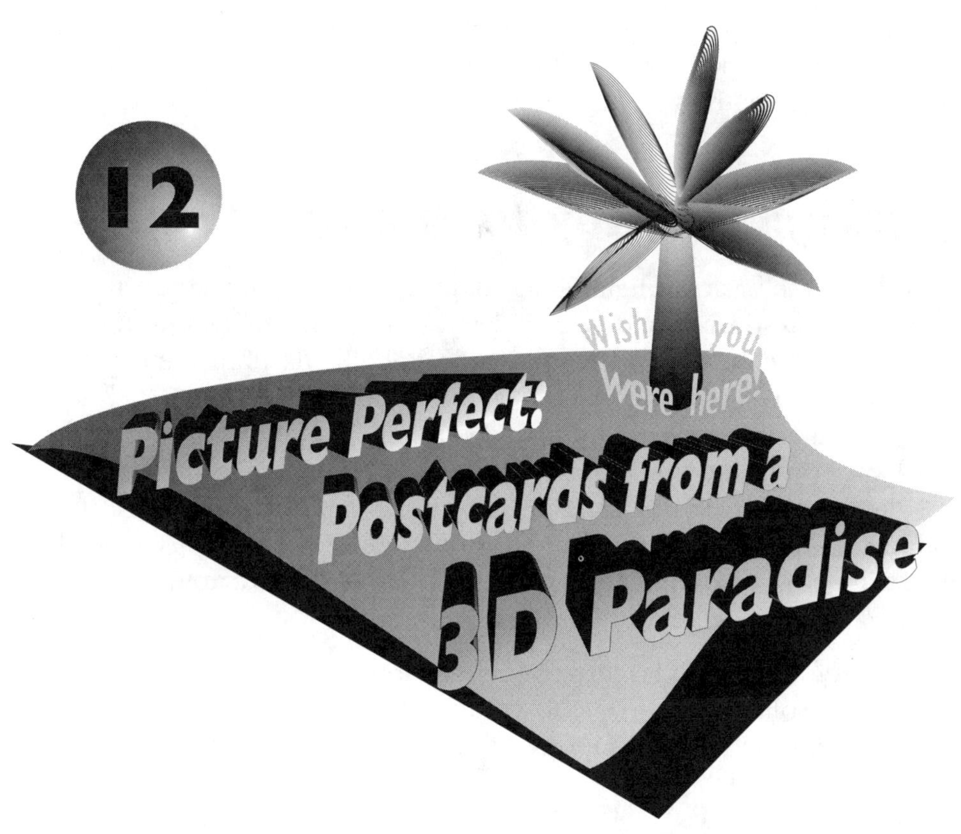

Introduction

The reason you have been spending so much time in 3D is to create more realistic models. A large part of the realism is the visualization of the finished model. The best part is that you get to show off your new skills with really professional results. To achieve these results you must think like a professional artist. With MicroStation you can create any scene imaginable. To make sure that your scene is convincing, you must think about the camera placement (see Chapter 7), scene lighting, the materials used in construction, shadows, and what kind of output are you generating (screen viewing, color printing, or video).

In this chapter, you will begin with lighting, then learn about textures and rendering output. Finally, you will have all the tools to impress even the most jaded of 3D natives.

Lighting: The Key to Photorealism

You explored the basics of lighting in Chapter 7. There you learned about the basic lighting source options: global and source. Global lighting affects the entire design cube. Global lights include ambient, flashbulb, and solar. Global lighting is used to simulate ambient light conditions, such as outdoor lighting. Source lighting includes point, distant, and spot lights. These source lights are used to simulate lamps, headlights, spot lights, or light sources with a defined origin. Various lighting techniques will be introduced using simple geometry. The complete use of lighting in a real project will be covered in the discipline-specific projects (Chapters 14, 15, and 16).

After the wireframe object is created, MicroStation can render the 3D object as if it were a solid object. The color of the object will vary from light to dark, depending on the orientation of the surface to the light source and the intensity of the light source.

To further enhance the reality of your scene you can assign texture maps to objects, levels, and level and color combinations. Texture maps are simple 2D images (of anything you scan or already have in the computer) that are wrapped around the 3D wireframe. The process is similar to applying wallpaper. Using this technique you can apply the image of water to a shape to make it look like a pond or lake.

To further enhance realism, MicroStation lets you use bump maps. Bump maps take texture maps and enhance them. The image is made to look bumpy, like a real surface, by using light and dark highlights on the texture map. The effect simulates dents and bumps.

MicroStation provides further tools for photorealism: reflection, refraction, and radiosity. These are available only if you use MasterPiece (Bentley's latest ray tracing and animation package for MicroStation, which is highly recommended), ModelView, or other third-party applications.

Reflections are maps of a reflected image, and with true ray tracing it does provide accurate reflections of objects in the model. Ray tracing is a method of tracing the path of a light ray from the light source to the reflective object, and then to the camera (what you see in the view).

Refraction is a bending of light waves as they pass through a transparent material whose density is different from air (or surrounding environment). The refraction control lets you simulate light passing through water or a curved glass.

Radiosity is a technique that simulates the way light reflects and bounces off objects. Thermodynamic formulas used to calculate heat transfer and dispersion are used to determine the amount of light an object will reflect. Radiosity produces stunning effects when used with scenes lit by reflected light. For example, if you had a white wall and a red piece of paper near it, you would see a red hue on the wall as light from the red paper hits the wall. Radiosity is not a standard MicroStation feature, but is available in ModelView and MasterPiece.

Choosing the correct lighting requires skill and experience. Nothing is more important to good renderings than lighting.

Global Lighting

Global Lighting is found under *Settings→ Rendering→ Global Lighting.*

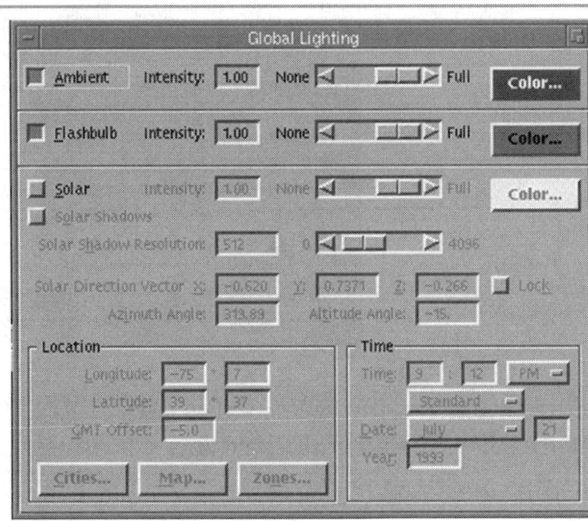

Global Lighting dialog box.

You learned about ambient lighting in Chapter 7. Remember that ambient lighting illuminates all surfaces in the model equally, so increasing its intensity will reduce the contrast and depth cueing of your rendering. Ambient light is good for simulating lighting indoors. No shadows are cast by ambient light.

Flashbulb lights are located right above the camera and have an intensity range of 0 (none) to 1.0 (full). The Flashbulb light can change its color and does not cast shadows. Flashbulb lights are good for looking at a rendering for the first time, before any light sources are placed. They also have the effect of producing highlights, visible to the camera. The flashbulb also has the effect of making objects in the foreground brighter than those in the background.

Solar lighting is a powerful lighting tool that provides accurate lighting conditions for any spot on the globe. For example, an architect can show his or her clients how the kitchen will look in the morning and show through which window the sun will set. All shadows and light angles will be accurate if the location of the site is input correctly, because you are positioning the sun in the sky.

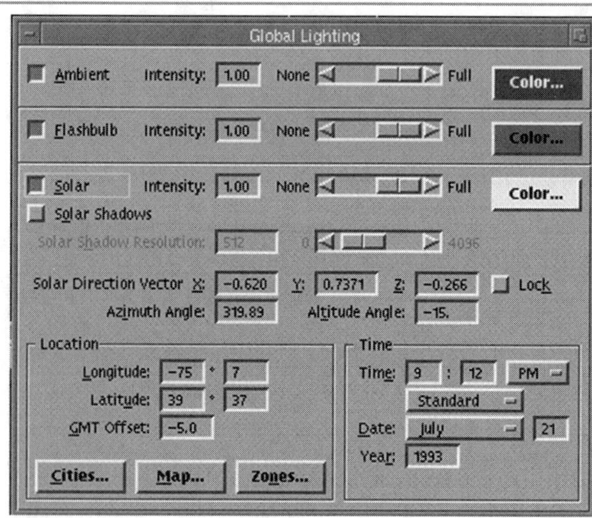

The Global Lighting dialog box with Solar lighting turned on.

Solar lights can cast shadows (only in Phong renderings) by turning on the Solar Shadows button. However, you must also turn on the Shadows button in the *Settings → Rendering → View Attributes* dialog box.

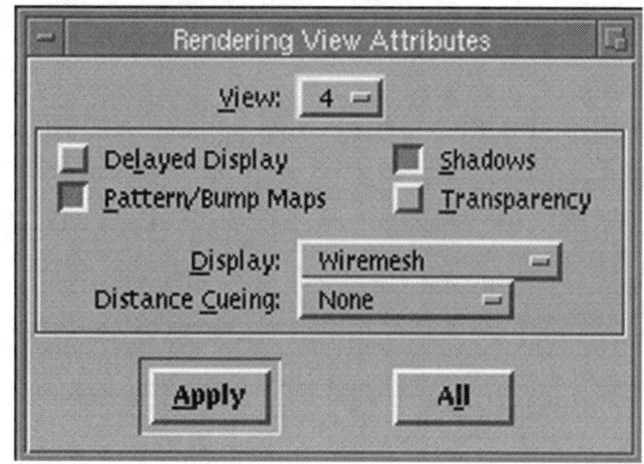

The Rendering View Attributes dialog box.

The intensity can be set from 0 (no light) to 1.0 (full brightness). The color can be changed by clicking on the Color button and then selecting the color of your choice.

Solar Shadow Resolution is the resolution of the shadow map generated by objects casting solar shadows. A lower value will have less definition and accuracy, but will greatly reduce rendering time.

Solar Direction Vector is the vector of the light from the sun. This is the direction in which the sun is shining. The vector is unitary and is defined by X, Y, and Z coordinates. Imagine a line drawn from the origin (0,0,0) to the coordinate. The length is one unit, but that's not important. What is important is the direction. The Lock button locks in the value from the X, Y, and Z coordinates. If the button is off, the Location and Time determine the angle of the sun.

Azimuth Angle is the angle of direction (0 to 360 degrees) of the sun. The azimuth angle is like the longitudinal coordinate. Zero or 360 degrees represents true north, 90 degrees is east, 180 degrees is south, and 270 degrees is west.

Altitude Angle is the angle from the horizon, ranging from 0 to 90 degrees. Just above 0 degrees indicates a rising or setting sun, and 90 degrees represents 12 noon.

The next two window panes also allow you to set the position of the sun. You just learned how to position the sun. Now you will specify the location on the planet, and the time and location of the sun is calculated for you.

The Location and Time window panes are grayed out if Lock is set.

Longitude sets the longitude at the model's location. The Longitude is measured from the prime meridian (which goes through the town of Greenwich, England) and is usually specified by number of degrees west of the prime meridian.

Latitude sets the latitude of the model's location. Latitude is measured north or south of the equator in degrees.

GMT Offset is the amount of time in hours offset from Greenwich Mean Time. This is the time difference in hours between Greenwich and the model's location.

There are three additional buttons along the bottom to help locate a model's location. These buttons are Cities, Map, and Zones.

Cities helps locate a particular city. The latitude, longitude, and GMT offset are given for a selected list of cities around the world. Highlighting a city and selecting OK sets the latitude, longitude, and GMT offset for that location.

Map opens a dialog box of the world's map. The map is used to set latitude and longitude by pointing and clicking on any location on the map. The GMT offset is not set by this method.

Zones opens the GMT offset by Time Zone dialog box. This sets the GMT offset by selecting a time zone.

The other window pane is for Time. This controls the time of year and day for a model's location. You can key in the time and select AM or PM. You can also select either Standard or Daylight savings time. The Date and Year options set the exact date of the rendering. For example, it might be interesting to see the shadows cast at Stonehenge on the summer solstice (June 23). First, you would need an accurate model of Stonehenge, oriented correctly, and then you could set the time and date and do a Phong shading to see how the shadows are cast.

Solar lighting provides realistic lighting conditions for realistic visualization.

Source Lighting

Source lights are, as the name implies, sources of light within the model, such as a lamp, computer screen, or headlight. There are three types of source lighting and these are point, distant, and spot lights. You learned the basics of these lights in Chapter 7.

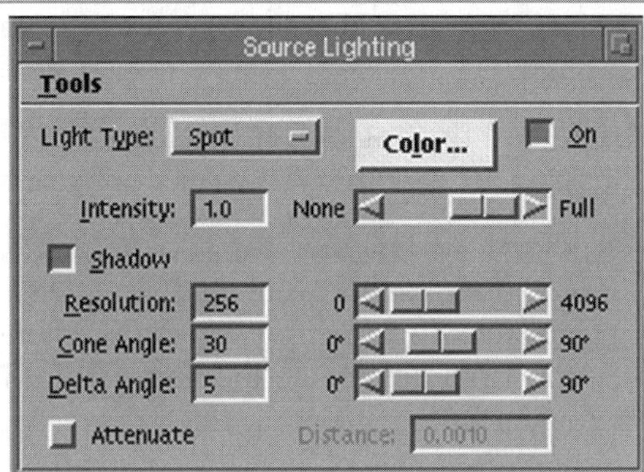

Source Lighting dialog box.

Remember that source lights are placed by cell. Attach the *lighting.cel* file if you have not already done so. The actual cell names are POINLT, DISTLT, and SPOTLT. The light attributes can be set using the data fields in the cell. You can change the color, intensity, and concentration using the cell or the Source Lighting setting box. If you use the data fields (and not the dialog box) then you must key in the following:

```
DEFINE LIGHTS
```

in the Command Window, for the new lighting attributes to take effect.

The Source Lighting dialog box is the best place to work from.

Light Type specifies the type of light to be placed: point, distant, or spot.

Color allows you to change the color of the light.

The **On** button turns the light source on or off if not selected.

Intensity controls the intensity of the source light. It can be controlled by inputting a number or by using the slide bar.

The **Shadow** button turns on the shadow map. Only distant and spot lights create shadows. Shadows are cast only in Phong renderings.

> **NOTE:** *You must turn on the Shadows button in the* ***Settings → Rendering → View Attributes*** *dialog box.*

Resolution sets the resolution of the shadow map. A lower value casts a less accurate shadow map while also reducing rendering time. Using larger values helps to increase the accuracy of the shadow map, but increases the rendering time.

Cone Angle sets the angle of a spot light. Use this variable to focus the spot light beam. This command works only for spot lights.

Delta Angle sets the falloff of the spot light. The Delta Angle of the spot light sets the width of the edge of intensity falloff of the beam. A 90-degree setting means that the intensity (greatest intensity at angle 0 degrees) decreases to 0 at 45 degrees to left and right of the focus point. A setting for Delta Angle of 60 degrees means a decrease to 0 at 30 degrees to the left and right of the focus point. This command works only for spot lights.

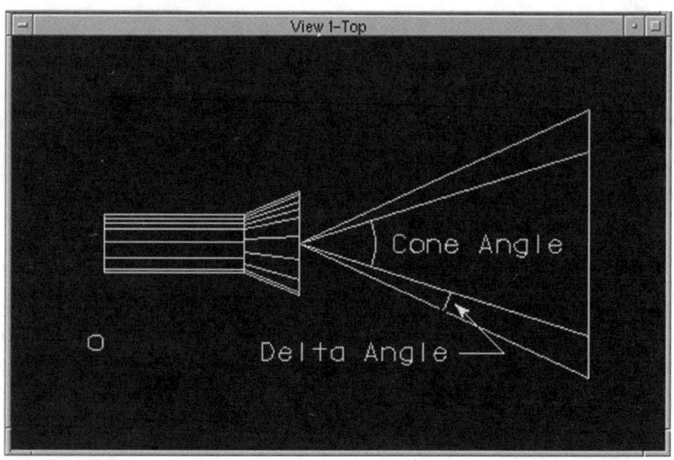

The Cone Angle and Delta Angle.

The **Attenuate** button controls the decrease in intensity over distance. Source lights attenuate (or decrease in intensity) by the inverse square of the distance, that is, if an object is 2 feet away from the light source then only 1/4 of the light gets to it. If an object is 3 feet away from the light source, then only 1/9 of the light gets there. MicroStation lets you set the one-half intensity distance, or the distance at which only half the light reaches. At twice the half-intensity light distance the intensity is zero.

Under the Tools pulldown menu in the Source Lighting dialog box you have the following commands: *Place New Lights*, *Edit Light*, *Scan For Light*, *Apply Values*, *Move Light*, *Target Light*, *Delete Light*, and *Clear Shadow Maps*.

Place New Light is used to place a light source cell with the current settings.

Edit Light allows you to change an existing source light setting.

Scan For Light lets you find the next source light in the model.

Apply Values applies the current settings to selected source light.

Move Light lets you move a light source, using the mouse or precision input.

Target Light enables you to select a new target element for a spot or distant light. If you change the target of a spot light, make sure that Cone Angle, Delta Angle, and Attenuation are set correctly for the new target.

Delete Light lets you delete a source light.

Clear Shadow Maps deletes existing shadow maps.

Point Lights

A point light is similar to a lightbulb, that is, light radiates out from all directions. Point lights do not cast shadows in MicroStation. These lights are useful in general lighting situations, for instance office lighting, interior lighting, lamp lights, and other point sources of light.

Distant Lights

A distant light source provides light in a unidirectional "ambient" fashion. Remember, ambient light sources illuminate all surfaces in the model equally. A distant light source illuminates all surfaces facing the distant light source equally, regardless of where in the design cube it is placed.

For example, a surface that is facing in the same direction as a light source, but is behind it, is also illuminated.

Distant light sources can cast shadows, if all the settings for shadows are set. These light sources are best used for modeling light sources that are not in the view, but the light they cast are. For instance, the light cast by the moon is best modeled by a colored distant light. Light coming from another room through an open door is another use of distant lights. There are many cases of this type of lighting, so you must ask yourself, where is my light coming from? Is it in my field of view ? If yes, then choose point lights, if no, then use distant lights.

Spot Lights

Spot lights are a lot of fun to use. Spot lights are conical light sources. Using the Source Lighting dialog box you can control the focus point and the edge fading. You can select any color you want for each spot light. Spot lights can cast shadows, and resolution of the Shadow Map is set the same way as before. The spot light works like a real spotlight, just like the ones used in theaters. Of course, they can be used in lots of different places. Any conical or focused beam of light is modeled with a spot light, such as headlights, flashlights, can lighting, flood lighting, and many others.

Shadow Maps

Shadow maps are placed in Phong rendered views, when certain light sources are used (solar, distant, and spot only). Shadow maps have their own resolution, depending on the number you set in the Source or Global Lighting dialog box. To turn shadows on you must depress the Shadows button in the *Settings → Rendering → View Attributes* dialog box. Shadow maps are also controlled through the *Settings → Rendering → General* dialog box.

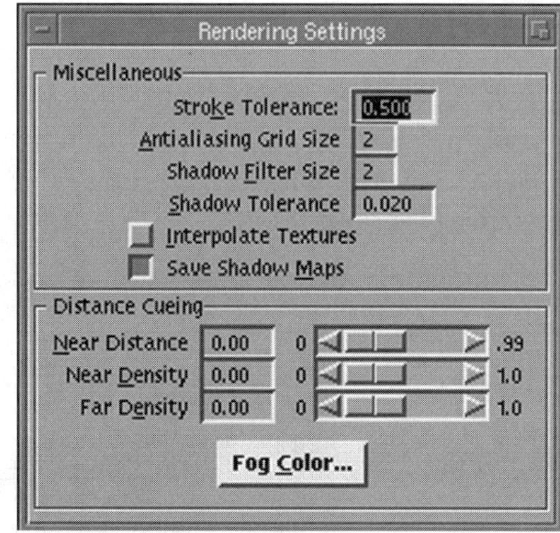

General rendering settings.

In this dialog box you have a couple of settings to control for shadows.

Shadow Filter Size controls the softness of the shadow. The range is from 0 to 15, where 0 gives extremely crisp-edged shadows and 15 gives fuzzy shadows, with softer edges.

Shadow Tolerance helps to prevent an object from casting a shadow upon itself. The tolerance is the fraction of maximum distance from the light source to the object. The default value is 0.02, which usually works to prevent an object from casting a shadow upon itself. Larger values are helpful when the incident angle of light on the element is low. Having larger tolerances generally causes inaccurate shadow mapping.

Save Shadow Maps saves the shadow maps the first time they are rendered. Any future rendering will use these saved maps and thus reduce time, because they will not have to be recreated.

The shadow maps are saved to disk in the format *name.l01* for a design file called *name.dgn*. The l01 refers to the shadow maps created by the first light source (that can produce shadows), and l02 to the second light source, and so on. If a light source is moved, then the shadow maps for that source are no longer valid; therefore, upon rendering, the shadow maps for that light source only will be recalculated.

Materials

Few things add as much realism to your images as textures. Making water look like water and copper piping look like copper piping is not simple. However, Safari Sam will guide you through it step by step, and you will be well rewarded for your efforts.

The Rendering Settings dialog box is found under *Settings* → *Rendering* → *General.*

The General Rendering Settings dialog box.

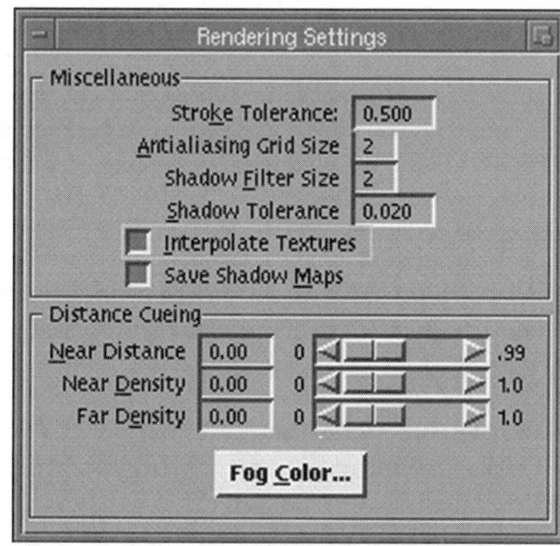

Stroke Tolerance is the maximum distance, in pixels, from the true surface to the surface mesh. This variable is used to determine the size of the polygons used to estimate the true surface, when it is rendered. The smaller the value, the more realistic the rendered mesh will be to the original surface. However, rendering time increases. The default value is 0.500.

Antialiasing Grid Size determines the pattern used for Phong antialiasing quality and number of passes. The grid pattern is calculated from the value. For example, a value of 2 yields a grid of 2 x 2, and makes 4 (2 multiplied by 2) passes. A grid size of greater than 3 is almost never needed and will significantly increase rendering time. Setting the grid size to 3 will require nine passes.

Interpolate Textures causes the surface color of two image patterns to be merged (or interpolated) between the two closest pixels (one pixel from each image pattern). This results in smooth transitions from one image to another. However, sometimes you want a good definition between two images; in that case, make sure this button is turned off.

The next issue to cover is the way MicroStation stores materials and their attachment to elements. The file used to store materials is the material palette (.PAL). Another file stores material attachment to elements, which is the material allocation table (.MAT).

The material palette (.PAL) file stores the definitions of a group of materials, for example, metals, carpets, or people. It is a relatively small file because it just stores the definitions of materials, not the actual materials themselves. MicroStation has many predefined palettes of images. They are classified into palettes of similar images. The palettes are Backdrop, Backyard, Carpet, Door&Win, Fabric, Flora, Glass, Granite, Homeofic, Marble, Masonry, People, Rugs, Surfaces, Tile, Vehicles, and Wood.

You can save your own images into a palette file using the *Settings* → *Rendering* → *Define Materials* command. Using the Material Name field, you can key in a new material and the adjust the pattern map settings, and finally, use the *Save As* command to create a new palette file.

The material allocation table (.MAT) file tracks which materials are attached to which elements, as well as which material palettes are used. Combinations of the levels and colors are used to define materials. A rose marble is attached to all elements that are on level 40 *and* color 3. The material allocation table is a text file and can be edited with a text editor.

Defining Materials

MicroStation has a sophisticated set of options to define almost any material. They are on the Define Material Dialog palette *(Settings* → *Rendering* → *Define Material).*

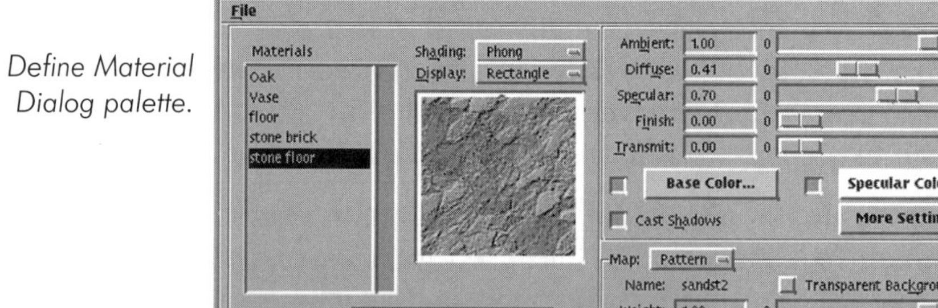

Define Material Dialog palette.

The left section of the palette organizes and allows previews of materials. The **Material** area of the palette lists all of the defined materials in the material palette (.PAL). Each material must have a unique name. To avoid conflicts, even materials in different material palettes should have unique names. Materials can be added, deleted, and renamed with the buttons at the bottom of the palette.

Shading and **Display** alter the preview area of the palette.

The settings on the upper right section of the palette control the way the material interacts with the lighting of the model.

Ambient controls the intensity of the material's ambient light. Zero is no ambient light, and 1 is full ambient light.

Diffuse controls the intensity of the material's diffuse light. Zero is no diffuse light, and 1 is complete diffuse light.

Specular controls the intensity of the material's specular light. Zero is dull, and 1 is shiny.

Finish sets the focus of specular highlights. Zero is a low polish (rough surface), and 1 is a high polish (very smooth). The specular and finish options work together to control a material's highlights.

Transmit sets the material's transparency (transparency must be turned on in the view being rendered). Zero is opaque (nontransparent), and 1 is totally transparent.

Base Color is used to define the overall color of the material. By default this option is off, and the color of the element in the model is used. When the base color button is selected, the Modify Material Color Dialog palette is displayed. Depending on your screen resolution, almost any color can be selected.

Specular Color functions the same as Base Color, but defines the specular color of the material.

Cast Shadows can be toggled on or off to change the material's ability to cast shadows. This can be useful in situations where a transparent material should not be casting shadows.

More Settings is a set of options for ModelView, MasterPiece, and other third-party rendering packages. They do not affect MicroStation's default rendering.

The lower right section of the dialog palette controls the texture of the material. The texture is defined by a pattern and/or bump map. Patterns are raster image files (usually *.TIF* files). They are usually created by scanning actual materials (such as marble or wood) or photographs (trees and people). Bump maps are black and white bitmap images (usually *.TIF* files as well). This bitmap is used to simulate the raised and lowered portions of real materials, such as carpet, stucco, and tile with grout.

Pattern Maps

Let's deal with pattern maps first.

Pattern map.

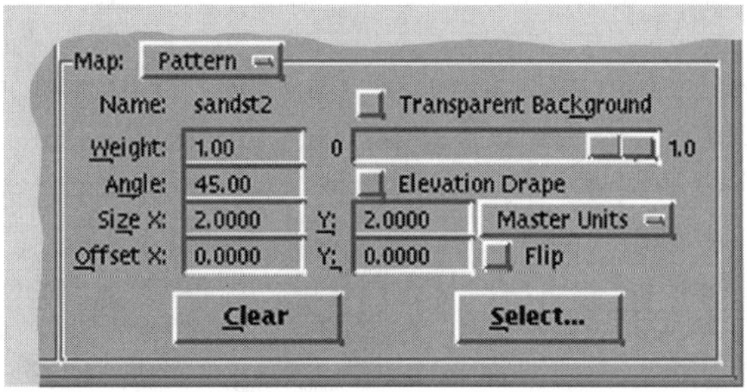

The **Select** button at the bottom of the palette opens a dialog palette to select the pattern map.

The **Clear** button removes and resets all of the pattern map settings.

Name displays the pattern map currently selected.

Transparent Background allows patterns such as people or cars to just display the person or car, not its background. The background is determined by the color of the upper left pixel. That color is usually white.

Weight determines to what degree the pattern map will override the diffuse material color: 0 is no pattern map, and 1 is total pattern map.

Angle will rotate the pattern map when it is applied to elements in the model. This is helpful in simulation of a brick or other pattern.

Elevation Drape "drapes" the pattern map over the view. A good example of this would be a real photograph mapped to a topographical surface.

Size X/Y makes it possible to tile a pattern onto an element. Different values can be put in the X and Y spaces. There are three settings for this option. The Surface option scales the pattern relative to the element. The Master Units option scales the pattern relative to the master units of the design file. Sub Units scales the pattern relative to the sub units of the design file.

Offset moves the origin of the pattern map in the X or Y direction. The value placed in the dialog area translates to a percentage of the pattern map's size.

Flip mirrors the pattern map on the element.

Bump Maps

Now let's deal with bump maps. Most of the options are the same as pattern maps. We will just cover the options that differ.

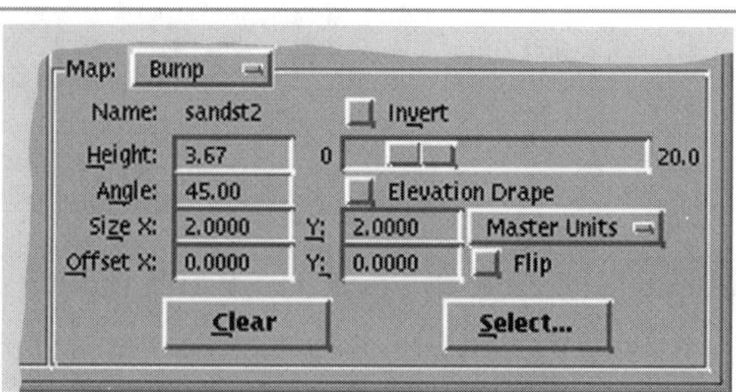

Bump map.

Height adjusts the height of the bump map. Zero flattens the map out, and 1 raises it.

Invert reverses the effect of the bump map. High areas (black) become low areas and low areas (white) become high areas.

Attach Materials

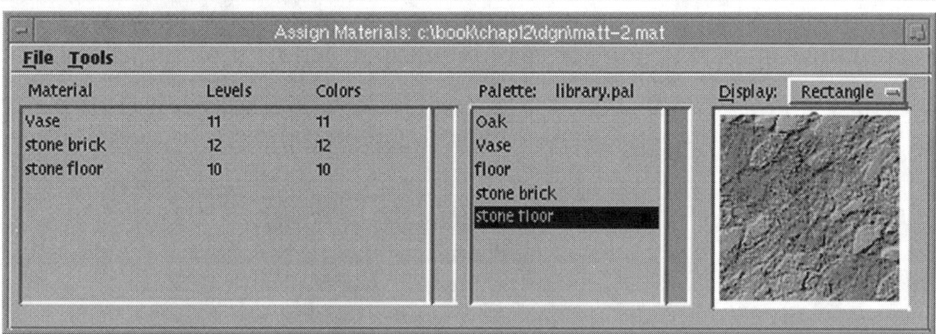

Attach Materials dialog palette.

Attaching materials is a relatively easy process in MicroStation. You can choose *Tools → Assign Materials By Selection* or use a dialog box by using the *Tools → Assign Material* command. Using selection is straightforward since you pick the object(s) you wish to assign the material to. Using the dialog box is just as easy; you just have to define which level and which color on that level will have the material attached to it. You can use the standard delimiters for level selection. For example, the following keyins for levels would be correct:

```
3-6,8
12,20
```

Refer to the levels keyin command (`on=` and `off=`) for the standard delimiters.

NOTE: *Regardless of how you assign materials, MicroStation only remembers levels and colors, so all elements on the specified level and with the specified color have the material assigned to them.*

After attaching the materials you want, select *Save Material Table* from the File menu of the Assign Materials dialog box. More examples of material mapping are covered in the projects section of the book.

Attach Material file options.

Attach Material tool options.

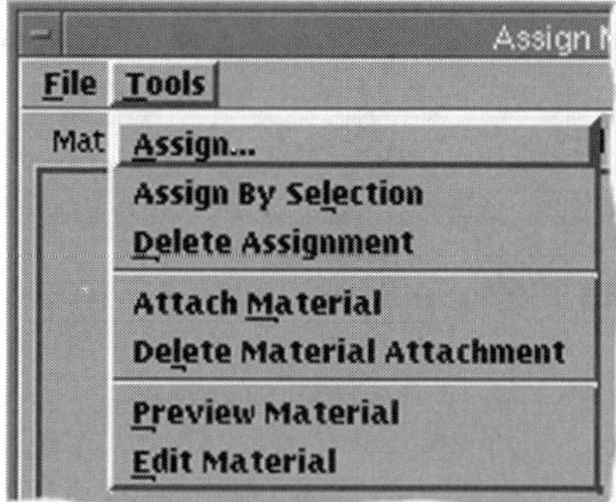

Rendering

Now that you have the tools to light the terrain around you, let's go the last step—setting up the rendering options and rendering it. In this section we will travel through tools that allow you to place a background behind your model (such as a sky), have distant portions of the model fade away, and even save images of the model as a postcard to send back home.

Backgrounds

Let's start with setting up a background. MicroStation has two main methods for establishing a background. The first is Active Background. It is basically a view attribute. Once it is set up, it can be turned on or off in each view. Active backgrounds take time to update. Because of this, it is usually turned off in all views except when rendering.

To set up the Active Background, key in

```
Active Background
```

in the Command window. This opens a dialog palette that allows you to select a image file to be used as the background. All views have the same background, and only one background can be selected at a time in a design file. A variety of image file formats are supported. Once the desired image file is attached, the background option must be selected on the View Attributes palette *(View → Attributes)*.

The image file selected as the active background is scaled or stretched to fill the view. It is also displayed parallel to all of the views. It will appear in the same orientation in an iso view as a front view.

If you just want an interesting background this works well. Sometimes it is undesirable to stretch or adjust the background. In this case you can import an image to use as your background. This can really tax your system, so use it sparingly. It is like putting a refrigerator in your backpack, but sometimes it is worth the effort.

Importing the image will allow you to maintain its proportions. The image becomes an element in the design file. It can be moved and scaled. A good technique is to place it on a unique level so that the level can be turned off when the image is not necessary.

> **NOTE:** *At times it is difficult to select an imported image. If you wish to modify an image, drag the selection tool to enclose the image. Once selected it can easily be modified.*

To import an image open the Select Image File dialog palette *(File → Import → Image)*. Once the image file is selected, MicroStation asks you to select the rectangle origin. This will be the origin point of the image. When the origin is selected, a dashed rectangle appears and MicroStation asks for the rectangle corner. The rectangle is constrained to the proportions of the image, but can be scaled to the desired size. When the corner is placed, the image is imported into the design file. This can be a valuable tool, but it can slow down the best of machines.

A better method is to place a shape behind the objects of interest and then map it with an image. The *Place Block* command is very helpful for this purpose. You would map the Block with the image of your choice. By altering the size of the image, using the Edit Materials dialog box, you can adjust the appearance of the image.

Distance Cueing

The next tool is *Distance Cueing*. This allows you to make the distant portions of your model fade into the distance, as if into a real atmospheric haze. There are three basic options for this tool: None, Depth Cueing, and Fog. They are controlled on the Rendering View Attributes dialog palette *(Settings → Rendering → View Attributes)*. More controls are found on the *Settings → Rendering → General* dialog box in the Distance Cueing window pane. This dialog box controls Near Distance, Near Density, Far Density, and Fog Color.

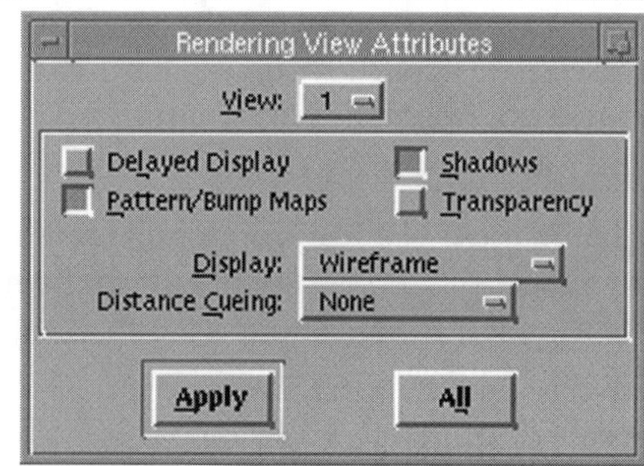

Rendering view attributes.

When Distance Cueing is set to None (default), all elements that are being lit will display.

Distance Cueing set to none.

The Depth Cueing setting gradually fades elements to black as their distance from the camera (or front clipping plane) increases.

Distance Cueing set to Depth Cueing.

Fog gradually fades elements to the fog color as their distance from the camera (or front clipping plane) increases. This can be useful in simulating atmospheric conditions or other sticky places. Every adventurer can find a use for this tool, whether you are under water, on the top of a mountain, or in the thick fog of London. Fog controls are also on the *Settings → Rendering → General* dialog box.

Near Distance sets the distance at which fading begins between the front and back clipping planes. If Near Distance is set to 0.5, then in the front half of the view volume there is no fading.

Near Density sets the fog intensity at the near distance.

Far Density sets the fog intensity at the back of the view.

Fog Color is used to select the fog color.

Distance Cueing set to Fog.

Saving Images

Well, now you have reached the summit of MicroStation's rendering. Why don't we save some images for those postcards? There are a number of options that can be modified when saving images, such as file format, type of shading, resolution, etc. The Save Image Dialog palette is opened from the Command window *(File→ Save Image As...)*.

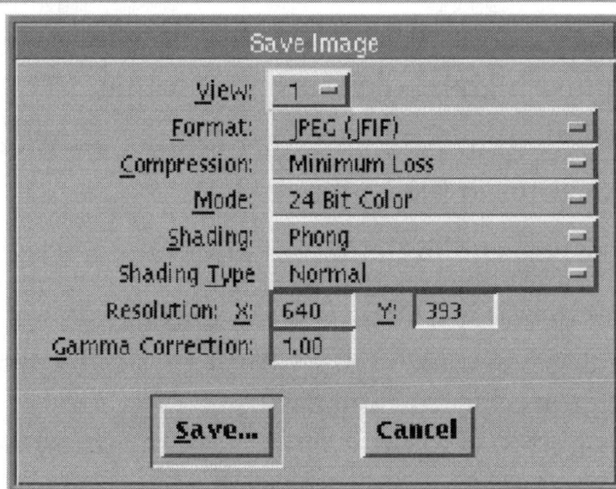

Save Image As palette.

Let's go through the options one at a time.

View sets the view that will be saved.

Format determines the file format that the image will be saved in. MicroStation supports a large number of file formats. The best format is determined by the purpose of the file and the platform (operating system) used. The file formats are listed below.

- ❒ CompuServe GIF (Graphic Information Format)
- ❒ IMG (8 and 24 bit versions)
- ❒ Intergraph's (24 bit)
- ❒ Intergraph's COT
- ❒ Intergraph's RGB
- ❒ Intergraph's RE
- ❒ Jiffy's JPEG (Joint Photographic Experts Group)
- ❒ Windows BMP (Bitmap)
- ❒ PCX (file format from Zsoft's PC PaintBrush program)
- ❒ PICT (Macintosh QuickDraw Picture file format)
- ❒ PostScript (Encapsulated PostScript, developed by Altsys Corp in cooperation with Aldus and Adobe for swapping high-resolution PostScript files from one application to another [.EPS])
- ❒ Sun Raster
- ❒ Targa (originally designed for video image storage, but now a popular high-resolution image format [.TGA])
- ❒ TIFF (compressed and uncompressed. Tagged Image File Format is a common file format [.TIF])
- ❒ Wordperfect WPG (WordPerfect's graphics file format)

Choosing the best format is a matter of taste and need. The biggest need is to be able to transfer files from one computer to another by diskette (fondly referred to as "SneakerNet"). The best compression and quality is found using the JPEG format. The best quality is either the Targa or TIFF format, but their file sizes are large. The GIF format is outdated, giving you moderate quality with larger file sizes.

Compression allows you to control the balance between compression and quality. This option only applies to the JPEG file format.

Mode controls the number of colors used in the saved images. Different formats allow different mode options. A greater number of colors will increase the file size of the image.

Shading defines the rendering method (see Chapter 7 for more information).

Shading Type defines whether normal, antialias, or stereo technique will be used to create the image.

Resolution X Y determines the size of the image in pixels. When one number is adjusted, the other is automatically adjusted to maintain the view aspect ratio.

Gamma Correction lightens or darkens the final image. The default value of this option is 1. Lower than 1.0 will darken the image. Higher than 1.0 will lighten it. The valid range for this option is 0.1 to 3.0. See the color calibration discussion later in this chapter.

That wasn't so bad, and now you know how to use this tool to produce your postcards. There is one last concept to deal with before we trek on to other features—quality versus file size. It is wise to consider the purpose of the image before creating it. If it will be printed on a gray-scale laser printer, color is a waste of time and resources.

The various file formats save information with different methods. JPEG is great for its file compression, but because it actually reduces the amount of information in the file, images can get slightly fuzzy (a nice effect sometimes). Targa on the other hand creates large files but they are of high quality.

The mode of the image file has a big impact on file size. It does not affect image resolution; however, the greater the number of colors, the more realistic the image.

The shading technique and shading type will not affect the file size, but will obviously increase the realism (quality) of the image.

The resolution of the saved file has a great impact on the file size. All images are saved at 72 dots per inch (DPI), which is the typical screen resolution. For example, if you want the image to be 8 inches wide, set the width to 576. If you want an image with a higher DPI, save the image larger in the X and Y directions and use a third-party image modification package (discussed later in this chapter) to reduce the image size and increase the DPI. This will produce an image with a higher quality (and a larger file size). Gamma correction has no effect on file size or quality, just appearance.

Output

So your postcard is ready to be printed. In this section we will trek through the jungle of methods to modify the postcard and output options. Output devices, color calibration, image modification packages, and service bureaus will be our stopping points on this journey.

Types of Output Devices

There are an ever increasing number of output devices. New technological developments and techniques are constantly driving this market. The best model printer at the beginning of the year may be a mid-range printer at the end of the same year. Carefully consider your requirements and options before buying. In many cases, it is advisable to get a low-end output device and use a service bureau for high-quality output (discussed later). The price per print will be high, but may be much less expensive than purchasing your own output device.

Let's talk about a number of the more popular systems for output. For the small sizes (8 1/2″ x 11″, 11″ x 17″) there is color laser, thermal wax, dye sublimation, and inkjet. On the large scale (up to E size) there are inkjet and electrostatic printers. And there are also slide imagers.

The *color laser* image is low gloss and resists scratching well. It is a process that uses heat to fuse solid particles of toner onto paper. A variety of papers can be used with this type of printer. Various color laser options are now available, starting from the desktop models for small workgroups or a single station (manufacturers like Hewlett-Packard and QMS have models for approximately $5000) to the large workgroup color laser copiers (CLC), such as those from Canon, Xerox, and Kodak. These larger CLCs are very expensive ($40,000 to $60,000), but very fast because they come with their own print server (a Fiery or Silicon Graphics machine).

Thermal wax printers provide good color saturation. They work well for presentation materials with solid colors, line work, and curves. Heat is used to transfer color from a transfer ribbon (also known as the "donor"). The transfer ribbon (which looks like a half-used roll of paper towels) is made up of three or four colors in a repeating pattern. The colors are CMY or CMYK (C, cyan; M, magenta; Y, yellow; K, black). As the cyan portion passes over the paper, small dots of cyan are transferred to the paper in the appropriate locations.

Dye sublimation printers offer continuous tone color, which provides near photo quality images. They do have difficulty with lines and arcs and are not as saturated as thermal wax printers. Like thermal wax printers, heat is used to transfer color from the transfer ribbon. However, dye sublimation printers do not use a set dot size. They can control the amount of color transferred and "mix" colors in the same area.

Inkjet printers are a less expensive output option. They spray a stream of ink through nozzles onto the paper. All colors are laid down at the same time, as the nozzles quickly scan back and forth across a page. Inkjets lack the color saturation and continuous tone qualities of other output technologies; however, the price per print and price of the printer make this a good entry-level option.

On the large scale (up to E size) there are inkjet and electrostatic plotters. The ability to produce large scale print (48″ x 34″) depends on the model. However, enlarging an image to this size reduces the quality of the image. "Aliasing," color shifts, and pixelization can occur, but viewed from a distance they are impressive.

The inkjet technology on large-scale plotters is the same as the inkjet printers, just on a larger scale. The carriage scans back and forth across the paper, spraying a stream of ink through nozzles onto the paper. The advantages of inkjet plotters are their relatively low cost (and getting lower) of ownership and operation for low to medium volume output (1 to 10 plots per day). They also can print over 16 million colors for large photorealistic plots. The drawbacks of inkjet plotters are that they are slow and need a couple of minutes of drying time after plot completion.

Electrostatic plotters use a specially charged paper. As the paper travels across the image head, the charge is reversed on the areas where toner is to be applied. The paper then travels across the toner fountain and opposite charged toner particles are fixed on the paper. The advantages of electrostatic plotters are that they are extremely fast (faster by far than any other output media) and the ink is dry upon plot completion. Electrostatic plotters can also plot color. The drawbacks are the initial cost of the plotter, starting around $12,000 for monochrome to $30,000 for color (these are not true color plotters, the colors are limited to about 256, depending on model and manufacturer). Furthermore, the paper is expensive, about $1 per foot at E size, and liquid toner is needed on occasion. However, if you have high volume plotting needs, there is no substitute.

Another valuable output option is the *slide imager*. It allows images to be "shot" onto regular slide film. When developed you have a high-quality, durable, and presentable image. However, without a slide projector, they are pretty useless.

Color Calibration

In basic terms, color calibration is the process of making what you see on your screen and your output look identical. This seems to be an obvious need and a useless process, but in fact it is very important. Scanners, graphic cards, monitors, and output devices are produced by different companies with different priorities. They describe color in different terms, and monitors and scanners (usually) use RGB (red, green, blue) and printers use CMYK (cyan, yellow, magenta, and black). Monitors usually display information at 72 to 90 DPI. Printers average 300 to 600 DPI. These and other differences often make images on the monitor and final prints look quite different.

Color calibration solutions can be software, hardware, or a combination of both. They can be inexpensive or expensive. The ideal solution for you will depend on your computer platform (Macintosh, DOS, Windows NT, UNIX, etc.), hardware, and requirements. These color calibration solutions are often offered by the companies that produce the output devices. Interestingly, the color of the monitor is usually adjusted, not the printer. This is because monitors are usually the weakest link in the chain, but the one we interact with most.

Most color calibration solutions focus on two main areas, white point and gamma. *White point* affects the intensity of colors and is measured in temperature on the Kelvin scale. Most monitors are set to 6500K; the higher the temperature, the whiter the whites become. Monitors usually have a high white point, making colors look brighter. This will make printouts look dark and oversaturated.

Gamma affects the contrast of the monitor. In a perfect system the gamma curve is flat or 1.0 (good luck). In this case the monitor is faithfully representing the image. A value higher than 1.0 increases contrast, darkening the image. A value lower than 1.0 decreases the contrast, lightening the image. Adjust your monitor first, then make a printout, and almost invariably you will have reduced the gamma correction. After an iterative process of gamma adjustment and printing, you can save this gamma correction for the monitor and printer.

Software solutions usually allow you to alter the white point and gamma of the monitor to match your output. Hardware solutions can be in line (connect to the monitor cable) or on screen (have an apparatus that attaches to the screen).

Color calibration is a somewhat complicated process and may need to be done for each different software package you use. It is hoped that in the near future this process will become invisible to users. As hardware and software standards are established and more sophisticated computers become available, this problem should decrease and one day disappear.

Third-Party Image Modification Packages

There is an increasing number of image modification packages on the market today. Each has it own unique features; however, they also have many of the same features. It would take an entire book to discuss even just a few of them in any kind of detail. We will discuss some of the more standard features and the ones most useful in finishing our postcards.

Some of the most useful features are those that allow you to alter the size and resolution of your images. Because MicroStation saves images at a resolution of 72 DPI, it is often helpful to enlarge an image. An image can be enlarged in two different ways. The first is to increase the actual dimensions of the image. The second is to increase the DPI. You can even use a combination of the two to achieve the ideal dimensions and resolution for the output you have chosen.

> **NOTE:** *Making multiple adjustments in dimensions and resolution will cause a degradation of the image. The computer must add or subtract information each time the image size is altered. Every time the image size is adjusted, the effect is multiplied. It is useful, but plan ahead and use it sparingly.*

These packages also allow you to place text on the images, such as a title, a note, or even a signature. Different fonts and styles can be used, as well as different colors.

Filters are another large set of features. A filter will alter an image. It might simulate a motion blur, a impressionistic painting, or even a bright sunlight reflection. Filters can be applied to a portion of the image or the entire image. There are hundreds of different filters. Every package has a different set of filters, and plug-in filters can be bought to supplement them. Many add-on or plug-in filters work with multiple packages.

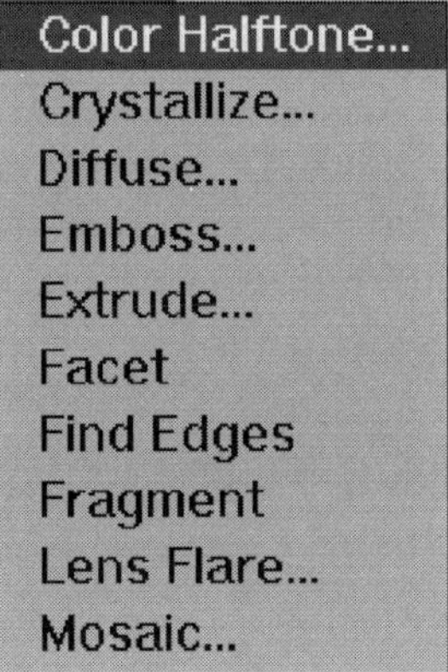

Image modification package filters.

Some of the more popular packages are PhotoShop, PhotoStyler, Picture Publisher, and Fractal Painter. Investigate the packages to find the one that best meets your needs. It should be noted that PhotoShop is the most widely used image editing package in the world, and Fractal Painter is probably in the top three for professionals. PhotoStyler may not be supported in the future because Adobe (manufacturer of PhotoShop) has purchased Aldus (manufacturer of PhotoStyler).

Service Bureaus

For those of you who are not ready to adventure into this area or can't afford the equipment, there is help. There are an increasing number of service bureaus popping up. They can modify your images and print them. They usually have the most recent software, hardware, and the skills to use them. If you wish to have nice color printouts, large-scale printouts, overheads, or slides, service bureaus can help. They can also scan pictures or materials (such as marble tiles or wallpaper for materials or photos of backgrounds).

Summary

Even though Safari Sam has just taught you about lighting and rendering in more detail, there is still much more to learn. Experience is the best teacher here. Safari Sam is confident in your skills. You can now explore all the variations you can think of and learn how each variation changes your image. Many more tricks of rendering are covered in the projects section of the book (Chapters 14 through 16).

References

For more information on the topics covered in this chapter, see the following references.

LeWinter, Renee, and Baron, Cynthia: Output Options for Artists. *Computer Graphics World*, January 1991, pp 34-40.

McMillan, Tom: The Promise of Portable Color. *Computer Graphics World*, September 1992, pp 30-36.

Morisson, Michael: *Becoming a Computer Animator.* Sams Publishing, Indianapolis, IN, 1994.

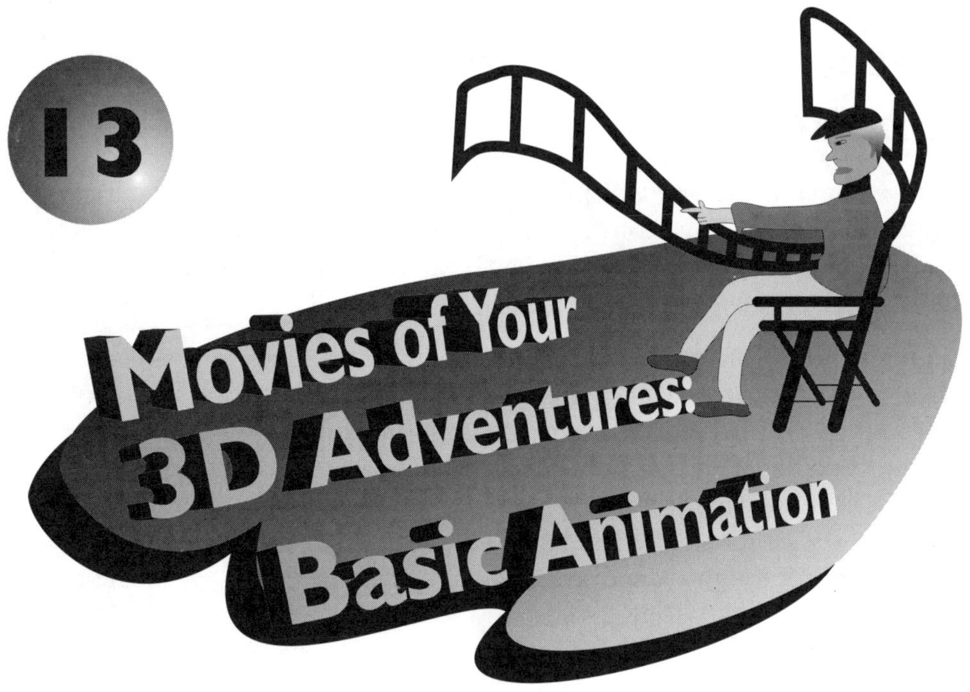

Introduction

The final product of your 3D study are movies of your adventure. You can now create 3D objects and assign realistic materials to them. Next Safari Sam is going to show you how to animate the camera so you can produce "flybys."

Animation is science unto itself and requires an entire book to explain it properly. Safari Sam is going to show you the basics. You may not be ready for Hollywood yet, but anyone with good basic 3D skills is on the right path.

A Background to Computer Animation

Computer animation is a recent phenomenon at the desktop level. Until recently, you could only get professional results using UNIX workstations and Amiga computers. Now, Bentley has brought the power of animation to the PC world.

Throughout history artists have learned various techniques and skills to illustrate what the artist sees in his or her imagination. Tools such as brushes and pens were used to bring the art to life, and many years were spent perfecting the skills before images of value were created. Computers help us close the gap in time, allowing us to produce valuable art relatively quickly. In the past the media used by artists were either 2D paintings or 3D sculptures. Although skilled artisans were able to excel, given their limited choice of media, they all suffered from the fact that the objects they created were fixed in time and space.

Computer graphic imagery (CGI) can represent any moment or sequence of moments in time and space. For example, most of us have seen animations of the "Big Bang" explosion depicting the birth of the universe. We see the universe develop to its current state, some 12 billion years later and a universe of information crammed into a few minutes. Alternatively, there are animations of electron and proton interactions where 1 second may represent 1-billionth of a second (a nanosecond). Computer animations open up a whole new universe of possibilities, not just to artists and movie makers, but to engineers, designers, architects, and the curious.

Let's first define the terms of our realm. *Animation* is the illusion of movement. The objects depicted in our animations change incrementally from one still image (we will refer to these still images as *frames*) to the next. If you play back the frames at the right speed, your eye perceives movement. We know from years of testing that a human's visual perception is limited to about 12 frames per second. The movement at 12 frames per second appears jerky and discontinuous, and adding more frames per second helps to smooth out motion. The film industry (movie) standard is 24 frames per second and the standard for animation and TV (in the United States) is 30 frames per second. Therefore, 30 separate pictures, each with the objects moved incrementally, are shown to our eyes every second, giving our minds the illusion of continuous movement.

You can see this quickly becomes a daunting task for our animator. A 1-minute animation requires 1800 frames (30 frames per second x 60 seconds per minute). A 2-hour animated feature film requires about 215,000 frames. A Pentium PC can usually render one frame of a complex animation every 15 minutes, so a feature film would take 53,750 hours, or 6.1 years, just to animate it, not create it.

Large animation projects such as feature films require a large team of artists. As a story is developed, rough sketches are placed on large boards and concepts are outlined in a process called *storyboarding*. This technique is the same as that used in comic books and strips. Each picture in the storyboard represents a key moment in the story, usually when some action takes place. These key frames were the responsibility of the chief artist, who usually produced the 1st, 30th, 60th, etc. frames of the animation. The remaining frames were done by staff artists called "in-betweeners," because they created the frames in between. The key frames contained the elements of motion or where the motion changes, like before a jump, at the top of a jump, and landing. The remaining process required inking and filming the celluloid pictures.

Computers are used to scan sketches. Scanning is the process of digitizing a picture. Each picture is broken up into rows and columns. For TV this would be 480 rows by 752 columns. The computer can then be used for creating models and beginning the animation process. Most animation packages are keyframe animation systems. Like the chief artist, you create the key frames and the computer creates the frames in between.

MicroStation is not a keyframe animator (however, MasterPiece, an MDL application, is a keyframe animator). You can, however, create flythrough animation with MicroStation.

Flythrough and Flyby Animation

Flythrough and flyby animation are when the camera "flies" through the model. The FlyThrough Producer is found in the *View* → *Render* →

FlyThrough menu. The elements of the model *do not* move in these animations, just the camera. The camera is moved by defining a path through the scene and telling MicroStation how many frames it takes to travel that path.

FlyThrough Producer is used to produce animations.

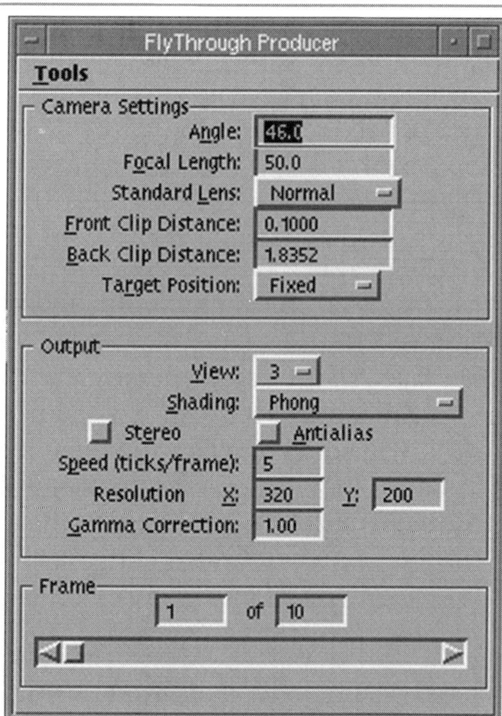

There are three window panes on this dialog box: Camera, Output, and Frame. The camera settings should look familiar to you since they are essentially the same as before (see Chapter 7).

Angle sets the angle, in degrees, of the lens's field of view. The Angle and Focal Length are tied together. Increasing the Angle decreases the Focal Length.

Focal Length sets the distance at which an image forms after entering the lens, while viewing a distant subject. The Focal Length is set in millimeters. These two variables allow you to create any lens you wish.

Standard Lens allows you to choose any of the standard lens types in MicroStation, such as Telescopic, Telephoto, Portrait, Normal, Wide, Extra

Wide, and Fisheye. See Chapter 7 for the angle and focal lengths of these lenses.

Front Clip Distance sets the Front clipping plane of the camera along the view Z-axis. Anything in front of this plane is invisible to the camera. If an element is clipped by the plane, then the camera will show only a portion of that element.

Back Clip Distance sets the Back clipping plane of the camera along the view Z-axis. This clipping plane is usually set behind the back wall of the model.

Target Position sets the definition of the target at which the camera is pointing. You have two options, Fixed or Float.

Fixed target means the camera is always pointing at the same object, while moving along its camera path. This option is useful for aerial site studies and requires less rendering time than Float.

The Float option makes the camera point in the direction tangent to the path. This option is helpful for walkthroughs, because the camera is pointing straight ahead at all times.

The Output controls the source view of the flythrough and the look of the animation.

View sets the view in which the animation will take place.

Shading sets the method by which each frame of the animation will be rendered. The options are as before: Hidden Line, Filled Hidden Line, Constant, Smooth, or Phong. Remember, the more complex a shading technique, the longer it takes to render each frame (and don't forget that each second of animation requires 30 frames).

Stereo allows you to create separate images for the left and right eyes. The dual-toned image gives the illusion of depth when viewed with the red-blue lens glasses provided.

Antialiased sets whether or not Phong-shaded images are antialiased. This command greatly increases rendering time.

Speed sets the playback speed per frame for FLI format files only. The fastest possible setting is 60 frames per second. Each tick is 1/60 of a second (commonly refered to as a field, instead of a tick). You have to set ticks per frame. The question, then, is are you going to have 1 tick per frame, 60 ticks per frame, or some number in between. A setting of 1 tick per frame yields 60 frames per second. A setting of 60 ticks yields 1 frame per second. Usually, the best setting is 5 or 6. Remember, Speed is only for FLI files.

Resolution XY sets the resolution of the animation in the X (columns of pixels) and Y (rows of pixels) directions. A higher setting increases the time for rendering substantially, as you will see later. *Pixel* is short for *picture element.* A pixel is a dot on your screen. Thousands of these dots are arranged in a grid fashion (with X and Y coordinates) on your screen. Image data arranged in this fashion is called *raster data.*

You may remember that MicroStation uses vector data (instead of coordinate or raster data) to keep track of pixels. This enables MicroStation to manipulate graphical elements with more speed and intelligence. Raster data formats are widely used in image editing and processing, since it is easier to acquire data from a scanner. Upon rendering, MicroStation maps raster data onto vector elements.

✔ **TIP:** *The first time you animate a model, use the FLI file format with speed set to 20 ticks per frame and resolution set low (320 x 200) to get a quick preview of the animation.*

The resolution is limited by the amount of memory on your video card. Since you are using the full color palette of MicroStation, the RAM in your video card can only store a limited amount of color information for higher resolutions. So the tradeoff is more colors and less resolution or fewer colors and more resolution. Better yet, buy a 4MB video card. After all, you are a CAD guru now, occupying your own little mountain in the 3D wilderness.

The resolution of standard VGA cards is 640 pixels horizontally by 480 pixels vertically. For NTSC compatible systems (like VHS tapes or TV) use 480 by 752. For higher resolutions 800 by 600 gives good results in both resolution, color, and speed of rendering. A standard of 320 by 200 is a good resolution for testing your animation. The number of colors is set when you record your animation.

Gamma Correction sets the brightness of the output images. The default value is 1.0, and the range is from 0.1 to 3.0. A value greater than 1.0 will lighten the image and a value below 1.0 will darken the image. See Chapter 12 on Rendering for more information.

The Frame window pane controls the currently viewed frame (the left field) and the total number of fields in the animation (the right field). The scroll bar shows the progress through the animation.

The Tools menu of the FlyThrough Producer dialog box is used to Define Path, Preview Camera, Preview View, and Record.

The Tools menu of FlyThrough Producer.

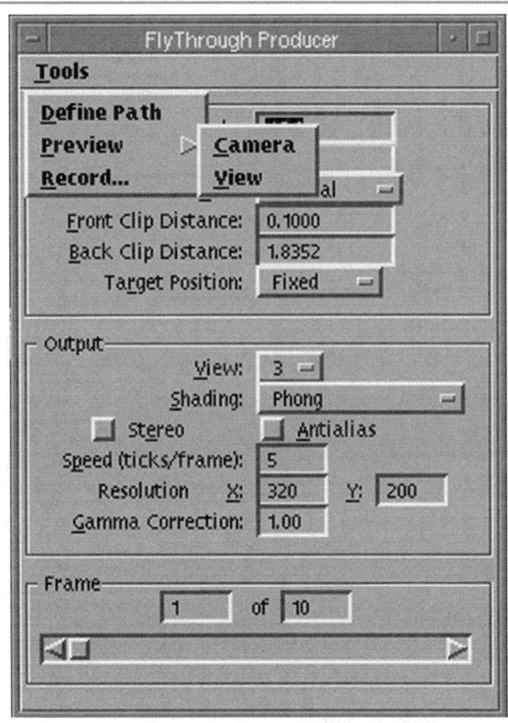

Define Path is used to define the path the camera takes through the model. The first step in the process is to place a line, line string, arc, ellipse, curve, or B-spline curve to define the path the camera is to follow. The path element will not be visible in the animation so long as it is a construction element (use *Element → Information* to find out). You must define the start point and the end point on the element.

A little study of kinematics (the physics of motion) is important to get a realistic flyby. If your camera is recording at 30 frames per second, then you want to travel a reasonable distance for your model in that time frame.

First, you must determine the length of the path element. Use the *Measure Length* command on the Measuring palette.

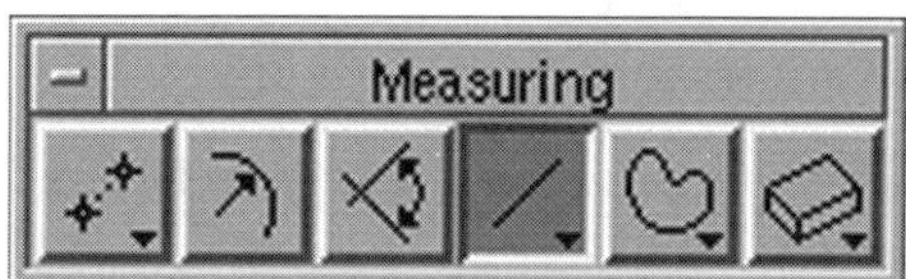

Measure Length command.

Let's say that the length of the total path was 30 feet. Now you must assume a reasonable velocity for your camera to calculate the total time of the camera on the path.

```
total time in seconds = (distance on path in units/
                         velocity in units/second)
```

How do you pick a rcasonable velocity ? Here are some rules of thumb:

❐ For walkthroughs a good velocity is 1 to 2 feet per second. At 1 foot per second the total time for the previous example would be 30 seconds.

❐ For aerial flybys, the best values are 10 to 20 units per second. This varies widely, depending on how close to the ground you are. At 10 units per second the total time for the previous example would be 3 seconds.

❐ Lastly, you determine the total time of the animation. Remember, the number of frames gets large very quickly. So keep it to a few seconds, unless you have nothing better to do with your machine for a couple of days. For a walkthrough 30 seconds is needed. If the animation is 30 seconds long you have (30 seconds x 30 frames per second) 900 frames of total animation. This would take a long time. For flybys you have (3 seconds x 30 frames/second) 90 frames, which is not too bad.

So, going slowly and carefully through a model take a lot more frames than a quick flyby.

Preview Camera shows the position of the camera, and its view volume at each point on the camera's path, without recording.

Preview View lets you see each frame in the display view without recording the animation.

The *Record* command allows you to record your animation.

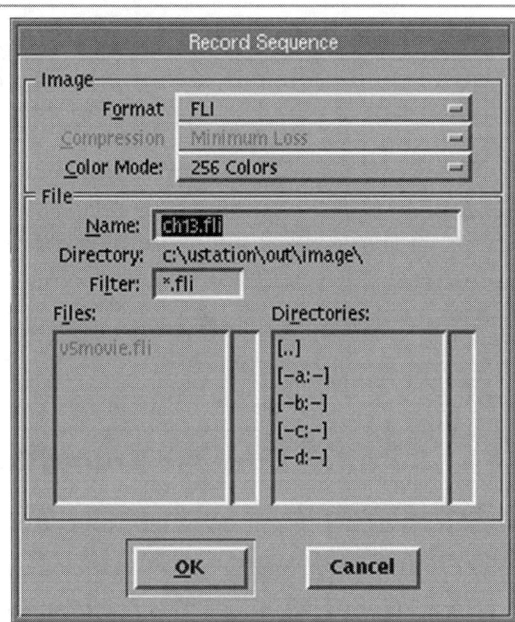

The Record Sequence dialog box.

Format allows you to choose the format of the output file. The FLI and FLC formats are the only true animation formats. The default color setting is 256 colors.

Color is encoded in bits for each pixel. If you set the resolution to 320 x 200, then you have (320 x 200) 64,000 pixels. The number of bits per pixel determines the number of colors. If there are 8 bits per pixel or 1 byte per pixel (8 bits = 1 byte) then you have 256 colors or shades of gray, because 2^8 is 256. Then 1 byte x 64,000 = 64KB of video RAM needed to display the image. Twenty-four bits can encode 16,777,216 colors (2^{24} = 16,777,216). Twenty-four bits = 3 bytes, so an image of resolution 1024 by 768 at 24 bits would require (3 bytes x 1024 x 768) 2.35MB. To do this rendering you would need a 4MB (or greater) video card. Try to calculate the hard limits of your video card. It is important to know your machine's limitations. It is possible to create images at a higher resolution, for playback on a better machine, but this is not recommended.

Other formats available are output as sequentially named single files. Many animation recording units (those that take computer files and dump them to VHS tape) use the Targa format. The MicroStation implementation of the Targa format is encoded by 24 bits per pixel. Review Chapter 12 for these file formats.

Compression sets the degree of compression in the JPEG format only. Minimum Loss gives the best image but the least compression.

Color Mode allows you to select the number of colors for your output image. FLI and FLC are defaulted to 256 colors. Other formats support either 8 bit or 24 bit color. Remember, rendering times increase greatly if you select 24 bit color (the CPU must process three times as much color information).

File lets you name the file(s).

Exercise 1: A Roller Coaster Animation

This exercise is based on the exercise done in Chapter 5, the creation of a roller coaster. Are you ready to ride it?

Select the **Inside3D → Chapter 13 → RollerCoaster Exercise** command.

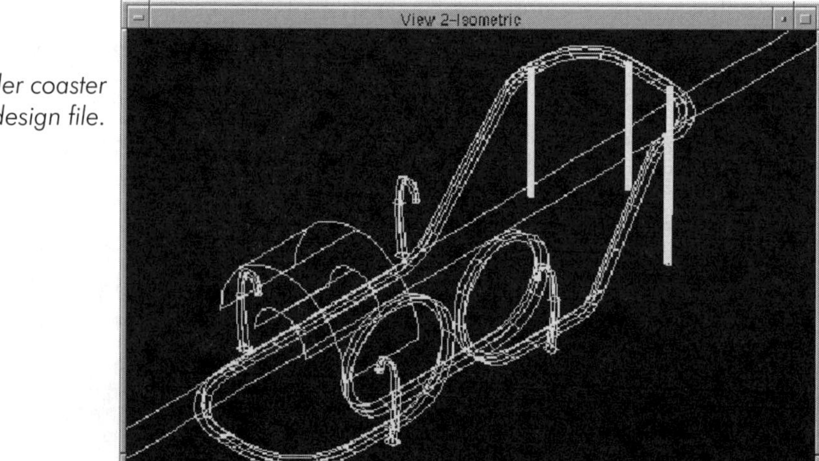

The roller coaster design file.

If you remember, you created a roller coaster from Tubular, Skin, and Offset Surface tools. Well, the animation is very easy.

1. Open the **View → Render → FlyThrough** dialog box and use the following settings. You must have at least 600MB of free space on your hard drive and a few hours of spare machine time to do this rendering, if you map the surfaces with like the *coaster.flc* on the CD-ROM. Reduce the resolution and the number of frames for smaller file size, or leave the surfaces unmapped. The actual file size will be a lot smaller but MicroStation needs the space for a swap file.

   Standard Lens = Wide

   Front Clip Distance = 0.1

   Back Clip Distance = 300

   Target Position = Floating

   View = 2

   Shaded = Phong

   Speed = 2

   Resolution = 640 x 480 (reduce to 320 x 240 for a one quarter reduction in file size

   Gamma Correction = 1

   Frame (right field) = 450 (for a 15-second ride; you can make this shorter, 150 frames or 5 seconds requires at least 215MB of free hard drive space to render)

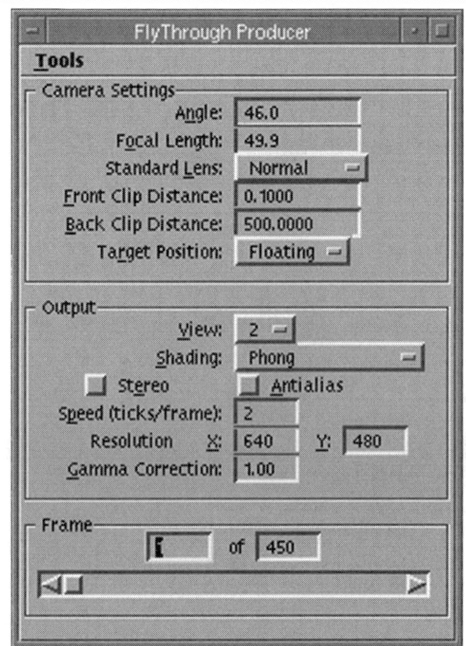

The FlyThrough Producer with the correct settings for the roller coaster.

2. Use the *Define Path* command and select the projected arc used to create the track with the *Offset Surface* command. Tentative point to the arc in the middle of the loading house as your start point, and data point to accept. Then tentative point to the same point (the roller coaster finishes where it starts) and your end point, and data point to accept.
3. Select the *Record* command from the Tools menu of the FlyThrough Producer and use the following settings:

   Format = FLC

   Name = roller.flc
4. Then select the OK button. Time to get some java.

You can use the **File → Display → Movie** command to view your completed FLC. The file can be enhanced even further by adding textures and lights. See how realistic you can make it.

Summary

Animation requires that you plan out the events in advance. You must think about how the camera is going to move and what you want to film. The animation process gives you endless possibilities with your 3D models. In addition to the models you create, the growing Internet community has resources where other models can found. Check out Appendix B for more information.

The animation process is much more complex than what we have covered here. Bentley's MasterPiece for MicroStation is an excellent way to continue with your animation career. MasterPiece is a true keyframe animator, with parametric actors and true ray tracing. You have now completed your 3D training. It's time to go on to some more complex projects.

Introduction

This introduction is by Bob Coffee and Carlos Soria of LPA, Inc. Architects, who provided master planning, architecture, landscape architecture, and interior design for the San Marcos Town Center project.

The 15,000 square foot San Marcos Library is one of four buildings designed by LPA as an integral part of the 200,000 square foot San Marcos Civic Center Complex. The other buildings in the complex include a four-story, 150,000 square foot City Hall, a one-story, 30,000 square foot Community Center, and a four-level, 600-car parking structure. The library is the smallest building within this 9-acre complex.

Extensive 3D modeling was used to study the massing and building form for each of the buildings within the Civic Center Complex. The proper scale of open space was as critical to the design as were any of the four individual buildings. The planning concept oriented the city hall, community center, and library to a central "civic garden." Computer modeling allowed the LPA design team to explore a variety of options in both building placement and building massing. Ultimately, exact alignment and visual relationships between the buildings become an important component of the human scale experiences. "Walking through" the design with the computer assured the designers that they were accomplishing their goals. These same simulations were invaluable in "explaining" this 3D spatial experience to the client and were used to assist architectural renderers in preparation of their renderings.

Conceived as a simple "loft for learning," the library building is one big room, like the original San Marcos Schoolhouse. Flexibility and a need for minimal staff supervision suggested a design with few interior partitions, providing an uninterrupted volume for stacks and reading spaces easily monitored from the circulation desk. Once again the specific 3D modeling of the library allowed the designers to demonstrate to their client that they had achieved the desired goal. The interior volume of the library was completely modeled to show all furniture, book stacks, and lighting techniques. The base for this computer model was also used to refine the exact placement of all furniture within the library for the Schematic Design Furniture and Fixture Plan.

The entire north wall of the library is glazed and provides the library patron a strong visual connection to the hills beyond. Conversely, the library becomes an "open book" and beacon to the passersby at night. The library staff had some concerns with direct solar light penetration along this north face. Once again, with the help of the computer model, the design team was able to evaluate light penetration based on time of day and year. Coupling this technical study with another study illustrating the visual effect of the lighted library at night as seen from the street, the "beacon" idea was embraced by the client, and their concerns were eliminated.

The LPA architects always discussed the library as a place of enlightenment, a place for the citizens of San Marcos to feed their dreams, meet their heroes, and learn the truth. They wanted the building to fit comfortably and appropriately in its context and to be an inspirational space. In this instance, they used the computer to share this ability with their client. This accelerated the design process and made it more efficient and enjoyable for all those involved.

Overview

The purpose of this exercise is explain some methods and concepts that will be useful in modeling, not just in architectural design, although the project is a building. We will spend some time on the evaluation of the purpose and requirement of individual models, actual modeling, and assigning materials.

This exercise will not go through the entire modeling process for the building. However, the parts not covered in the exercise will be provided on the CD-ROM so a complete model will be the final product. The emphasis is on concepts and methods, not individual tools and commands.

Let me introduce you to the library. Below are the floor plan and elevations that show the basic spaces and rooms of the library.

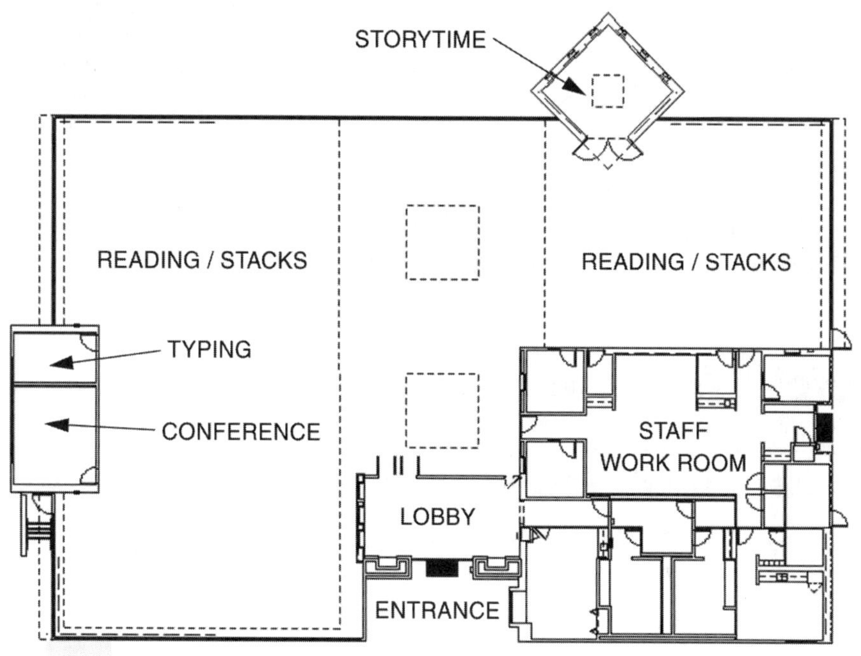

Floor plan of the San Marcos Town Center Library.

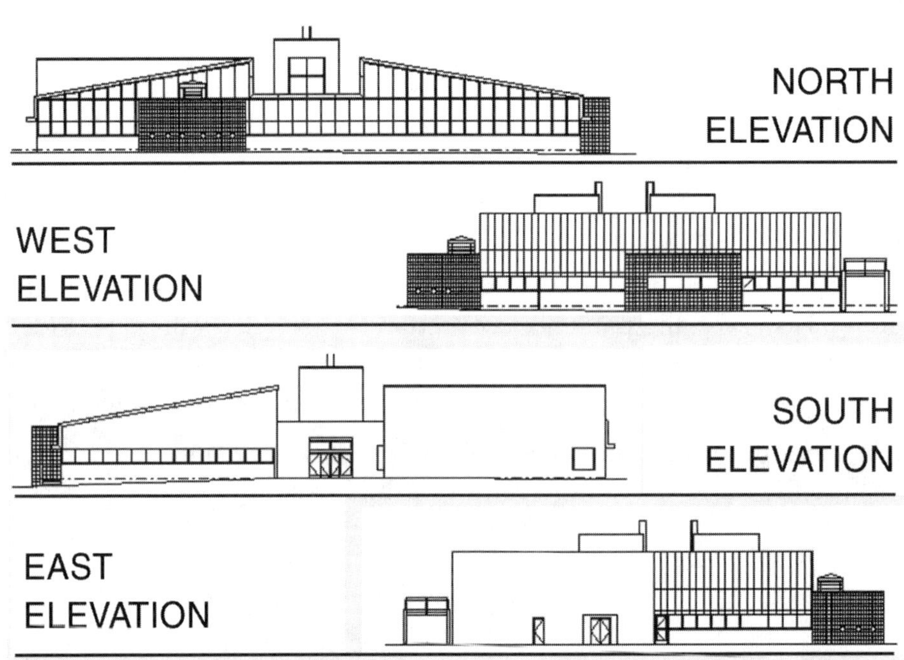

Elevations of the library.

One of the main purposes of this exercise is to produce a model that faithfully represents the real building.

Quick Volume Study

Exercise 1: Quick Volume Study

1. Open the Quick Volume Study design file **(Inside3D → Chapter14 → Quick Volume Study)**.

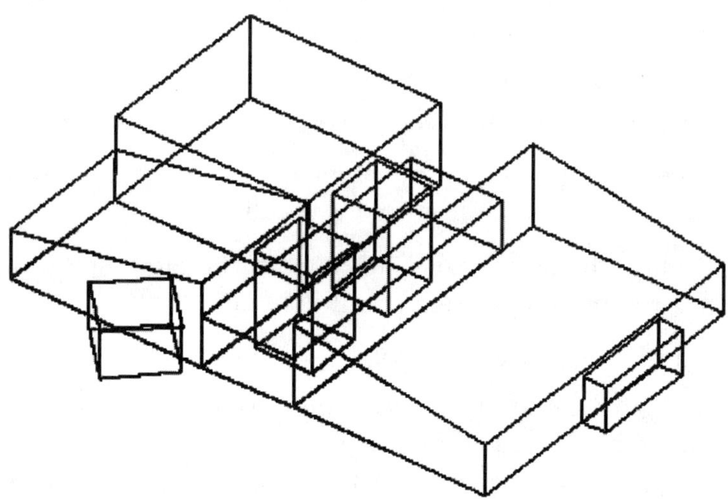

Roof plan of the library.

The roof plan of the library is already attached to the design file. You will now create some volumetric shapes on top of the roof plan. The roof plan is used rather than the floor plan because it better represents the exterior shape of the building. The heights at key points of the building have been included in the design file.

Let's start with the tall towers in the center of the library.

2. Change active level to 11 and color to 1.
3. Activate the *Place Slab* tool and change the Type setting to solid.
4. On the upper tower, tentative point to the top left corner and place a data point. Repeat the process on the top right corner and then the bottom right corner.

   The basic shape of the slab has now been established. To define the height you will use precision input. The precision input will be relative to the last data point entered (the bottom right corner).

   ```
   DL=0,0,30
   ```

The first tower is now created and in place. You can copy it to place the second tower.

5. Activate the *Copy* command.
6. In the Iso view, select the tower at one of its bottom corners.
7. Place the copy by placing a tentative point on the lower corner of the tower in the corresponding location.

Now let's place the slabs that will represent the Storytime room and the Conference room at the west side of the library.

8. Activate the *Place Slab* tool.
9. Using the shape of the Storytime room, in the roof plan, place a slab with a height of 14 feet, 0 inches.
10. Do the same for the conference room. It has the same height.

Well, it looks good so far, but you must add a few more volumes before you are finished. Let's place the center volume of the library and the southeast volume.

11. Turn on level 62 in all the open views. The three points needed to construct the center volume are now called out. Use them to create a slab with a height of 15 feet, 0 inches.
12. Turn level 62 off and 61 on in all open views. Use the new set of callouts to create a slab with a height of 25 feet, 0 inches.

The only volumes that you must still create are the sloped portions. You will create a 2D shape representing the section of the slope volume. Once that is done, you will use it to project the volumes.

13. Turn level 61 off and level 60 on in all open views.
14. Activate the *Place Shape* tool and use the displayed information in the design file to place the shape.
15. Mirror the new shape over to the other side of the building. Set the key point devisor to 2 and use one of the towers to define the mirror point.
16. Project the first shape south 100 feet, 0 inches.
17. Project the second shape south 44 feet, 0 inches.

Now you have a complete volume study.

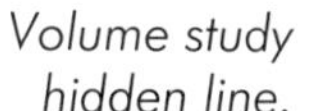

Volume study hidden line.

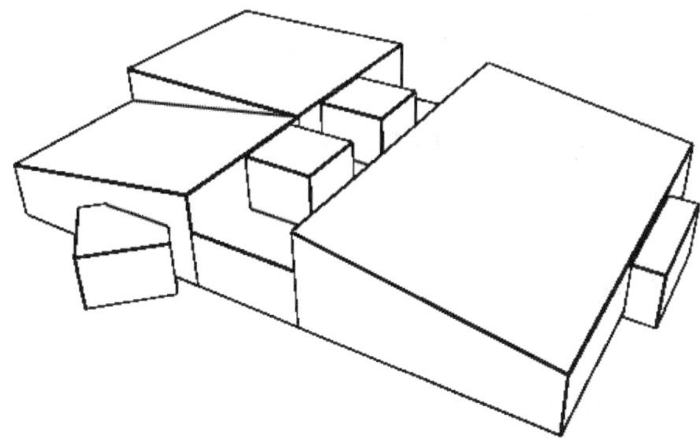

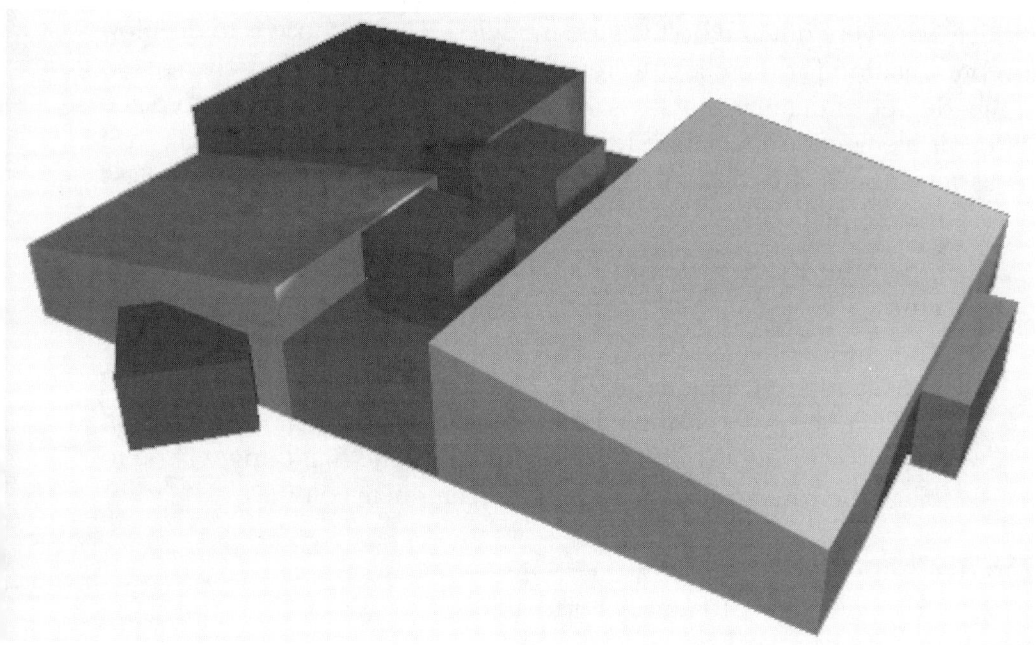

Volume study Phong shaded.

The clients love the design, but as usual they want just a bit more, some materials and windows and simple things like that. Fortunately you are working with MicroStation, and it is here to make you and your project look good.

Your volume model will not be useful in creating a more realistic model, so you will leave it behind. The best place to start the new model will be with the exterior of the building. You will place the walls and glazing (windows) and then the roof.

The Exterior

Exercise 2: The Exterior

In this exercise you will construct the exterior of the library. The walls, windows, doors, and roof will take shape. We will try to achieve a faithful model. Not every nut and bolt will be modeled. The purpose is to develop a representation for the client to review. Things are bound to change, so a high degree of detail is not advisable at this point. It is important to place different elements on separate levels with separate colors. Materials will be assigned later; if things are on incorrect levels, you might have a glass floor and a carpeted roof.

1. Open the Exterior design file **(Inside3d → Chapter14 → Exterior)**.

   The orthogonal and Iso views are active and the floor plan is already referenced into the file. The points on level 63 will be used as reference points during the exercise. You will begin construction at the lower right portion of the building. The first thing will be to place the four walls that form the rectangular volume.

The Rectangle

2. Change level to 10 and color to 10.
3. Activate the *Place Block* tool rotated.
4. Tentative point to point 1 and place a data point.
5. Tentative point to point 2 and place a data point.
6. For the third point of the block, enter the following precision input:

`DL=0,0,25`

7. Repeat the process using points 1 through 4 to create all four walls.

Now that you have the walls in place, let's add some detail.

8. Activate *Place Block* rotated.
9. Tentative point to point 5 and place a data point.
10. Tentative point to point 6 and place a data point.
11. Enter the following precision input:

    `DL=0,0,6.75`

    `(6.75 is the same as 6:9, both equal 6 feet, 9 inches.)`
12. Now, in the Iso view, move the new block up 2 feet, using the following precision input:

    `DL=,,2`
13. Use the *Copy Parallel* tool to copy the block inward 9 inches (or 0.75).
14. Use *Construct Solid of Projection* to project the larger block 1.75 (1 foot, 9 inches) and the smaller block 2 feet.
15. Activate the *Construct Difference Between Surfaces* tool. Subtract the smaller block from the larger to form the window.
16. Now Activate the *Group Holes* tool. Identify the wall as the solid element and the inner block as the hole.
17. Delete the outer block.

Now, it's time to place the mullions and the glazing.

18. Change the level to 14 and the color to 14.
19. Place a block on the inside of the window, away from the wall.
20. Copy it using precision input (`DL=0,0:2,0`) (2 inches) and change its level to 15 and color to 15. This will be the glazing.
21. Copy the original block parallel 2 inches (or 0:2).
22. Use the *Group Holes* tool and select the original block as the solid and the block copied parallel as the hole. This will form the mullion.

23. The last step is to project the mullion. Activate the *Construct Solid of Projection* tool and project the shape 6 inches.

View of window.

Now you have the first (and most complicated) window created and you can see into the library. Let's place another window.

24. Activate *Place Block* rotated.
25. Tentative point to point 7 and place a data point.
26. Tentative point to point 8 and place a data point.
27. Enter the following precision input:

    `DL=0,0,6.75`

28. Now, in the Iso view, move the new block up 2 feet, using the following precision input:

    `DL=,,2`

29. Create another duplicate block by coping it using precision input (`DL=0,:2`). This will be the glass, so change the level and color to 15.
30. Create a duplicate block by coping it using precision input (`DX=0`). You will use this to form the mullion.
31. Now activate the *Group Holes* tool. Identify the wall as the solid element and one of the blocks as the hole.
32. Change the level and color of the duplicate block to 14.

33. Activate the *Copy Parallel* tool, set the distance to 2 inches, and copy the block inward.
34. Use the *Group Hole* tool to subtract the inside block from the outer.
35. Then project it inward 6 inches.

Well that will suffice for this side of the entry, let's go to the other side and begin construction on it. You will start by placing some shapes to represent the walls and then place some windows.

The Wall

1. Change active level to 10 and color to 10.
2. Activate the *Place Shape* tool.
3. Tentative point to point 9 and place a data point.
4. Enter the following precision input:

   `DL=,,25`
5. Tentative point to point 10 and enter the following precision input:

   `DL=,,15`
6. Tentative point to point 10 again and place a data point.
7. Tentative point to point 9 and place a data point.

This finishes the construction of the south wall. Now let's construct the wall near the entrance.

8. Activate the *Place Shape* tool.
9. Tentative point to point 9 and place a data point.
10. Enter the following precision input:

    `DL=,,25`
11. Tentative point to point 11 and enter the following precision input:

    `DL=,6,25`
12. Tentative point to point 11 again and enter the following precision input:

    `DL=,6`
13. Tentative point to point 9 and place a data point.

Now you will place the windows on the south wall.

14. Activate the *Place Block* rotated tool.
15. Tentative point to point 10 and place a data point.
16. Tentative point to point 12 and place a data point.
17. Enter the following precision input:

    `DL=,,4`

18. After the block is created, move it with the following precision input:

    `DL=,,4`

19. Use the *Group Holes* tool to subtract the new block from the wall.

To construct the mullions for the windows you will project them from the floor plan.

20. Change the active level and color to 14.
21. Activate the *Construct Solid of Projection* tool and make sure it is set to solid.
22. Activate the distance option and set it to 4 feet, 0 inches.
23. In the Top view project the mullions upward along the entire southwest wall. Do not project the glass; you will construct the glass out of one element.
24. Change the level and color of the projected mullions to 14.
25. Activate the *Place Slab* tool and set it to solid.
26. Tentative point to point 10 and place a data point.
27. Tentative point to point 12 and place a data point.
28. Enter the following precision input:

    `DL=,.:6`

29. Enter the following precision input:

    `DL=,,:2`

30. Copy the slab just created up 3 feet, 10 inches (`DL=,,3:10`).

Now you have all of the mullions in place. The next step is to construct the glass and then move the entire window structure up 4 feet into place.

31. Change the active level and color to 15.
32. Activate the *Place Block* tool.
33. Tentative to point 10 and enter the following precision input:

    `DL=,:2`
34. Tentative to point 12 and enter the following precision input:

    `DL=,:2`
35. Enter the following precision input:

    `DL=,,4`
36. In the Front view place a fence just large enough to include the window structure. Activate the *Move* command, make sure Fence-Inside is set, and move the window up 4 feet (`DL=,,4`).

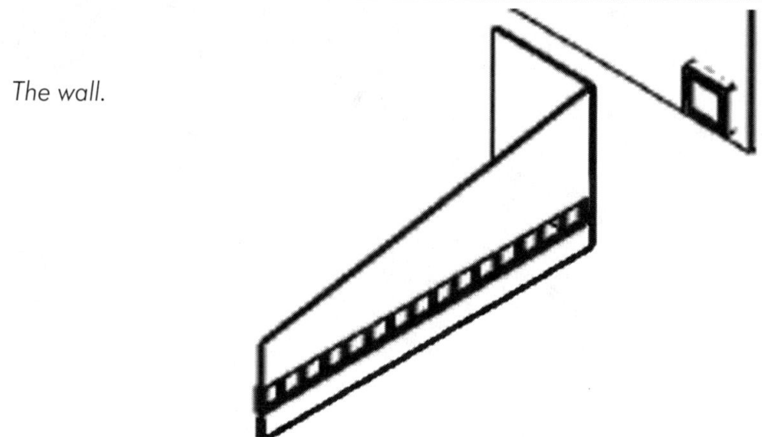

The wall.

Now you have the concept of creating the walls and windows. You don't need to spend the time going through the process of constructing each and every mullion. The remaining portion has already been constructed for you. It is stored in the cell library for Chapter 14.

37. Open the library **(Inside3d → Chapter14 → Cell Library)**.
38. Make windows the active cell and set the active scale to 1 and the active angle to 0.
39. Tentative point to point 10 and place the cell.
40. Drop the cell.

The Entrance

The entrance of the library needs some work, so let's head over there and get the construction started. You will begin with the doors.

1. Turn on the south elevation (it has been attached for you). Window into the entrance in the Front view.

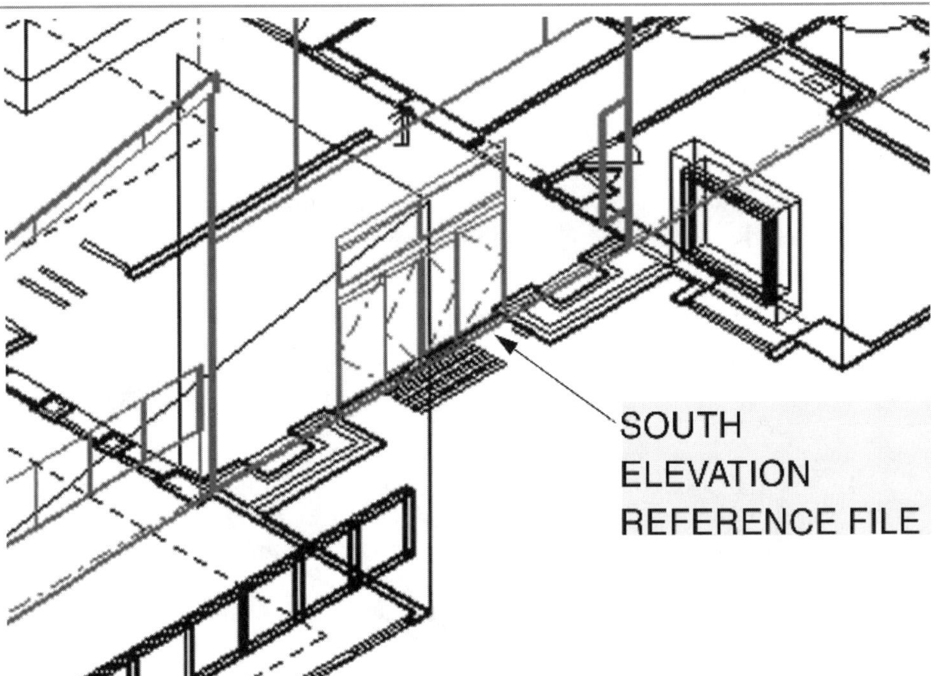

View of south entrance elevation.

Let's begin with the windows above and then move to the doors.

2. Copy all of the blocks that define the windows and structure above the doors into the file using precision input:

   `DX=0`

3. Copy the outer block that defines the windows back 2 inches (`DL=,:2`) and change its level and color to 15 (glass).
4. Change the active level and color of the remaining blocks to 14 (aluminum).

5. Form the window frame by using the *Group Hole* tool to subtract the inner block from the outer block.
6. Project the window frame back 4 inches using the *Project Surface or Solid of Projection* tool (set to solid).

Now that the windows and structure are in place, it is time for the doors.

7. Copy the inner and outer blocks that define the leftmost door.
8. Copy the outer block back 2 inches (`DL=,:1`) and change its level and color to 15 (glass).
9. Change the active level and color of the first two blocks to 14 (aluminum).
10. Use the *Group Hole* tool to subtract the inner block from the outer block and project it back 1 inch.

That forms the first door. To construct the left-center door, you will copy and modify the one just constructed.

11. Turn off the display on the south elevation reference file.
12. Place a fence large enough to enclose the door just constructed (make sure the views' clipping planes do not include the north side of the building).
13. Copy the fence using the following precision input:

    `DL=3:4,:2`

Because the center doors are smaller than the outer doors, you must use a fence stretch to modify the new door.

14. Place a fence just large enough to include the right side of the center door.
15. Activate the *Stretch* command and stretch the right side of the door to the left 10.5 inches:

    `DL=-:10.5`

The left doors are now constructed. You will mirror the right doors into place.

16. Place a fence just large enough to include both the doors.
17. Set the key point divisor to 2.

18. Activate the *Mirror* command and set the Mirror About setting to vertical. Copy and Fence should also be activated.
19. Tentative point to the center of the uppermost portion of the entry and place a data point.

The right doors should now be in place and the entry is starting to take shape. To finish, you must complete the construction of the walls that surround the doors.

20. Set the active level and color to 11 (Brick).
21. Activate the *Place Shape* tool.
22. Tentative point to point 14 and place a data point.
23. Enter the following precision input:

    `DL=,,15`

24. Tentative point to point 13 and enter the following precision input:

    `DL=,1,15`

25. Tentative point to point 13 and place a data point.
26. Tentative point to point 14 and place a data point.
27. Copy the shape just created left 8 feet, 10 inches:

    `DL-8:10`

28. Mirror these two shapes to the right side of the entry.
29. Activate the *Place Block* rotated tool.
30. In the Iso view, use the two outside shapes to place a block to form the top of the entry.

The next step is to create the soffit or ceiling above the doors.

31. Turn off the display on the floor plan reference file.
32. Turn off level 63 in the Iso view.
33. Keep the *Place Block* rotated tool active.
34. Tentative point to the uppermost left side of the structure above the doors and place a data point.
35. Tentative point to the corresponding point on the other side of the entry and place a data point.

36. Enter the following precision input:

```
DL=,-2:8
```

The last part of the entry is the front portion or face.

37. In the Iso view, use the *Place Shape* tool and travel around the entry to form this portion.

The Roof

Now for the good stuff. The roof is an amazingly easy but important aspect of this building. The design and copper material make the roof a powerful feature.

1. Attach the chap14 cell library.
2. Make roof the active cell and set active angle to 0 and active scale to 1.
3. Tentative point to point 10 and place the cell.
4. Drop the cell.

What you now have is the profile or section of the roof and the linestrings to project it along.

5. Activate the *Construct Tubular Surface* tool. Set the Type to solid and the Define By option to section.
6. Data point on the linestring that follows the slope of the roof to define the trace curve.
7. Data point on the profile of the roof to define the section.
8. Repeat the process on the other sloped roof.

The profile of the roof is projected along the linestring, forming clean corners. What an easy way to create such a complicated roof. The other portions of the roof were created with concepts already covered. They are contained in the chap14 cell library, so let's finish the construction of the project and move on to materials.

9. Attach the chap14 cell library.
10. Make finish the active cell and set active angle to 0 and active scale to 1.

11. Tentative point to point 10 and place the cell.
12. Drop the cell.

Materials

Exercise 3: Materials for the Library

Because of the time you have spent putting things on specific levels, applying materials will be simple. The materials that are needed have already been created and are stored in the material palette called library.

1. Open the Assign Material dialog palette.
2. From the assign material palette, open the material palette *library*.
3. Select Plaster as the active material.
4. Make the active level 10.
5. In the Top view turn off all levels except 10.
6. Perform a fit.
7. Use the selection tool to select all of the elements on level 10.
8. From the Assign Materials palette select Assign by selection.
9. Plaster has been assigned to every element that was selected.
10. Reset the selection tool (select nothing).
11. Repeat the process for all of the other levels. The following list presents the levels and the materials that are related with them.

Level/Color	Material
10	Plaster
11	Stone Brick
12	Tinted Glass
13	Copper
14	Aluminum
15	Clear Glass

To make your model a bit more exciting, a modeled site is provided. Reference the file *Site.dgn.* Materials will also need to be attached to the levels in the site plan.

Level/Color	Material
50	Slab
51	Concrete
52	Grass
53	Arcade White
54	Blue Paint
55	White Paint
56	Stone Floor

At this point you have a very convincing model to show to your clients. You can communicate a number of concepts at once, not to mention impress your clients.

If you spend some time setting up lighting and backgrounds, you will get some impressive renderings.

Summary

For more views of the San Marcos Town Center Library, see the enclosed CD-ROM. You can play with these files to learn more about the effects you can achieve to amaze your clients and friends.

Introduction

This is the final part of the journey. You have been taught the basic skills to conquer the 3D wilderness. Now it's time to put those skills to work. In this exercise, you will create a simple 3D mechanical part. You will learn some shortcuts and some pitfalls to avoid. For the most part, you have worked with the 3D tools one by one. After you graduate from wilderness camp and venture out on your own, you will learn which tools to use to achieve the best results. Often, the order in which you create your models is critical to their success.

The tools used in this exercise represent only one path that you can take to reach the goal. Enough of this standing around the campfire and chatting, let's get to work.

Creating the Part

To create this part, you will be using a variety of 3D tools. The cylindrical parts will be circles that have been projected using the *Construct Surface or Solid of Projection* tool. That same tool will be used to create the body of the part by projecting a complex shape. The boss feature will be created by projection also. The radius on the boss will be accomplished by using the *Construct Fillet Between Surfaces* tool. Once the part is complete, you will assign materials and do some renderings for the folks back home.

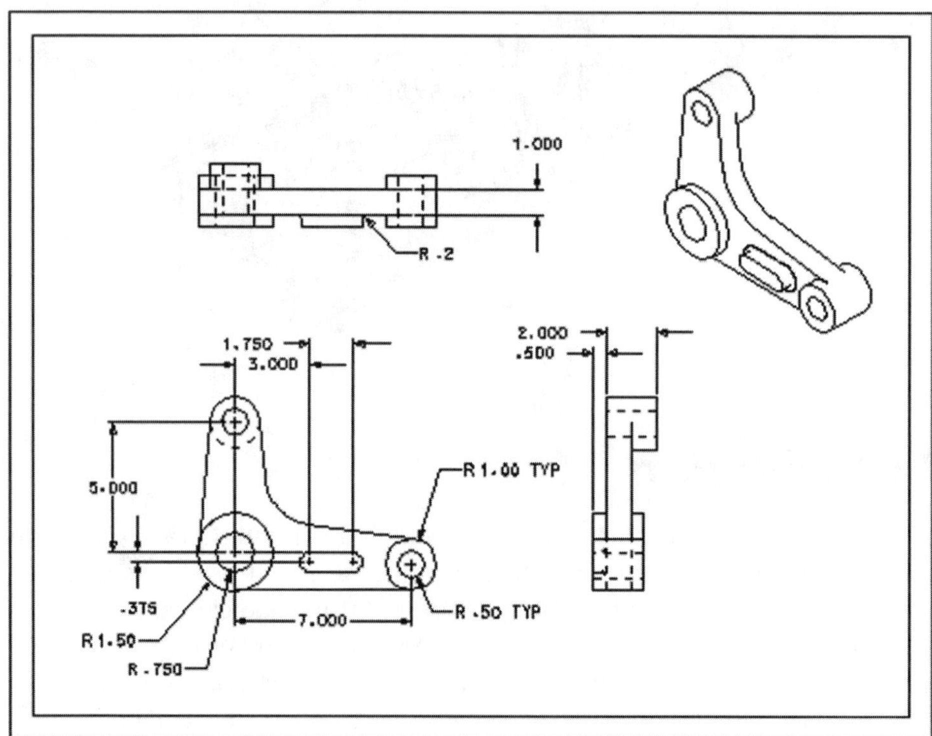

This is the part you will be creating.

For this exercise, you will need to follow a leveling schema. This schema will assist you in visually identifying which elements are profiles and which are surfaces.

Profiles are the elements that you will project into surfaces. They will follow this schema:

Level=5

Weight=2

Color=2

Surfaces will follow this schema:

Level=10

Weight=1

Color=1

1. From the Command window, select **Inside3D → Chapter 10 → Start Exercise**. I have created the first circle for you, and your Element Attributes have been preset by your trusty guide. (Every little bit helps when you're out in the wilderness.)
2. The circle in the file is the large circle of the part, the one with the 1.5 radius.
3. Select the *Place Circle By Center* tool. You will now create the circle inside the large circle.
4. In View 3, tentative and data point in the center of the 1.5-radius circle. Then key in the following:

   `DX=.75`
5. Now you will create the circle to the right of the two circles. Tentative point in the center of the circles and key in the following:

   `DX=7,-.5`

   `DX=1`
6. You now need to create another circle inside the circle you just created. In View 3, tentative and data point in the center of the circle on the right. Key in the following:

   `DX=.5`
7. Now for the circle on the top: In View 3, tentative point in the center of the large circle and key in the following:

   `DX=,5,-.5`

   `DX=1`

This creates a circle 5 inches up on the Y-axis and -.5 inches on the Z-axis. Notice that this circle is not on the same plane as the first four circles. The profile shape for the body of the part will be on this plane.

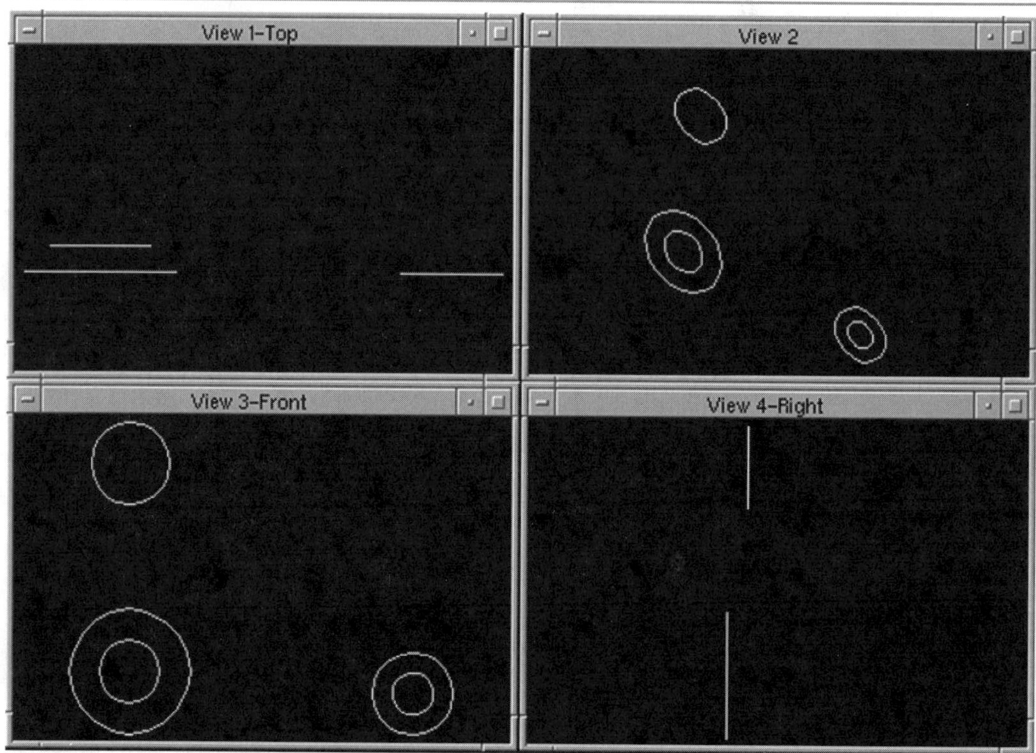

Your views will look like this.

8. You have one more circle to create. This is the circle in the center of the last circle created. Tentative and data point in the center of the circle on top, and then key in the following:

```
DX=.5
```

9. The next step will be to create the profile shape of the body. To do this, you will create a series of lines and arcs. Then you will chain them together. But first you need to set the active depth so that the lines and arcs are created on the same plane as the last set of circles you created. Select the *Set Active Depth* tool from the 3D View Controls subpalette.
10. You are going to set the active depth for View 3. Data point in View 3.
11. Now move your cursor into View 1 and tentative and data point on the last circle you created. (That would be the geometry closest to the top of View 1.)
12. To create the profile, you will be snapping to geometry that is on a different plane. To ensure that the geometry you create remains on the correct plane, turn on the Depth Lock. Select from the Command window **(Settings → Locks → Toggles)**. The Depth Lock is on the bottom of the palette.

You are going to create a shape that will look like this.

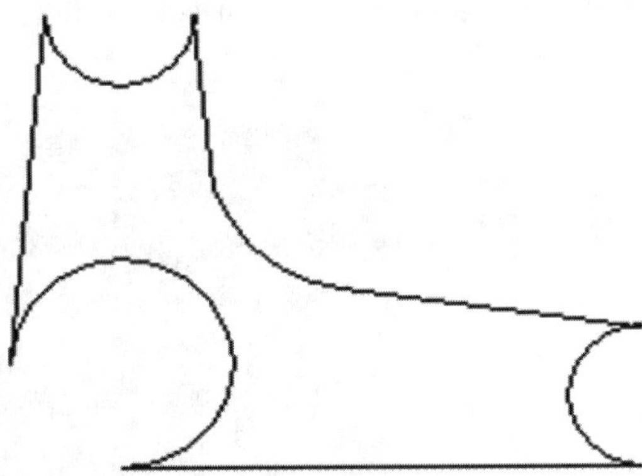

13. Select the *Place Line* tool. In View 3, tentative and data point on the left side of the 1.5-inch radius circle.
14. Now tentative and data point on the left side of the 1-inch radius circle above the circle you just selected. Then press Reset.

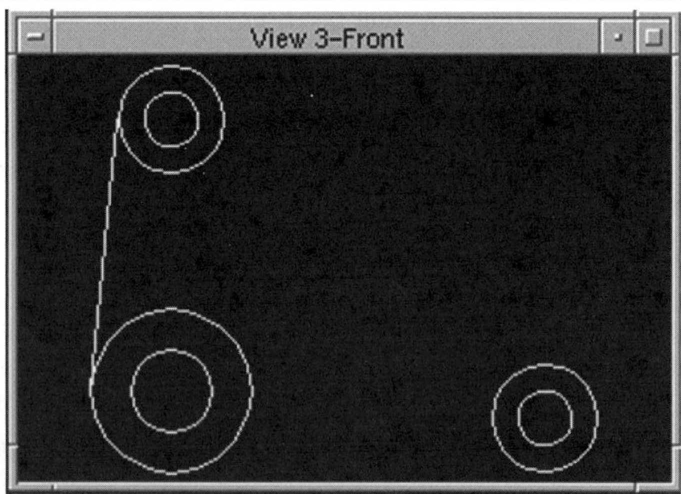

View 3 with circle created.

15. Now tentative and data point on the right side of the 1.5-inch radius circle.
16. Tentative and data point on the right side of the 1-inch radius circle above the circle you just selected. Then press Reset.

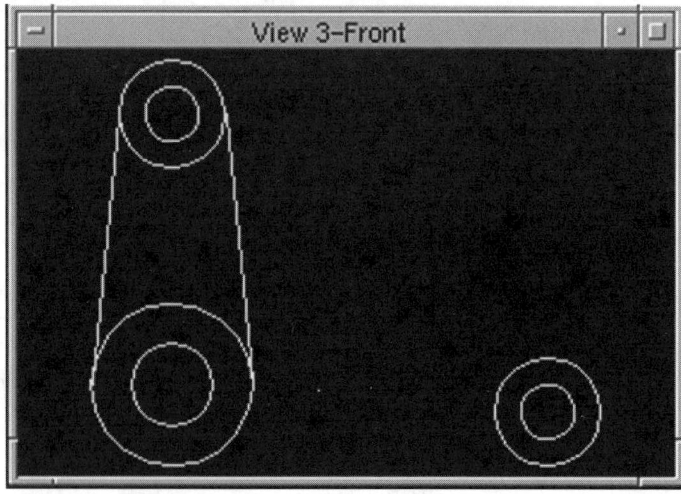

View 3 should look like this now.

17. Tentative and data point on the top of the 1.5-inch radius circle.
18. Tentative and data point on the top of the 1-inch radius circle to the right of the circle you just selected. Then press Reset.
19. Tentative and data point on the bottom of the 1.5-inch radius circle.
20. Tentative and data point on the bottom of the 1-inch radius circle to the right of the circle you just selected. Then press Reset.

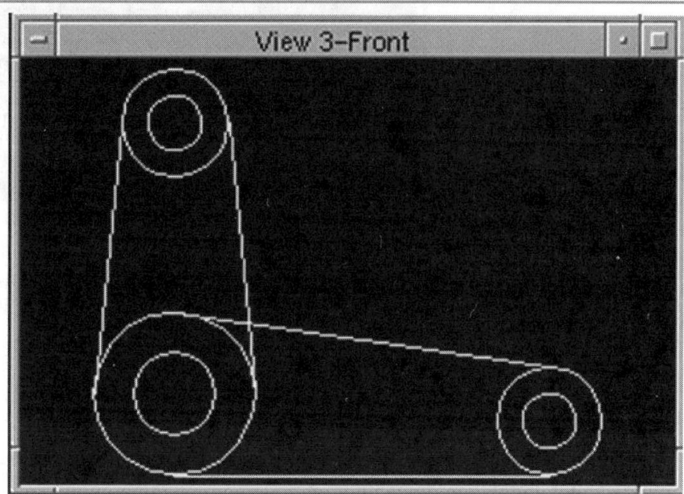

The part is almost completed.

21. Now you need to fillet the two lines that connect on the right side and top of the 1.5-inch radius circle. Select the *Construct Fillet and Truncate Both* tool.
22. Set the radius for this command to 2.
23. Data point on the line that connects to the right side of the 1.5-inch radius circle.
24. Data point on the line that connects to the top of the 1.5-inch radius circle. Then data point to accept the results.

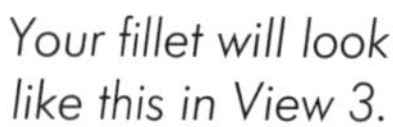

Your fillet will look like this in View 3.

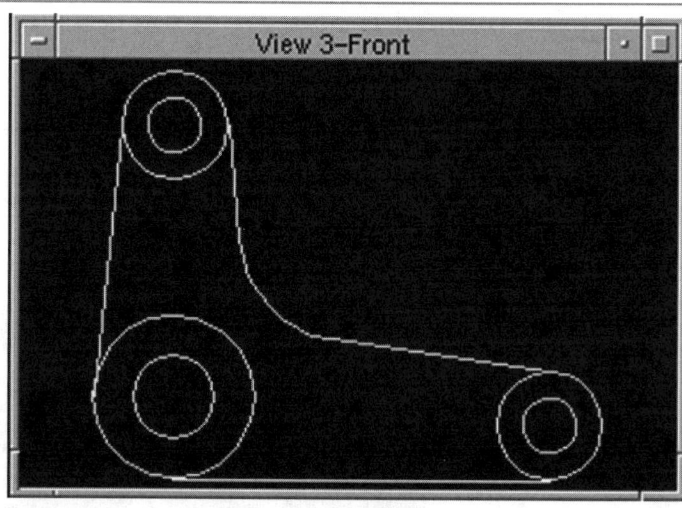

Now you are going to create three arcs. These arcs will be used to connect the lines and create a complex shape.

25. Select the *Place Arc By Center* tool.
26. In View 3, identify the first arc endpoint by doing a tentative point and a data point on the left side of the 1-inch radius circle at the top of the view.
27. In View 3, identify the center of the arc with a tentative point and a data point in the center of the circle in the top of the view (you started on the left side because arcs are created counterclockwise).
28. Now in View 3, identify the second arc endpoint by doing a tentative and data point on the right side of the circle.
29. Now for the next arc. In View 3, tentative and data point on the top of the 1-inch radius circle on the right side of the view.
30. In View 3, identify the center of the arc by selecting a tentative and a data point in the center of that same 1-inch radius circle.
31. Complete the second arc endpoint by doing a tentative and data point on the bottom of the 1-inch radius circle.
32. Now for the last arc. In View 3 again, tentative and data point on the bottom of the 1.5-inch radius circle to start the arc endpoint.
33. In View 3, tentative and data point in the center of the 1.5-inch radius circle to identify the center of the arc.

34. To identify the arc endpoint, in View 3 tentative and data point on the left side of the 1.5-inch radius.

If the circles were removed, the lines and arcs would look like this.

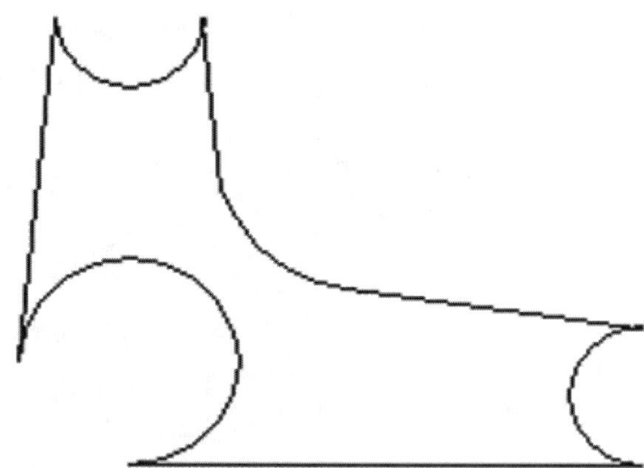

35. Now that the lines and arcs are in place, you are going to create a complex shape from them. Select the *Create Complex Shape* tool.
36. Set the method to Automatic. In View 3, data point on the line that connects the 1.5-inch radius circle and the 1-inch radius circle on the right side of the view.
37. To indicate the direction you want the shape to be created, data point to the left of where you selected the line.

NOTE: *The direction in which a shape is created will determine the surface normals. In this case, you want the surface normals to be pointing toward the front of the part.*

38. Continue to data point until the shape is closed.
39. Now it's time to project the circles. But, before you do that, you need to use the *Group Holes* command so the inside circles project as holes. Select the *Group Holes* tool from the Chain subpalette.
40. You are prompted to identify the solid element. In View 3, data point on the 1.5-inch radius circle.
41. You are prompted to identify the hole element. In View 3, data point on the .75-inch radius circle inside the 1.5-inch radius circle. Data point once more (away from any elements) and then press Reset to accept.

42. Repeat this procedure for the other two sets of circles. The outside circle is the solid, and the inside circle is the hole.
43. Now you are ready to project the circles into cylinders. Select the *Construct Surface or Solid of Projection* tool. I know that this is still new to you, so here's a clue. The tool is located on the 3D Free-Form Surface subpalette.
44. Make the tool settings match those displayed in the image below.

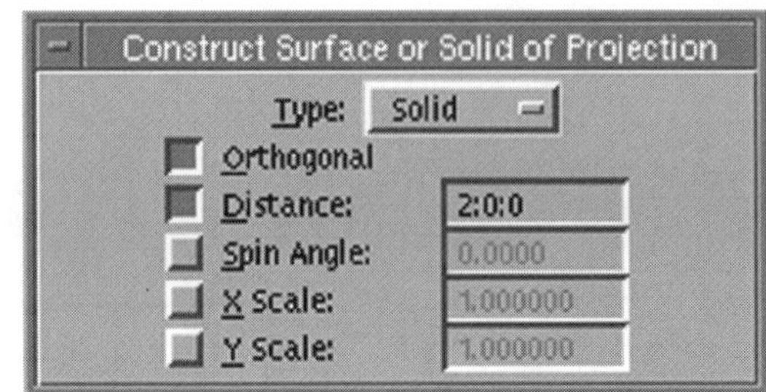

Set your tool settings to match these.

45. In View 3, data point on the .75-inch radius circle (that's the one inside the large circle).
46. Move the cursor into View 1. Now move the cursor toward the top of the view so the circle is projecting in the positive Y-direction. The cylinder should be projected toward the back of the part.
47. Once you have the projection going on the correct direction, data point to confirm the height.

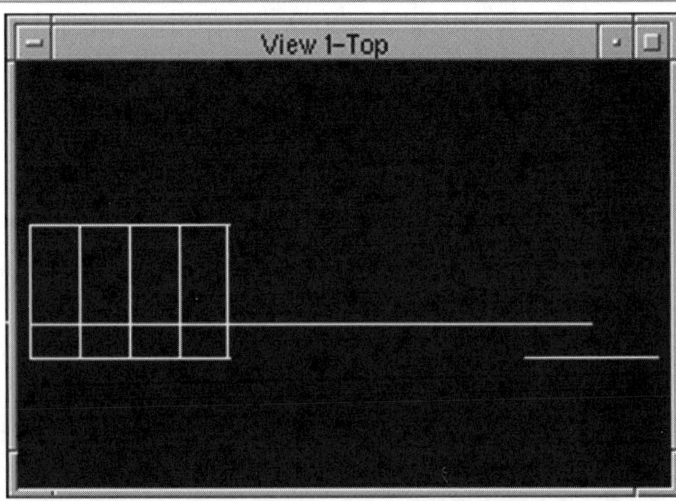

In View 1, the projection should look like this.

48. Repeat this process for the other two sets of circles. Remember that the top set of circles are not on the same plane as the other two sets. They will project .5 inches further back.

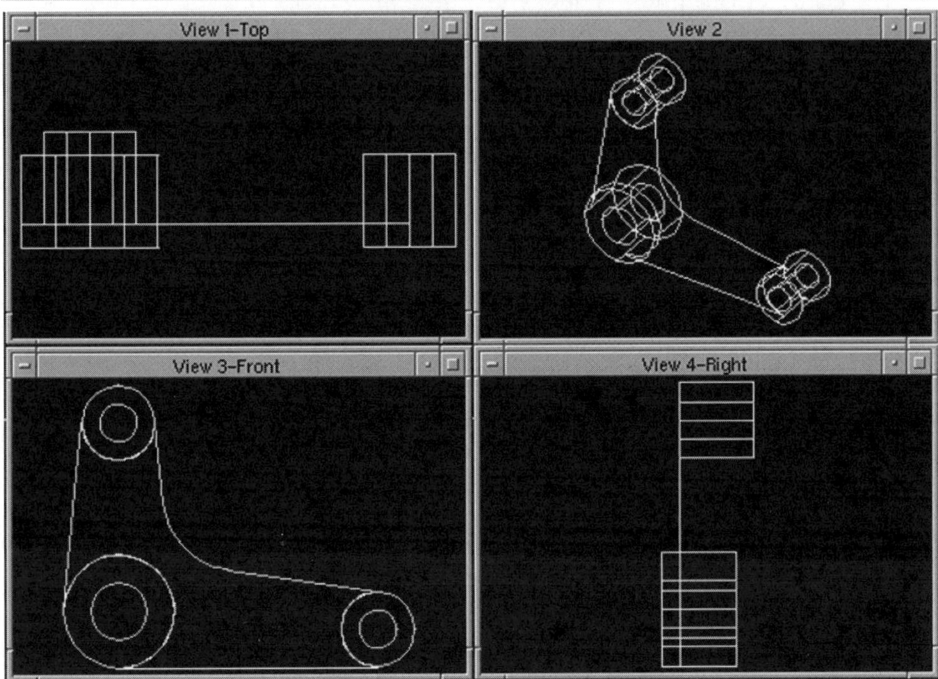

Your views will look like these, after the circles are projected.

49. It's time to project the body of the part. The thickness of the body will be 1 inch. Set the distance to `1`.
50. In View 3, data point on the complex shape.
51. Move the cursor into View 1. Again, you want to project the body toward the back of the part. Data point to accept the results.
52. For the boss feature on the front side of the body, you will need to change your active depth and turn off the Depth Lock. Select the *Set Active Depth* tool.
53. Data point in View 3 to select it as the view you want to set the active depth for. Then key in the following:

    `DX=,,.5`

This keyin sets the active depth to be the same as the front of the two lower cylinders.

54. To create the boss, you will create the two lines. Then use arcs to connect the lines. Select the *Place Line* tool.
55. Tentative point in the center of the 1.5-inch radius circle. Then key in the following:

    `DX=3`

    `DX=1.75`

56. Select the *Copy Parallel by Key-in* tool. Set the Distance to be `.75`.
57. Make sure you have Make Copy button selected. Data point on the line you just created.
58. Move the cursor below the line and data point to create a copy of the line. Press Reset to cancel the command.

The two new lines will look like this in View 3.

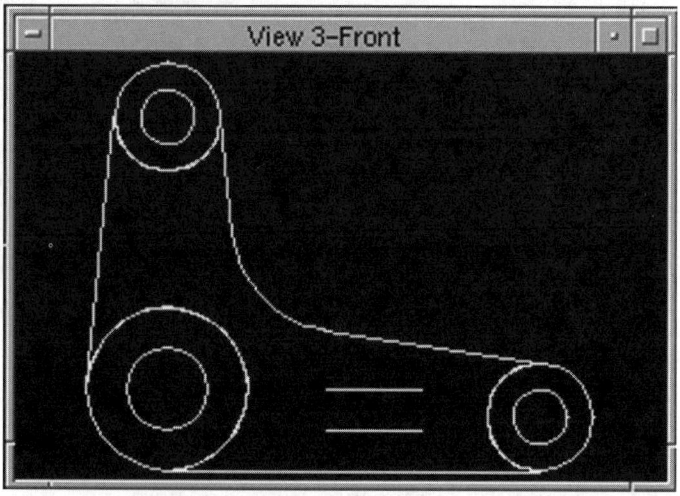

59. Select the *Place Arc By Center* tool. Did you remember to turn off the Depth Lock? If not, do it now. Make the settings match those in the following image.

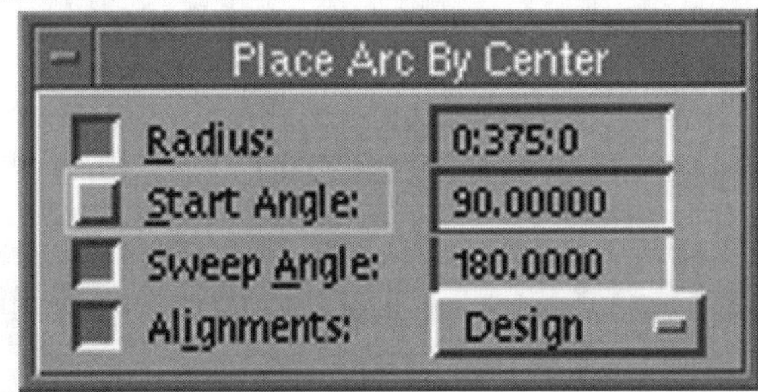

These are the settings for the Place Arc By Center command.

60. In View 3, tentative and data point to the left end of the top line. Notice how using the Design Alignments setting produces orthographic drawing aids. These will assist you in placing the other end of the arc.
61. In View 3, move the cursor counterclockwise until it snaps into alignment with the end of the arc connecting to the other line.
62. Repeat this process for the arc at the other end of the boss.

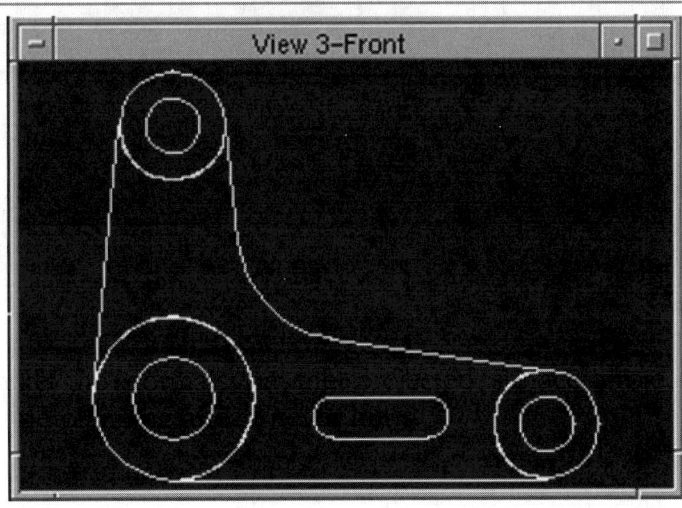

The finished boss feature will look like this.

63. The lines and arcs need to be chained together to make a shape. Select the *Create Complex Shape* tool.
64. When you chain the elements together, go in a clockwise direction. This will make the normals on the projected surface project outward. A little planning ahead never hurts.

65. Select the *Construct Surface or Solid of Projection* tool. Set the distance to 1.
66. In View 3, data point on the complex shape. Then project the shape toward the back of the part.

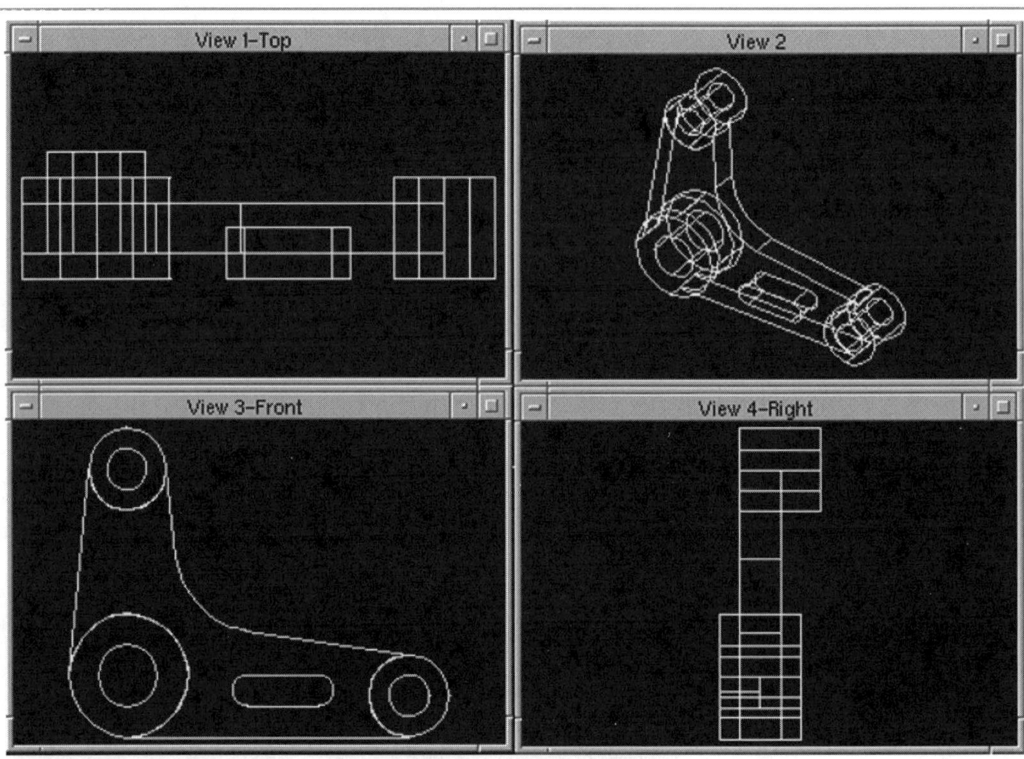

Your views should look like this.

67. To make the fillet on the boss feature, you will use the *Construct Fillet Between Surfaces* command. Select the *Construct Fillet Between Surfaces* tool. Set the Radius value to .2.
68. In View 3, data point on the boss feature. Notice that the surface normals are pointing outward. This will construct the fillet on the outside of the surface.
69. In View 3, data point on the body of the part. Again, notice that the surface normals are pointing outward. Data point to accept the surface.

70. The surface is trimmed, and the fillet now appears. Data point to accept the results. Congratulations! The part is now complete.
71. In the beginning of this exercise, I mentioned a leveling schema. If you haven't been following it, now is the time to make sure that the profiles and surfaces have the correct element attributes. The most important is the correct level assignment. After you are done, turn off level 5 for all views.

I'm convinced that looking at a wireframe all day has to be bad for your brain cells. Some material mapping and rendering is a great way to end the day.

1. From the Command window, select **Settings → Rendering → Assign Materials**.
2. From the Assign Material palette, select **File → Open Palette**.
3. From the Open Palette File palette, select the file called *metal.pal* and select the OK button.
4. Now select the material called Aluminum.
5. You are going to assign this material to your part. From the Assign Material palette, select **Tools → Assign By Selection**.

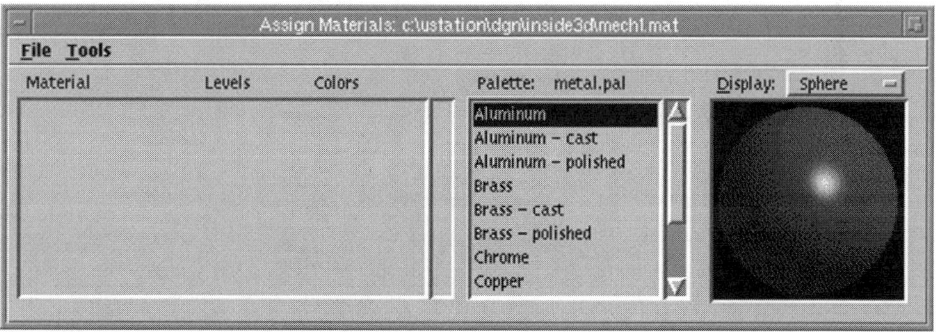

The Assign Materials palette.

6. You are prompted to identify an element. Data point on the body of the part and data point to accept it.
7. Now do a Phong render of View 2. Again, I have taken the liberty of preparing the file ahead of time. I placed three light sources on level 61. Your rendering will turn out much nicer this way. Feel

free to adjust them any way you would like. Experiment with the materials too. Job well done, traveler, you deserve time to play around.

Summary

You have applied the skills you've learned throughout this book to this mechanical project. If you feel proud of yourself, you are justified. The shortcuts and tools that you've learned about will carry you a long way in the real world.

Introduction

Safari Sam taught about free-form surfaces back in Chapter 9. You will use those tools to create an object created almost entirely from free-form surfaces and curves—a violin. The violin will be full sized. It's time to become a luthier (violin maker); wouldn't Stradivarius have been jealous of you working in MicroStation?

Stradivarius' "La Messie" violin, built in 1716.

First, a quick review of the history of the violin. The modern violin originated in the small Italian town of Cremona, where the giants of violin making came from. Nicholas Amati is regarded as the father of the modern violin, and in his shop he taught his sons, Antonio Stradivari, Andrea Guarneri, and Francesco Ruger. In addition, Cremona was home to the Bergonzi family, also great luthiers. The whole world came to Cremona to purchase the finest stringed instruments (violin makers of that time also made violas, cellos, and string basses).

To this day violins created in Cremona from 1650 to 1750 are the most prized of all instruments. The instrument shown above (La Messie or *The Messiah*) was made in Stradivarius' most prolific year. He made 15 violins in 1716. When he finished making La Messie, he realized that it was as close to perfect as an instrument could be. Therefore, he never strung it and kept it above his workbench until the end of his life. He was afraid that the sound of the instrument could not match its beauty, and he did not want to find out. It was the only instrument he would never sell. The instrument was only played once for 10 minutes in the late 1800s by the nephew of the great French violin maker, J.B. Vuilliame. The sound was said to match its beauty.

Today, the violin is in the Ashmolean Museum of Oxford University, in an environmentally controlled case and available for viewing. The instrument is regarded as *the* perfect violin, and one of the world's greatest works of art. Now, you are going to build another one.

The Violin

Here are the vital statistics for the La Messie violin back.

Length = 35 cm

Maximum width for top half = 17 cm

Maximum width for bottom half = 20.5 cm

Minimum width between middle of f-holes = 11.5 cm

Width between top of f-hole curves = 16 cm

Width between bottom of f-hole curves = 18.5 cm

The information for the violin was gathered from books referenced at the end of this chapter and one of the authors' own violins. Using simple dimensions and construction lines, an outline for the violin was established. Working Units are cm:mm: 1000 PU per mm.

1. Select the **Inside3D → Chapter 16 → Violin 1** command.

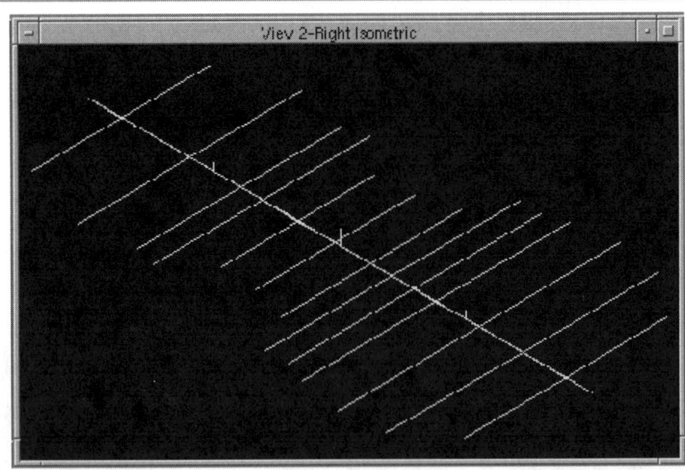

The construction elements for the top and bottom of the violin.

The construction elements will be used to create the B-spline curves, which will then be used to create the B-spline surfaces. In addition, the natural symmetry of the violin will be used to speed the modeling process. Not all components of the violin will be modeled. Safari Sam wants to leave some excitement for you.

Let's begin by noticing that all the construction components are on level 1, except for the central line, which is on level 2. So it would be a good idea if you started drawing your B-splines on level 2. The *Settings → Rendering → View Attributes* is set to Wiremesh.

Exercise 1: Building the Top and Bottom

1. Use the Snap points to create the three main curves in the Top view. Select the *Place B-spline Curve* command with the Catmull-Rom and Placement settings. The three curves are points 1-5, 1-5, and 1-7. Make a copy of the curves and put them on level 7. Stay on level 2 to build the top and bottom.

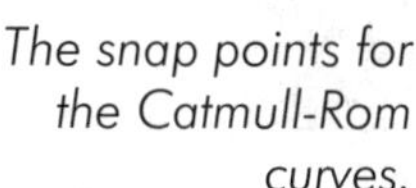

The snap points for the Catmull-Rom curves.

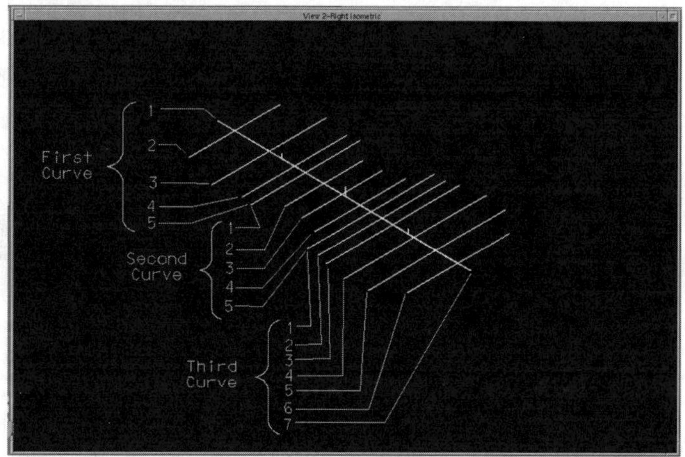

2. Next create another Catmull-Rom B-spline in the Front view, using the four snap points indicated, to create the longitudinal curve of the top.

Snap points for the longitudinal curve in the Front view.

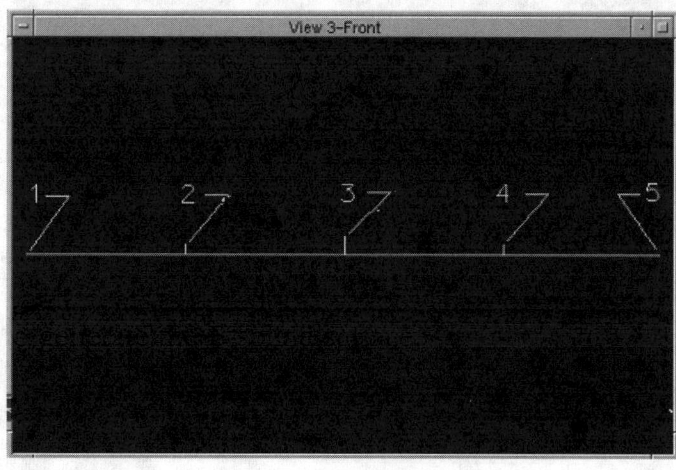

3. Turn off level 1. Now select the *Create Complex Shape* command, with Manual set, select the three curves and the longitudinal B-spline, and data point to accept the shape. This complex shape will be used to generate the B-spline surface.

Create Complex Shape command.

4. Select the *Create Surface By Edge* command on the Free Form Surfaces palette.

Create Surface by Edges.

5. Select the complex shape you just created and data point to accept. A smooth surface should now be created between all curves.

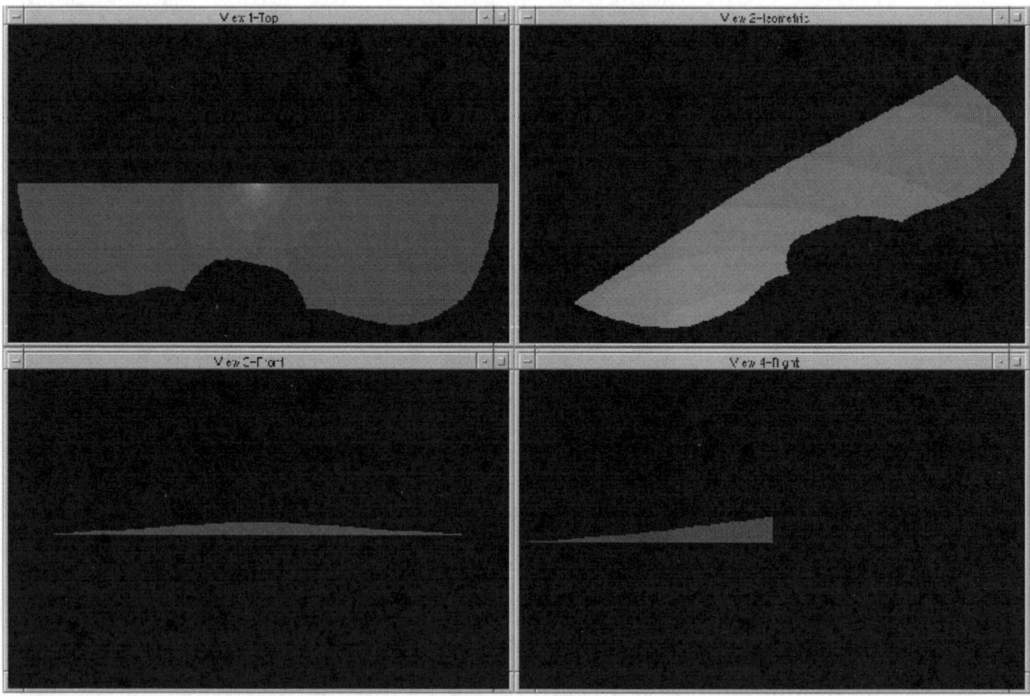

The Phong shaded view of the surface.

6. Use the *Mirror Element* command to create the other half of the violin back. Use Horizontal and Make copy as your settings. Snap to the endpoint of the central line as your rotation point. You want

to copy the complex shape, not the B-spline surface. The Shape is now a 3D shape.

7. To create the bottom of the violin, begin by drawing a line in the Top view. Snap to the endpoint of the central line. Key in the following:

   ```
   DX=,,-3.7
   ```

   You are using this keyin since the distance between the top half and bottom half is 37 mm.

8. Use *Mirror Element* again in the Front view, with Vertical and Make Copy set. Use the midpoint of the new line as the rotation point, and data point to accept the lower copy. Use the *Mirror Element* command again to create the second horizontal half of the bottom of the violin. Your model should look like the following figure.

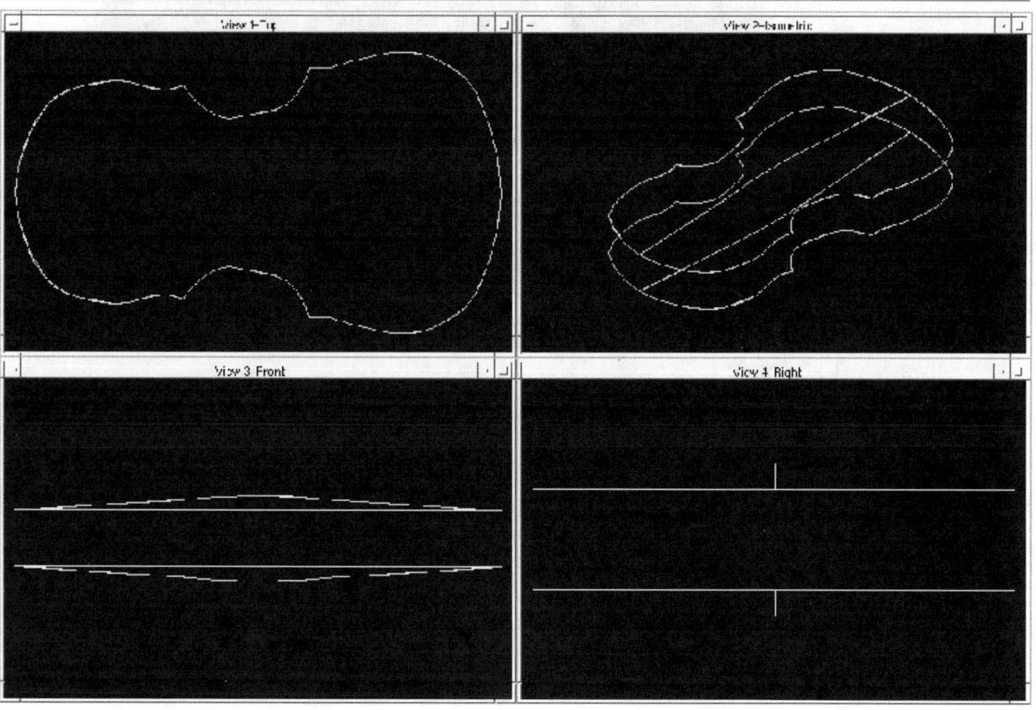

The top and bottom half of the violin.

Exercise 2: Creating the Sides of the Violin

Creating the sides of the violin is just as easy. The sides are between the top and bottom of the violin.

1. Change to level 7 and turn levels 2 and 1 off. You should have the three curves you created in the beginning. These curves will be used to create a solid of projection. Select the *Create Complex Chain* command and select the three curves to make one element out of them.

Iso view of curves for top and bottom edges of the side.

2. Draw a vertical line in the Front view at one end of the curve and key in the following:

   `DX=,-3.7`

   This is the distance between the two halves of the violin.
3. Select the *Create Solid of Projection* command and, in the Front view, select the complex chain curve and then select the bottom endpoint of the line just created to complete the command.
4. Mirror the side of the violin in the Top view to create the other side of the violin.

Exercise 3: Creating the F-holes

The f-holes are the holes in the top of the violin, through which the sound comes out. They are on either side of the violin and about halfway down lengthwise.

1. Change to level 10 and turn off levels 1 through 9. You will be placing a spiral to create the f-hole.
2. The end of the short line is the origin of one spiral, and the endpoint of the longer line is the origin of the other spiral. Select the *Place Spiral* command from the Curves palette. The settings are Type: Archimedes; Initial Radius: 0.05; Final Radius: 1.5; and Angle: 470 degrees. Place the spiral using the snap points shown (1-3).

Snap points for Archimedes spiral.

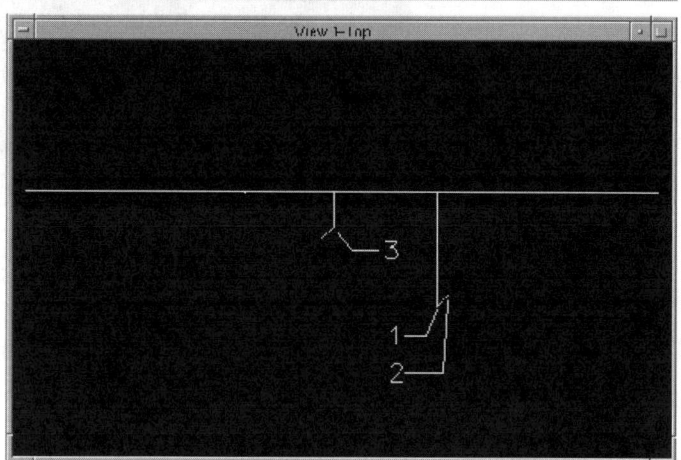

3. Use the *Modify Element* command to move the end of the spiral to the end of the short line and data point to accept.

The Modify Element command.

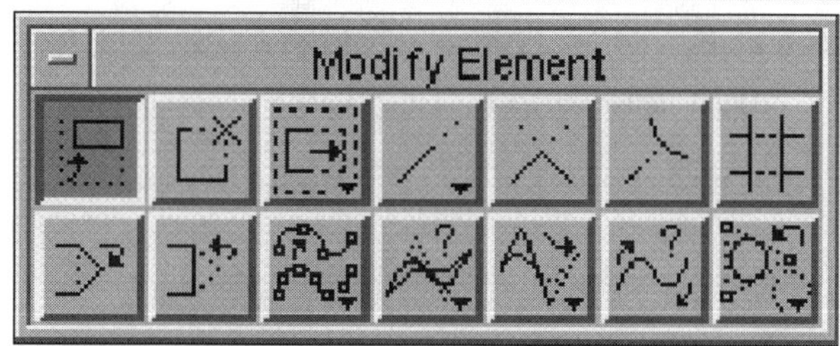

4. Select the *Mirror Element* command, with Vertical and Make Copy turned on. Mirror the element again using Horizontal, but Make Copy should be off. Finally, move the origin of the copied spiral to the end of the short line on the shorter of the two lines (point 1).

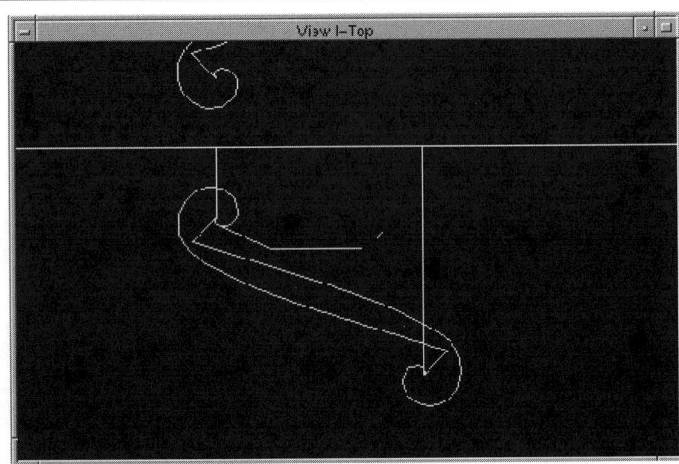

F-holes created and snap point.

5. To create a shape, use the *Chain Complex Shape* command with Manual set.
6. Use the *Move* command and in the Front view move the f-holes up 10 mm. Key in the following:
   `DX=,1`
7. Finally, turn on level 2 and use the *Punch Surface Region* command to create the f-holes in the top of the violin.

Exercise 4: Creating the Finger Board

The finger board is a piece of ebony wood, where the violinist places his or her left fingers to play the strings.

1. Change to level 11, and turn off levels 1 through 10. In the Right view, use the *Place Arc By Edge* command to create one arc at either end of the line, using the snap points shown for each arc.

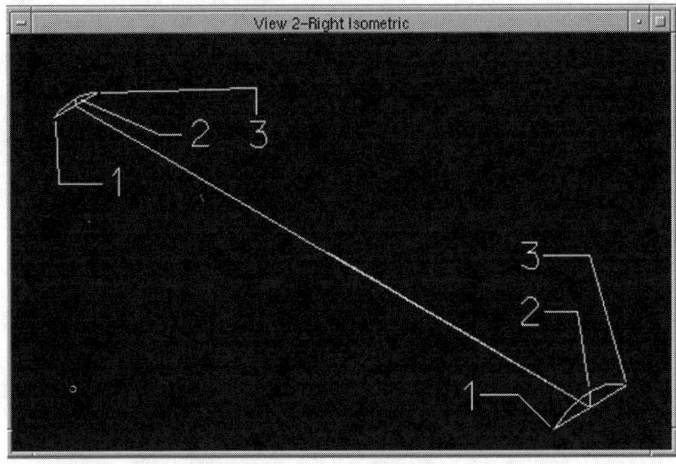

The arcs and the snap points used to create them.

2. Copy the larger arc in the Right view up by 4 mm, and the smaller arc up by 3 mm. Key in the following for the *Copy* command:

   ```
   DX=,0.4
   DX=,0.3
   ```

3. Draw lines between the ends of the original arcs and their copies, creating two shapes at either end. Select the *Create Complex Shape* command with Manual set and turn each set of two arcs and two lines into one shape.
4. Use the *Create Skin Surface* command to create the finger board. First, select the line in between the two shapes, then select the larger arc shape and then the smaller one, and data point to accept.

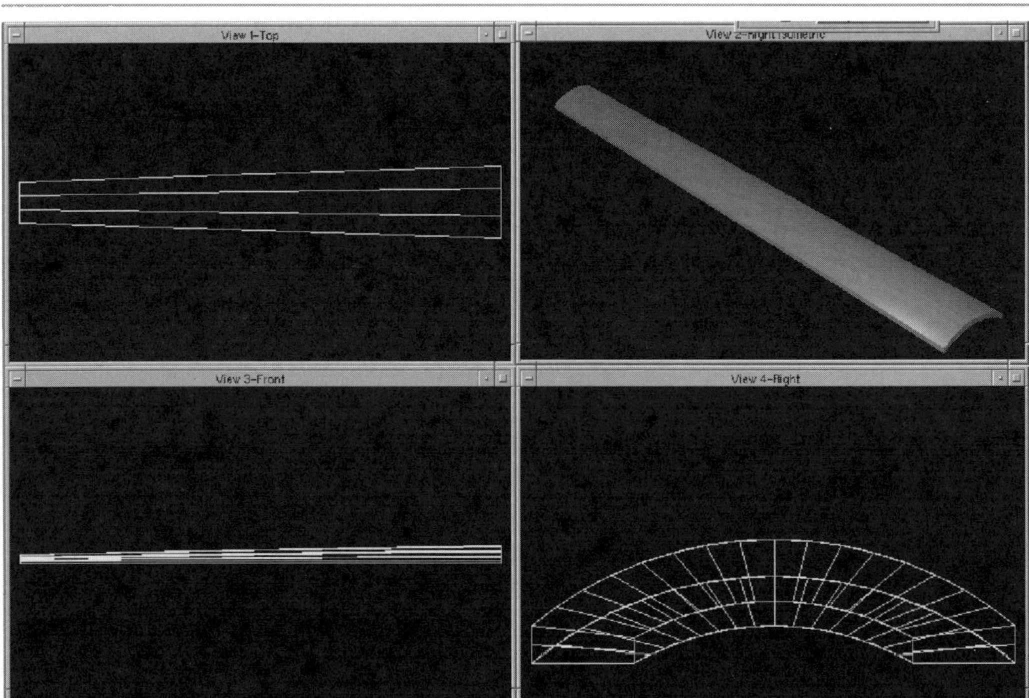

The finger board.

5. The finger board support is built in the exact same way, using two curves and a trace curve. The construction elements are on level 12. Turn off levels 1 through 11.
6. Use the *Place Arc By Edge* command in the Top view to create the first arc (the one attached to the body of the violin). Use the snap points indicated (points 1-3). Use the same command to create the second arc using the second set of snap points (1-3).

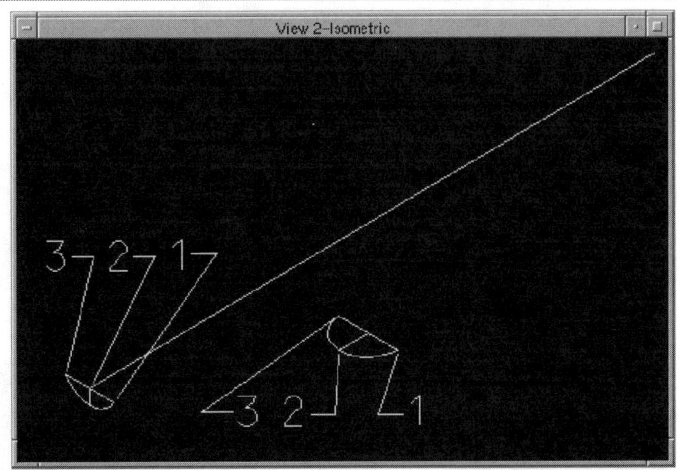

Snap points for the arcs for the finger board support.

7. Use the *Place Line* command in the Top view to create the trace curve. Snap to the top (middle of the arc) of the first arc drawn and key in the following:

   `DX=,,2.8`

8. In the Front view, select the middle of the second arc drawn and key in the following:

   `DX=10`

9. Now use the *Extend 2 Lines to Intersection* command to join the two newly created lines.

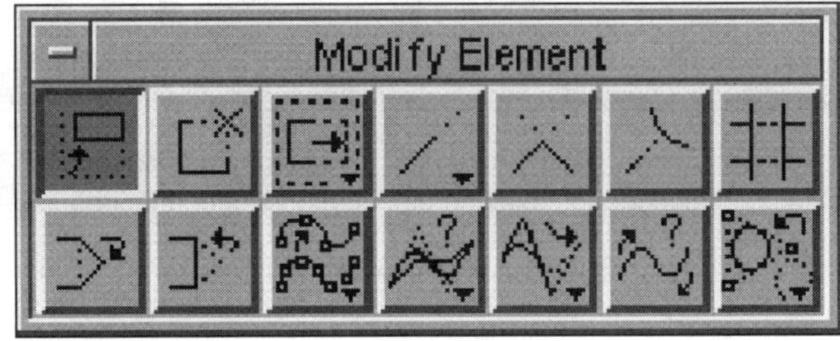

Extend 2 Lines to Intersection command.

10. Next select the *Fillet Element* command with a radius of 1.5 and Truncate set to Both and fillet the two lines. Use the *Complex Chain* command to turn the two lines and fillet into one element.
11. Use the *Create Skin Surface* command to create the finger board. Use the complex chain curve as the trace curve, select the larger arc and the smaller arc, and data point to accept. Notice that there is a gap between the body, the finger board, and finger board support. This can be filled using the *Blend Surface* command.

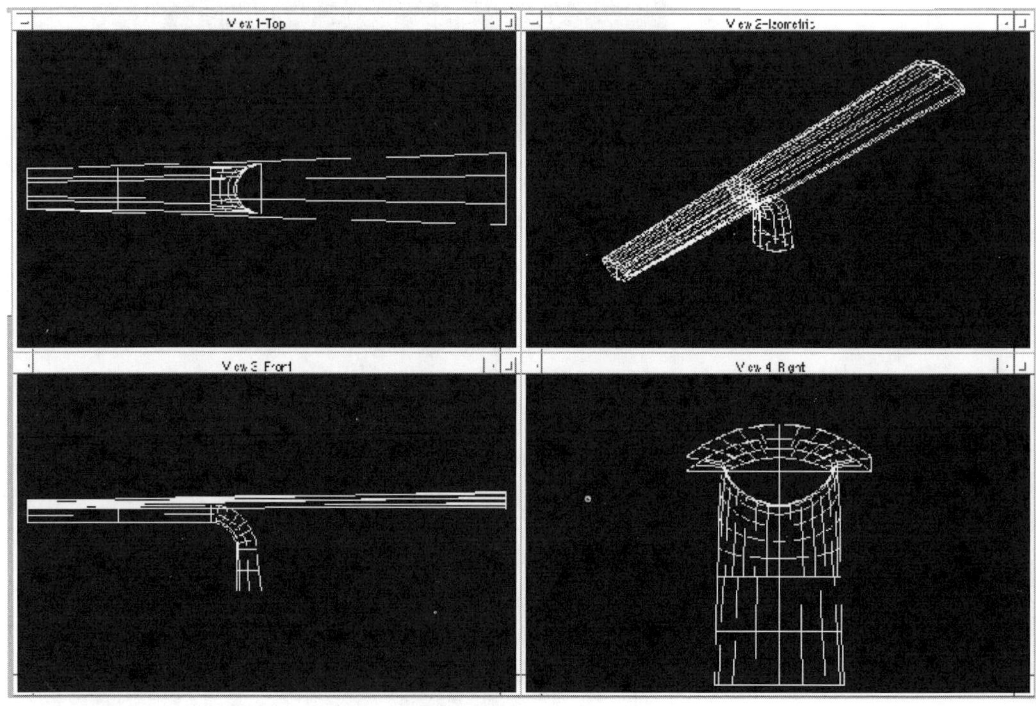

Finger board and support for the violin.

Adding Textures for Detail

At this point you can stop modeling and begin working on rendering the violin. The bridge (between the f-holes), the scroll, and the tail piece will be left for you to finish. This gives you the opportunity to build from scratch, by doing a little research. Look at the reference section for more information. At this stage you can begin to assign materials. A special TIF file is provided on the enclosed CD-ROM of both the La Messie violin and the wood from La Messie, so that you can map your violin with this wood. You must use Define Material to use this TIF in MicroStation. After defining your material, you should save the palette as violin or lamessie, or whatever you like (but save it).

1. Begin by turning on level 2. Levels 1 and 3 through 11 should be turned off.
2. Select the **Settings → Rendering → Assign Materials** command. Open the palette you just created with the La Messie wood in it (or you can use the MicroStation *wood.pal* file for more wood textures).

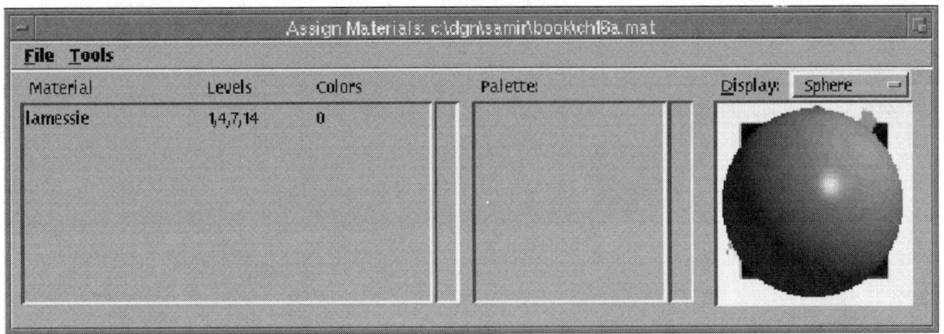

Assign Materials dialog box.

3. Use the *Assign* command and assign the La Messie material to level 2, color 0. Then *Assign* it to level 7, color 0. Then map it to level 12, color 0. This maps the body, sides, and finger board support. Make the finger board black or very dark gray. Use Phong rendering to see the (near) finished product.

Summary

Your 3D adventure is almost over. We hope these exercises helped you understand the design process better. There is still much work that can be done in these files, and many more details can be added. As a hearty adventurer you should be able to handle these problems easily. Being able to think of the complete process is important. Perhaps the most important thing to remember is that 3D elements are built from simpler 2D and 1D elements; therefore, when you look at a 3D object, try to picture the simple 2D elements that make it up. Then even a complex object becomes easier to draw. Finally, remember to break up your 3D object into smaller parts that can be modeled easily and then joined later on.

References

Hill W.H., Hill A.F., Hill A.E.: *Antonio Stradivari: His Life and Work.* Dover Publications, New York, New York, 1963.

Silverman, William A.: *The Violin Hunter.* Paganiana Publications, Neptune City, New Jersey, 1981.

Introduction

Now that you have climbed to the top of the 3D mountains, it's time to stop for a moment and enjoy the view. After a while, you look down the other side of the mountain. To your amazement, you see there was a much easier way to get to the top. That shortcut to the top is called MicroStation Modeler. If you are planning extended trips into the 3D wilderness, you would be well advised to invest in Modeler. As others who have come before you, you may be asking, what would Modeler give me that I don't have in MicroStation? Well, my weary friend, let me tell you about this mother of all power tools.

The story begins with the fact that Modeler is an MDL application. That means it integrates seamlessly into MicroStation. This MDL application was developed by the same good folks who brought you MicroStation, Bentley Systems, Inc. Modeler was created for the serious 3D users. Most folks associate solid modeling with the mechanical environment. Although Modeler is very well suited for that task, you would be short changing yourself to put that kind of limit on it. I have seen architects and civil engineers do some powerful things with it.

Differences Between Modeler and MicroStation

What makes Modeler different from MicroStation? There are several significant things. One of them is the solid modeling engine it uses. The engine is the ACIS solid modeling engine. This engine was developed by Spatial Technology, Inc. and is licensed to many other developers.

Another difference is the way Modeler stores information about 3D geometry. MicroStation stores the information about, for example, a slab, by the slab's X,Y,Z points that define the slab. Modeler, on the other hand, stores the slab as a parametric slab, with parameters that define the slab. The parameters of the slab would be the values used to define it, such as the length, width, height, and origin. Why is this better? If you wanted to edit the parametric slab, all you would have to do is edit one of the values (parameters) that define the parametric slab. For the MicroStation slab, you would have to modify the X,Y,Z points.

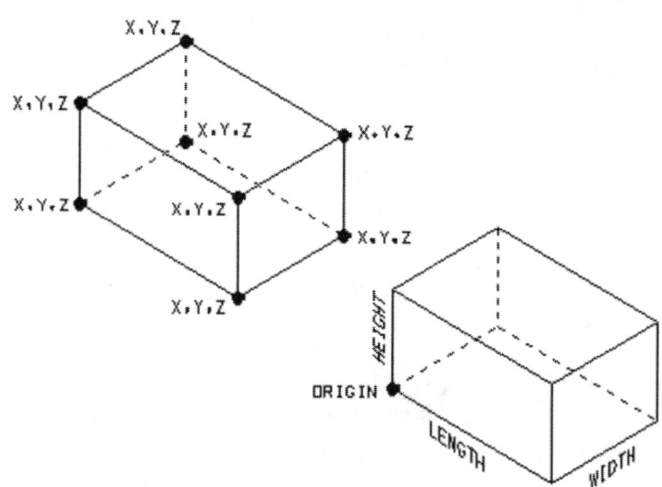

A MicroStation slab versus a Modeler parametric slab.

Modeler is a feature-based solid modeler. A feature is 3D geometry related to a parametric solid. For example, a radius on a slab is a feature. That radius (feature) has values (parameters) that were used to define it. Those values can be edited at any time and will update the radius on the slab. Can you see how this would save you hours compared with working with just MicroStation commands?

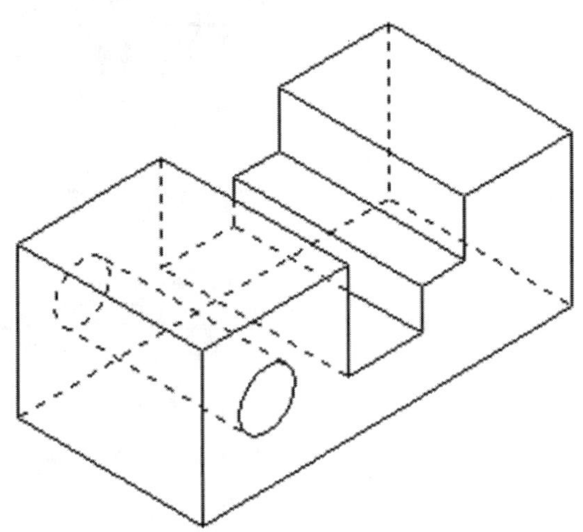

This is a parametric slab. The hole is a feature. The cut through the slab is also a feature. The values used to define all three can be edited to change the appearance of the object.

These are the values that define the parametric slab.

These are the values that define the hole. The hole is a feature.

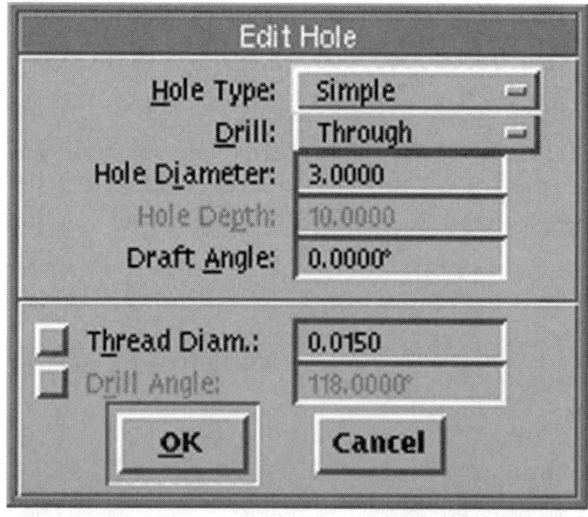

These are the values that define the cut through the slab. The cut is also a feature.

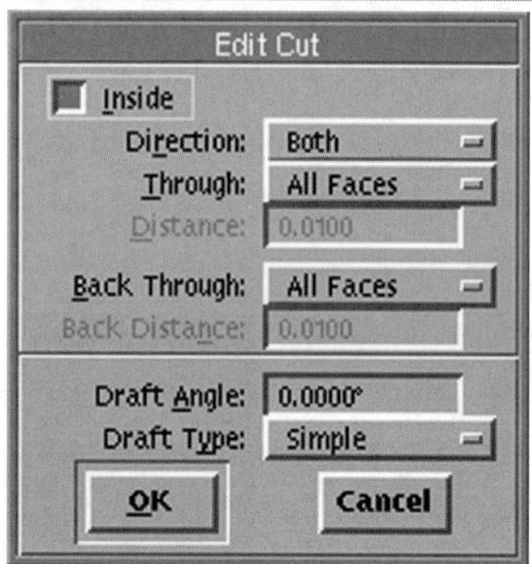

To demonstrate the power of feature-based geometry, I will edit the cut feature. I will change the direction, through type to blind, and distance.

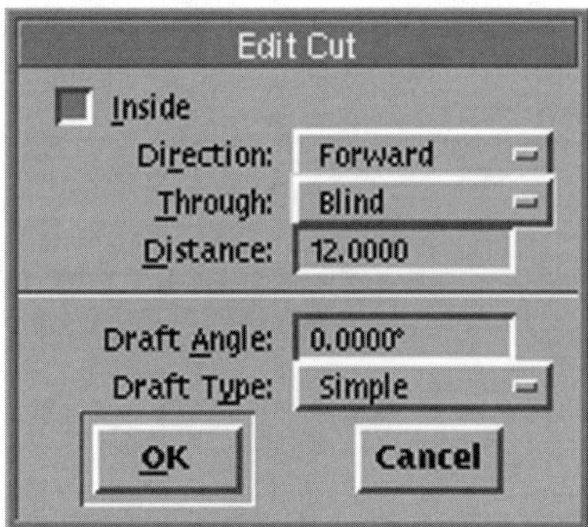

These are the changes being made to the feature.

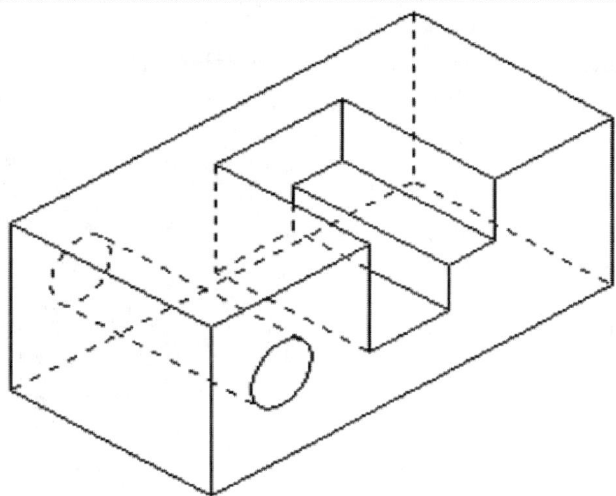

The results of the changes made to the cut feature.

If the previous example didn't get you excited, I know this one will. This example includes two demonstrations in one. First, I will show you how easy it is to create a hole through a radius on a slab. Then I will edit the radius value. Modeler makes this exercise a walk in the park.

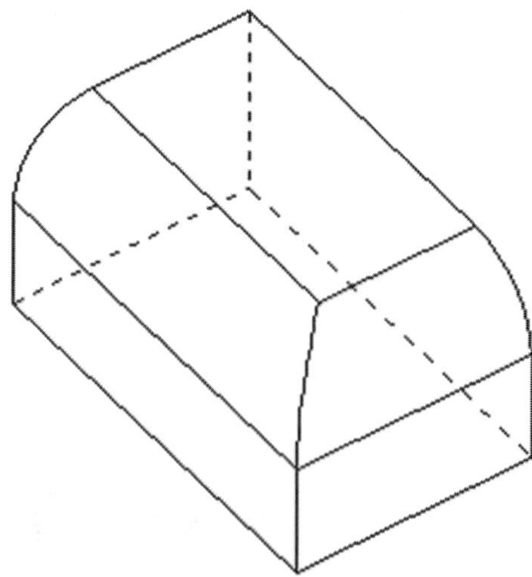

The slab with a radius along two edges.

Now I will create a hole in the slab. The hole will be through the blend of the two radiuses' edges. Try that without Modeler.

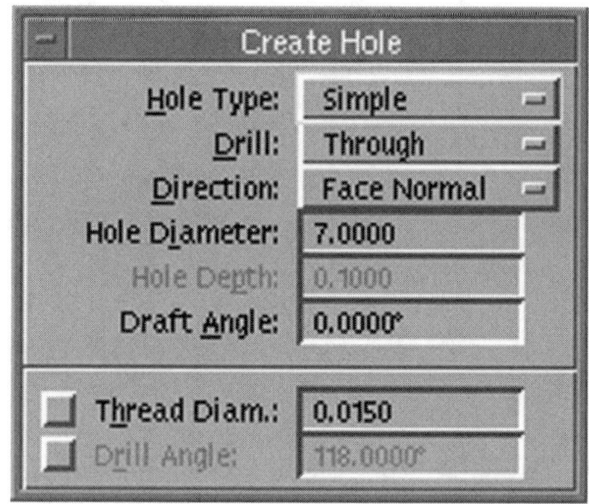

The Create Hole settings palette.

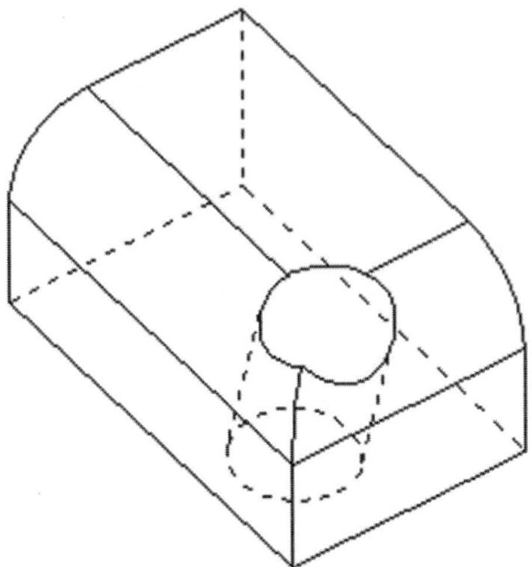

The slab with the hole through it.

Now it's time to change the hole type from a simple hole to a countersink hole.

The Edit Hole settings palette.

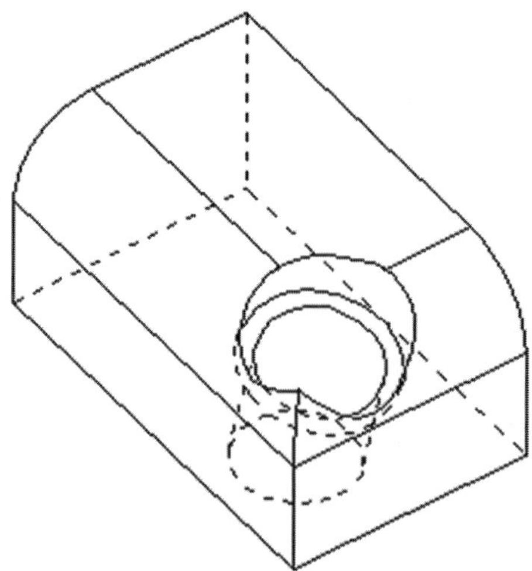

The results of changing the hole type from a simple hole to a counterbore hole.

Modeler Features

What you have seen so far is only a taste of what Modeler can do. Again, anybody planning an extended trip in the 3D wilderness could justify the need for this MDL extension. Safari Sam doesn't leave home without it. The following is a list of just some of the other tools and abilities Modeler comes with:

- It has a complete set of Boolean tools.
- It has the ability to suppress features for viewing and data export to finite element analysis programs.
- It can import Romulus files.
- It can import and export ASIC SAT files.
- It can export sterolithography (STL) files.
- Modeler has an enhanced Drawing Composition that provides bidirectional associativity. This means you can change a dimension in the sheet view and the part will change, and vice versa.
- You can use dimension-driven cells as profiles.

- The *Sketch Profile* tool allows you to create profiles with constraints predefined.
- It can create assemblies from existing parts. The parts can be attached to each other by joints. The joints control the type of movement and the degrees of freedom.

Summary

Modeler can dramatically extend your capabilities in the 3D world. Because it is a feature-based solid modeler, has parametric capabilities, and is an MDL extension, it makes a valuable addition to your 3D armor. For more information about Modeler, contact Bentley Systems, Inc.

The New Interface

The next release of MicroStation will be MicroStation 95. Some of you may have heard it called Version 5.5. In this new release, you will see many new changes. The most prominent will be the interface. MicroStation is a feature-rich program, and the number of commands can overwhelm a user. MicroStation 95 is the next level in ease of use. As an adventurer, I know that you are not easily frightened. The new interface streamlines the many commands that are in MicroStation. There are also many new tools that will assist you in the 3D wilderness. Some the most beneficial tools are view controls. Navigating in the 3D wilderness has become incredibly easy with MicroStation 95.

This tale of MicroStation 95 will start with the new interface. As the story progresses, you will hear about the new view controls and then the new tools. All of the skills that you have developed in Version 5 will allow you to appreciate MicroStation 95 all the more. First, we present the new interface.

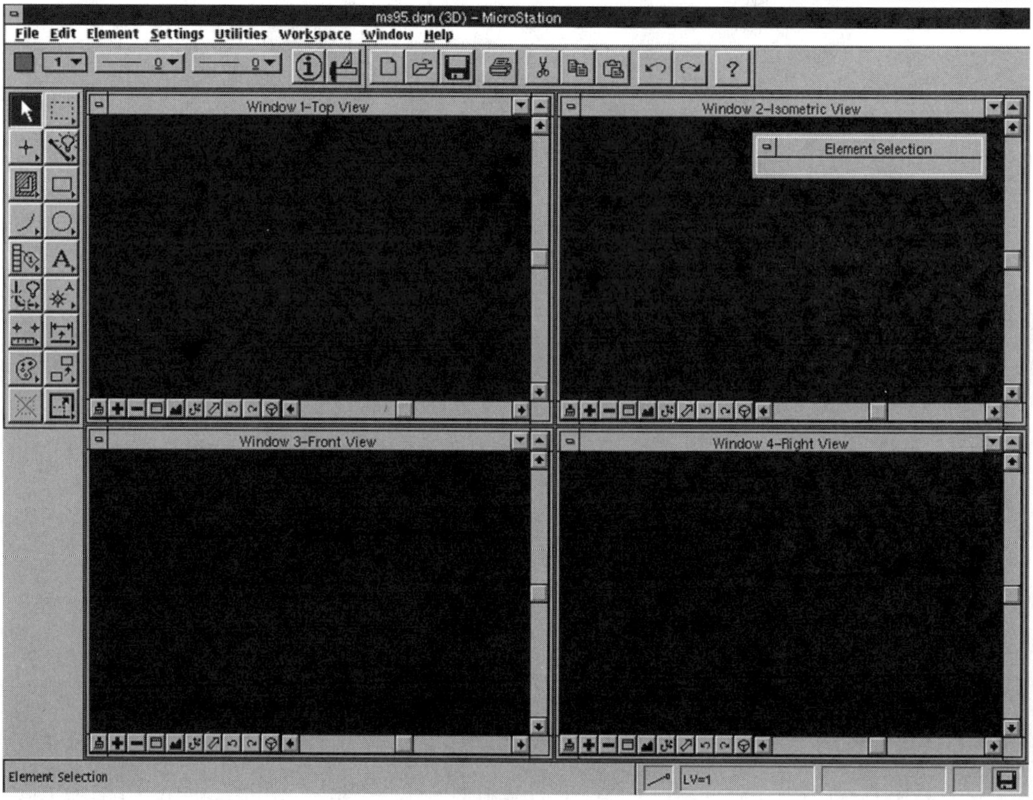

This is the new look and feel of MicroStation 95.

In the new interface you will notice that at the bottom of the frame of each view there is a series of tools. Theses are the view controls for that particular view. Also included on the view frame, there are scroll bars. If your first thought is, "I do not have enough screen real-estate now," don't panic. You can turn off the view controls and the scroll bars. As your guide, I would highly recommend that you first give this new view layout a try. Let's take a closer look at these tools and see what's changed.

The Update tool.

The Update View control, if selected, will update the view to which the tool is attached. Once selected, you can select any other open view to update. The control settings palette will display a button option to update all views.

The Zoom In tool.

The Zoom In control will still zoom in, but how it does it is the great part. First of all, once you have selected the Zoom In control, the tool settings palette will provide you with the zoom ratio option. The focus will jump to the tool settings palette. This means all you have to do is type in the new ratio value. You don't have to press ESC or data point on the text field. Then you will see a rectangle appear in your view. This rectangle will take on the shape of the view. What is inside the rectangle is what you will see when you zoom in. No more zooming in and zooming out to see what you want.

The Zoom Out tool.

The Zoom Out control, once selected, allows you to change the ratio. It will zoom out by that ratio in the view that you selected the control from. You can zoom out in any other view just by doing a data point in that view.

The Window Area tool.

The Window Area control has a new intelligence. Once you start to define the window area, the shape of the area will be confined to the shape of the view extents. You don't have to guess anymore what will be displayed in your view. You can see just exactly what will be inside the view.

The Fit View tool.

The Fit View control operates much the same as in Version 5. If you are working in 3D, there are two additional settings for this command. You have the option to *Expand Clipping Planes*. This will reset the clipping planes to fit all of the elements in the design file, into the view. The other option is to *Center Active Depth*. This setting will reset the Active Depth to a midpoint between the front and back clipping planes. These can be very useful, once mastered.

The Rotate View tool.

The Rotate View control is the Rotate View palette, except better. When working in 3D, you can rotate the view "dynamically." What does that mean? Once the control is selected, and a view chosen, by moving the cursor around, you will see a 3D display cube dynamically rotate around. Once you have rotated the cube to the position you want, just data point. If you are a Comprehensive Support Program (CSP) subscriber with Bentley, then you might recognize this command from the View Controls palette delivered with the Quarterly Vault.

The Pan tool.

The way the Pan View control works is totally new. How it reacts depends on if the view is an isometric (or custom) or one of the standard views. If the view is standard (or a 2D design file), you data point at the location you want to move from. Then you will see an arrow, with the tip of the arrow head at the end of the cursor. Now all you have to do is data point at the location you want the first point to be. The arrow dynamically showing the first point selected, and the arrow head pointing to the new location makes it child's play. In 3D, when you data point in the view you want to pan in, a cube is displayed. You can just dynamically drag it around with the cursor.

The View Previous tool.

This is the View Previous control. Not much change, except it is located at the bottom of each view frame.

The View Next tool.

This is the View Next control. Think of this control as the opposite to the View Previous control. Again, it's the new location that makes it handy.

The Change View Perspective tool.

This is the Change View Perspective control. This control allows you to change the perspective of a view dynamically. By selecting a data point in a view, you can control the degree of perspective by moving the cursor. This type of view manipulation is usually combined with camera settings.

Dockable Tool Boxes

Looking at some of the other new interface items, you will notice what looks like an icon button bar along the top of the screen. Actually, what you are looking at are two new tool boxes. Oh, did I forget to tell you, tool palettes are now know as "Tool Boxes." The faces are the same, just the names are changed. One of the new features with tool boxes is that you can "dock" them. Allow me to explain; if you drag a tool box to the edge of the screen, it will become part of the application window, that is, it will integrate itself into the border. The following images will show you what I mean.

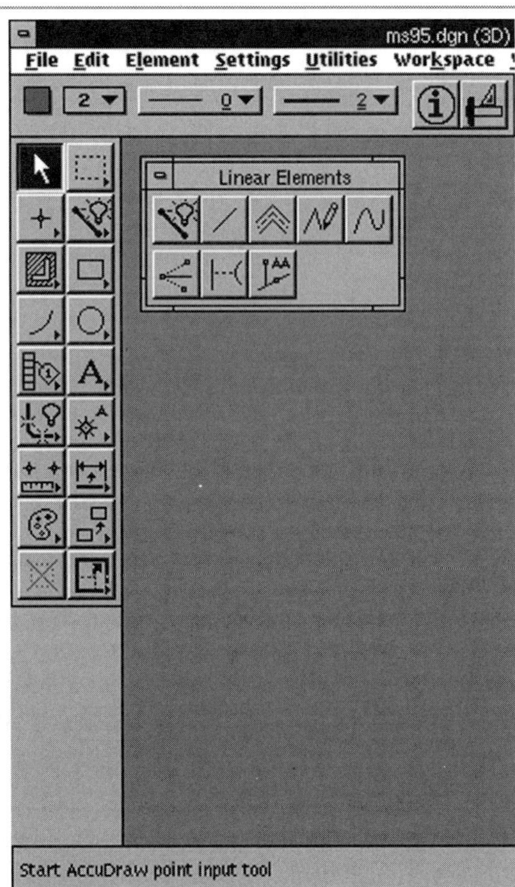

In this image, you can see the Main palette docked and part of the MicroStation application window. The Linear Elements palette is to the right, and not docked.

In this image, you can see the Linear Elements palette is now docked to the MicroStation application window.

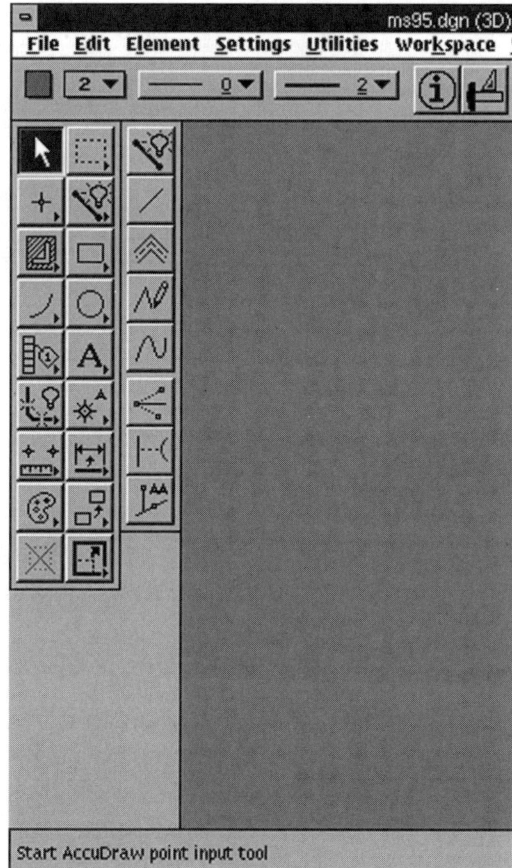

You can dock a tool box to either edge of the application window.

Resizing the Tool Box

In MicroStation Version 5, if you didn't like the shape of a subpalette, you had to customize one of your own. Now with MicroStation 95, you can just drag the corner of the palette and resize it the way you want. The subpalette, as it was called in Version 5, is now known as a "child tool box." The images below demonstrate some of the possible shapes of the Manipulate Element child tool box.

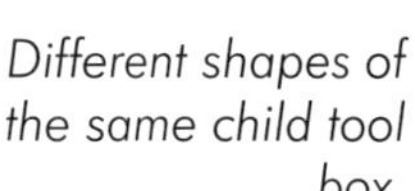

Different shapes of the same child tool box.

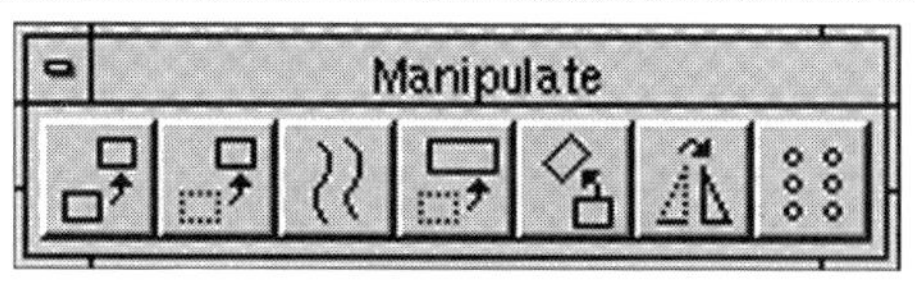

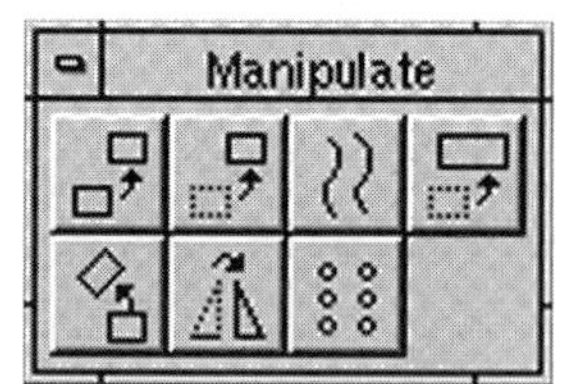

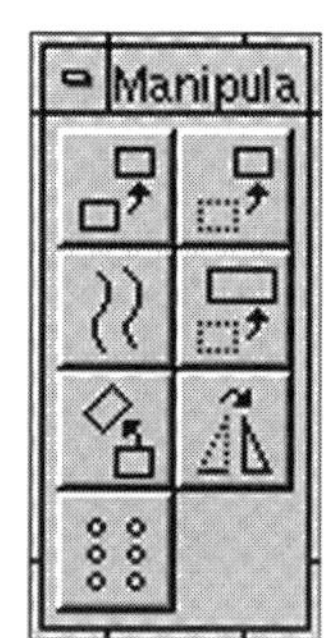

AccuDraw

One of the most productivity enhancing features of MicroStation 95 is AccuDraw. AccuDraw is designed to assist you in drawing creation. This is accomplished by two methods.

First is the drawing plane. This is a visual drawing aid that appears on the screen. This drawing aid indicates which drawing plane you are working on. With AccuDraw, there are four planes: the view, top, front, and side. The drawing aid can be rotated between these planes by a single-letter shortcut keyin. These shortcut keyins are part of the AccuDraw features. The shortcut keyins are user definable, one or two character keyins available while the AccuDraw window is open.

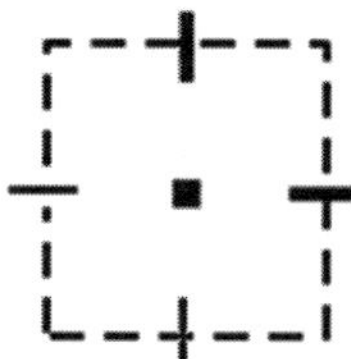

This is what the drawing aid looks like on your screen when in the rectangular coordinate system.

This is what the drawing aid looks like on your screen when in the polar coordinate system.

The second method used is the AccuDraw window. Once AccuDraw has been activated, the AccuDraw window will appear. The focus will automatically go to this window whenever a data point is used to start a command. For example, if you were to place a line, once a data point has been defined, all you need to do is move the cursor in the direction you want. AccuDraw will allow you to lock on the X and Y axes, without using the axes lock setting. Moving the cursor along the X axes, the AccuDraw window will display the distance values as you go. At this point, you can key in values for the X, Y, and Z axes. You don't need to use the `DX=`, `DL=`, or `XY=` precision input keyins. You can also tentative point on a certain point, and lock the end of the line on that axis by using the shortcut keyins of `X` or `Y`.

This is the AccuDraw window.

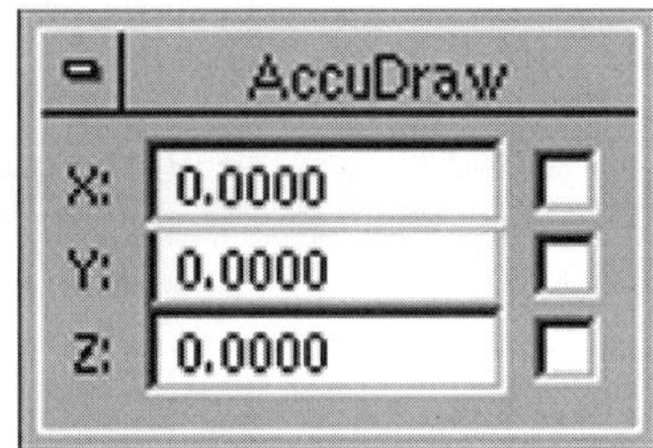

AccuDraw has other setting also. These settings are accessible from the AccuDraw Settings palette.

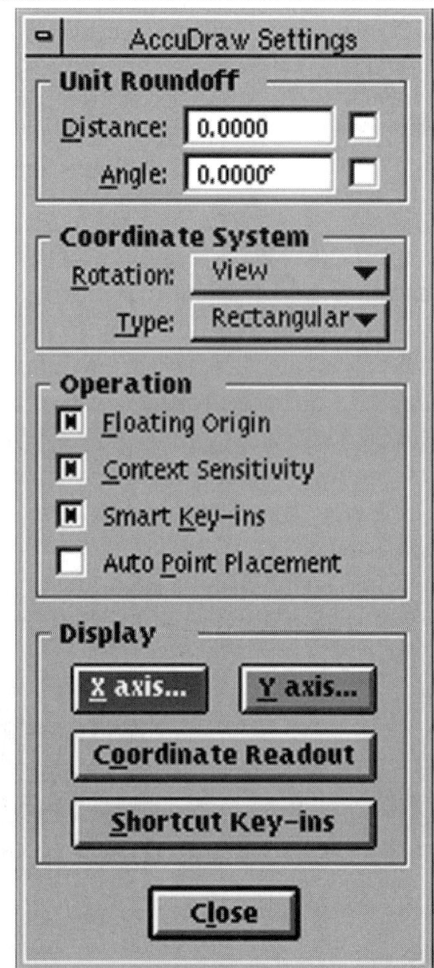

The AccuDraw settings palette provides numerous options to control AccuDraw.

AccuDraw will revolutionize the way you work in 2D and 3D. What I have told you about is the only the tip of the iceberg. This feature alone is worthy of a chapter. Once you start working with it, your productivity will soar.

SmartLine

There is a new way to place a line in MicroStation 95—the *SmartLine* tool. This new tool is designed to work in conjunction with the AccuDraw feature. This single tool allows you to place lines and arcs, and it automatically creates a complex chain out of the lines and arcs. Combined with the AccuDraw drawing plane, this new tool will simplify the creation of basic lines and arcs. When placing a line using the *SmartLine* tool, you will have an option of which kind of vertex you want. The options are Sharp, Rounded, or Chamfer. Because this is MicroStation, you can change the type of vertex on the fly. Also, the new enhanced *Modify* tool will let you modify the type or value for the SmartLine vertex. So if you really want a rounded corner with a 2-inch radius, instead of chamfered corner, you can just change the type and values. No more deleting and redrawing.

What's New with the 3D Tools?

MicroStation 95 introduces several new 3D tools. Some of the existing 3D tools have new settings. The location of the tools has changed very little. It won't take you long to get familiar with the lay of the land. Here is an overview of the new 3D tools in MicroStation 95.

Extract Surface Rule Line

Extract Surface Rule tool.

The *Extract Surface Rule Line* tool will allow you to extract a rule line from a surface. Now, you may be asking, "When would I use this tool?" Let's say that you have a B-spline surface that has been modified. This B-spline surface has curves or bumps in it. Using this new tool, you create lines that are not part of the B-spline surface, but reflect the contour of the B-spline surface. These new lines can be used for any number of different things.

Extract Surface Trim Boundary

The Extract Surface Trim Boundary tool.

The *Extract Surface Trim Boundary* tool is used to construct a linestring or B-spline curve from a surface boundary. For example, if you have a B-spline surface with a hole through it, and you want to create a linestring from the boundary of the hole, this is the tool you will use. You have the option of the type of element to create from the boundary. You can create a linestring or B-spline curve.

Modify B-spline Surface

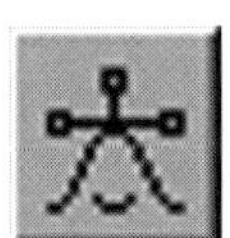

The Modify B-Spline Surface tool.

The *Modify B-spline Surface* tool is used to change the shape of a free-form to a B-spline surface. The way it allows you to modify that surface is unique. When you select the B-spline surface, it breaks the B-spline into sections. At the point where the surface is selected, a graphical handle appears. This handle is used to modify the B-spline surface. The handle has weighted curves that indicate the range that the surface can be modified. Once mastered, this tool will provide you with greater control of how you modify B-spline surfaces.

Changes to Existing 3D Tools

Place B-spline Curve

This tool has new tool settings. If you place the B-spline curve using the Through Point method, you now have the option to switch between producing an open or closed curve.

Reduce Curve Data

The *Reduce Curve Data* tool now has a setting that allows you to make a copy of the original.

Convert Element to B-spline

There is a name change. The tool setting named *Closed to Surface* is now *Convert Element to Surface.*

Punch Surface Region

This tool also has a new tool setting. You now have the option to specify the cutting direction. The new direction options are Forward, Backward, and Both.

Summary

MicroStation 95 is well worth the upgrade. The 3D tools will allow you to scale the 3D mountain even faster.

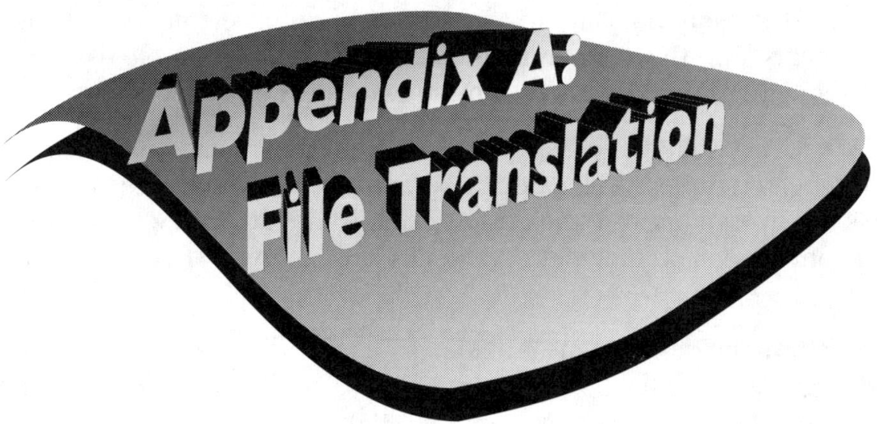

MicroStation and AutoCAD File Translation

Introduction

With version 5.0 of MicroStation, AutoCAD binary DWG drawing files can be opened and saved directly, without intermediate file translation from DXF or IGES. MicroStation still provides support for DXF and IGES file formats. However, the ability to export and import these different file formats is still available.

The Basics

The DXF format is the most widely used translation format. The file format was created by Autodesk. Since DXF and DWG files have the same entities, they are converted in exactly the same way, and the same settings are used for both. Hence, what you learned in doing DXF translations is still applicable in version 5.0 of MicroStation, and for the purposes of this discussion we will consider them (DXF and DWG) to be the same. You should remember that conversion from DWG or DXF to MicroStation DGN format cannot be exact since elements are defined in different ways.

In general, MicroStation has more information about its elements and has more elements, and these differences can cause accuracy problems. Let's look at some the differences between the two drawing formats.

AutoCAD: The design plane of DWG files is stored in memory as floating point numbers (numbers with a decimal place), giving you a large number of levels (or layers), because the design plane is infinite. In DWG files the naming conventions of linetypes and blocks can be up to 31 characters in length.

MicroStation: The design plane is made up of integers (2^{32} = 4,294,976,296 per axis), giving us only 63 levels in which elements can be created, because our design plane (or cube in 3D) is limited. The naming conventions in MicroStation for cells are limited to six characters.

AutoCAD DWG files do not support the following MicroStation features:

- Multi-line text
- True ellipses, elliptical arcs, curves, and B-splines
- True complex shapes and complex chains
- Compound custom line styles
- Cell libraries
- Color tables
- Line weight for elements other than polylines

AutoCAD polyline entities are the closest thing to these MicroStation elements. Although AutoCAD uses blocks in place of cells, each block is stored in a separate file, since there are no block libraries.

The AutoCAD color table can be modified but not saved; the MicroStation color table can be modified and saved. Furthermore, the default color tables are different, that is, color 1 (white) in MicroStation is not the same as color 1 in AutoCAD (red). Only polylines in AutoCAD can have line weights (thickness).

Basic Terminology

Entity, Element: The basic unit of graphical data in DWG and DGN files.

Receiving Application: The application MicroStation is exporting to.

Sending Application: The application that created a file that MicroStation will import.

Translation Choices

Each translation is unique. You must first determine which factors apply to your translation.

1. What is the sending and receiving application?
2. What type of data is being translated?
3. Are you translating vector or raster data?

The best way to convert between MicroStation and AutoCAD is by opening and saving DWG files directly. Do *not* use the DXF or IGES intermediate formats.

The elements in one drawing system may not have exact equivalents in another drawing system. Furthermore, elements that are common to each system may be used, and viewed, differently by each system.

Hence, the goal of translation is to get as much commonality as possible between the two systems, not absolute identical conversion. The level of perfect translation is dependent on the factors outlined before. So, it is best to learn what works best and minimizes data loss.

✔ **TIP:** *Work with smaller drawings to begin with and test elements you commonly use, or will use in future drawings.*

Settings Files and Log Files

Settings files help make file translation as accurate as possible. Once you have determined what the best settings are for your project, you may save them in a settings file, which uses the SFO extension. There is a setting file for each type of translation such as DWGIN.SFO for importing DWG files, and DWGOUT.SFO for exporting DWG files. Using these settings files ensures that consistent file translation occurs every time.

It is useful to use several settings files for the different types of translations and projects. For instance, you may want to have different settings files for import and export of different projects, sending and receiving applications, and clients and projects.

A log file is a simple text file that keeps track of and saves the translation process, such as the statistics, error messages, and everything displayed in the Status window. The log file is saved in the same directory as the file being translated, with the same name and the LOG extension. You can use the *File→ Display→ Text* command to view the file after the translation process.

General Translation Procedures

Selecting a DWG file is done in the same ways you do a DGN file. Saving information in one format or another is done by using the *Import* and *Export* commands from the File menu.

For more information, see your MicroStation reference guides.

Translation Tables

The translation table defines the way DWG or DXF formats and codes are converted to MicroStation elements. These tables can be edited using a text editor. The files are found in the USTATION\TABLES\DWG directory. The files have the TBL extension.

Summary

In general, the ability to transfer information from one package to another is easy. However, as your files become more complex (and nothing makes them more complex than creating 3D elements), the translation procedure becomes more complex. The best method of translation is to understand what you have created in MicroStation (or what you have in terms of an AutoCAD file) and what you want in your DWG or other type of file. Selecting the most generic options leads to easier translation, but the nature of the elements are compromised in that they can be altered to fit the needs of the translation.

The Internet

The Internet is a common network of computers around the world. The Internet found its origins in 1969 when the Advanced Research Projects Association (ARPA; a government advanced research entity) and several large universities decided on a protocol for common communication between large computers. The idea was to enable researchers across the country to share data easily and quickly. The Internet is not owned by anyone and there are over four billion addresses available (2^{32} = 4,294,967,296 addresses). Using an Internet service provider like CompuServe, America Online, NetCom, Primenet, and many others gives you the ability to "get" an Internet address when you log on.

Today the Internet represents the largest revolution ever, the revolution of ideas and communicating them. The revolution began about 10 years ago when the personal computer was introduced. Within a few years almost everyone had one in their home. They have become information appliances. The ability to get questions answered easily was never so widespread in human history. Today we can connect to news services, follow or trade stocks, read professional journal excerpts, search libraries and museums, and many other avenues exist.

The Internet can be used to send electronic mail (usually text messages, but some systems let you attach other files) or e-mail. A typical e-mail address could be:

```
username@mynet.com
```

File Transfer

Another function of the Internet is to transfer files. Many tools can be used to transfer files. The original method is called File Transfer Protocol (FTP). Using ftp you can logon to a remote computer using a guest ftp session. You can then change to the directory you want to and then "get" or "send" files to or from the remote computer. Better methods were developed to be more friendly, like Gopher, which lets you use a tree structure to maneuver about. Because of its menu-driven interface, it is the preferred method of navigating the Internet when downloading or uploading files.

Using the Telnet command you can logon to a computer as a user, thus giving you more control. You must have a user name and password to logon, though many systems have guest logon with your Internet address as your password. When you Telnet to a site, it's like connecting to a bulletin board system (BBS).

World Wide Web

Bentley System's Home Page

The latest method of communication on the Internet is the World Wide Web (WWW). The Web is a collection of text documents, videos, graphics, and sound. There are many sites on the Web, referred to as *home pages.* WWW addresses usually use the HyperText Transfer Protocol (http), an example of one is:

```
http://www.bentley.com/
```

This is the address for Bentley Systems. The Bentley home page has links to information regarding its products. There are places for you to leave messages for product managers. Customer support is given to help you find software updates, third-party developers, or technical notes about Bentley products (primarily MicroStation). There is also a section on the corporate information about Bentley and their corporate vision. Even a list of employment opportunities.

Above all Bentley's home page provides links to other MicroStation sites around the world. This feature is called the MicroStation MAINline. The MAINline acts as a jump off point for exploring the MicroStation world on the Web. There are four main categories to explore: Third-Party Software, Systems that run MicroStation, MicroStation at Universities, and MicroStation Images. The Bentley WWW site is a great place to learn about what is offered on the Web for MicroStation. Any questions in regards to Bentley's Web site can be directed to:

```
webmaster@bentley.com
```

Other helpful Web sites are found at the following addresses:

```
http://www.intergraph.com/
http://ftp.ncsa.uiuc.edu/Mosaic/
http://akebono.stanford.edu/yahoo/
http://gnn.com/gnn/gnn.html
http://sunsite.unc.edu/boutell/faq/www_faq.html
http://voyager.via.net/netbiz/artist.html
http://www.netsurf.com/nsd
```

Accessing MicroStation Information

There are many other forums for exchanging MicroStation information. On the major on-line services there are MicroStation forums. On CompuServe you need to type:

```
GO MSTATION
```

For America Online (AOL), type MSTATION for the MicroStation forum. These forums are available to CompuServe or AOL users only, and it is not available to Internet users. Much helpful information is distributed through these forums.

Another aspect of the Internet is the Usenet service. Usenet is a collection of news groups, each dealing with a specific subject. The information posted in a news group is provided by all the users of the Internet. Anyone can post information or reply to information in any news group. Examples of these news groups are:

- `sci.space.shuttle` (space shuttle issues and updates)
- `alt.sports.football.pro.buffalo-bills` (Buffalo Bills updates)
- `comp.sci.cad.microstation` (MicroStation news group; not available on all news servers)
- `alt.cad.intergraph` (Intergraph news group)

There are literally thousands of news groups and many dealing with computer issues. Exploration is a necessity.

The Internet allows you to also get in touch with specific entities at Bentley. Here are some helpful addresses:

- `support@bentley.com` (support for CSP partners; allows you to get any question about MicroStation answered)
- `bugs@bentley.com` (address to send bug reports to, available to all users)

A large problem with the Internet is that addresses change. Hence, keeping up to date is important. It is hoped that these addresses will be valid by the time you read this.

Index

B

C

D

G

K

L

M

N

O

Q

R

T

More
OnWord Press Titles

Pro/ENGINEER and Pro/JR. Books

INSIDE Pro/ENGINEER, 2E
Book $49.95 Includes Disk

Pro/ENGINEER Quick Reference, 2E
Book $24.95

Pro/ENGINEER Exercise Book, 2E
Book $39.95 Includes Disk

Thinking Pro/ENGINEER
Book $49.95

INSIDE Pro/JR.
Book $49.95

Interleaf Books

INSIDE Interleaf (v. 6)
Book $49.95 Includes Disk

Adventurer's Guide to Interleaf Lisp
Book $49.95 Includes Disk

The Interleaf Exercise Book
Book $39.95 Includes Disk

The Interleaf Quick Reference (v. 6)
Book $24.95

Interleaf Tips and Tricks
Book $49.95 Includes Disk

MicroStation Books

INSIDE MicroStation 5X, 3d ed.
Book $34.95 Includes Disk

MicroStation Reference Guide 5.X
Book $18.95

MicroStation Exercise Book 5.X
Book $34.95 Includes Disk
Optional Instructor's Guide $14.95

MicroStation 5.X Delta Book
Book $19.95

MicroStation for AutoCAD Users , 2E
Book $34.95

Adventures in MicroStation 3D (5.5)
Book 49.95 Includes CD

MicroStation Bible
Book $49.95
Optional Disks $49.95

Build Cell for 5.X
Software $69.95

101 MDL Commands (5.X)
Book $49.95 (4.x)
Optional Executable Disk $101.00
Optional Source Disks (6) $259.95

Managing and Networking MicroStation
Book $29.95
Optional Disk $29.95

Windows NT

Windows NT for the Technical Professional
Book $39.95

SunSoft Solaris Series

The SunSoft Solaris 2.* User's Guide
Book $29.95 Includes Disk

SunSoft Solaris 2.* for Managers and Administrators
Book $34.95

The SunSoft Solaris 2.* Quick Reference
Book $18.95

Five Steps to SunSoft Solaris 2.*
Book $24.95 Includes Disk

One Minute SunSoft Solaris Manager
Book $14.95

SunSoft Solaris 2.* for Windows Users
Book $24.95

The Hewlett Packard HP-UX Series

The HP-UX User's Guide
Book $29.95 Includes Disk

The HP-UX Quick Reference
Book $18.95

Five Steps to HP-UX
Book $24.95 Includes Disk

One Minute HP-UX Manager
Book $14.95

Other CAD

One Minute CAD Manager
Book $14.95

Manager's Guide to Computer-Aided Engineering
Book $49.95

Fallingwater in 3D Studio: A Case Study and Tutorial
Book $39.95 Includes Disk

Geographic Information Systems/ESRI

OnWord Press Distribution

End Users/User Groups/Corporate Sales

OnWord Press books are available worldwide to end users, user groups, and corporate accounts from your local bookseller or computer/software dealer, or from HMP Direct/SoftStore: call 1-800-223-6397 or 505-471-8822; fax 505-471-4424; write to High Mountain Press Direct/SoftStore, 2530 Camino Entrada, Santa Fe, NM 87505-8435, or e-mail to ORDERS@BOOKSTORE.HMP.COM.

Wholesale, Including Overseas Distribution

High Mountain Press distributes OnWord Press books internationally. For terms call 1-800-4-ONWORD or 505-471-2600; fax to 505-471-4424; e-mail to IPGORDERS@HMP.COM; or write to High Mountain Press, 2530 Camino Entrada, Santa Fe, NM 87505-8435, USA. Outside North America, call 505-471-4243.

Comments and Corrections

Your comments can help us make better products. If you find an error in our products, or have any other comments, positive or negative, we'd like to know! Please write to us at the address below or contact our e-mail address: READERS@HMP.COM.

OnWord Press
2530 Camino Entrada, Santa Fe, NM 87505-8435 USA